T0321365

Biohythane

Biohythane
Fuel for the Future

Debabrata Das
Shantonu Roy

Published by

Pan Stanford Publishing Pte. Ltd.
Penthouse Level, Suntec Tower 3
8 Temasek Boulevard
Singapore 038988

Email: editorial@panstanford.com
Web: www.panstanford.com

British Library Cataloguing-in-Publication Data
A catalogue record for this book is available from the British Library.

Biohythane: Fuel for the Future

ISBN 978-981-4745-29-1 (Hardcover)
ISBN 978-981-4745-30-7 (eBook)

Printed in the USA

Contents

Foreword

The nonjudicious use of fossil fuels and anthropogenic activity has led to the emission of greenhouse gases in the atmosphere. This has caused great harm to global weather patterns and has raised global temperatures. The need of the hour is to find a suitable alternative to fossil fuel–based economy so as to prevent the ill effects of global warming.

Extensive research work has been done on harnessing renewable energy sources (solar energy, wind, hydropower, biomass, and geothermal energy). This requires the development of an integrated energy system comprising innovative and highly efficient energy conversion technologies. The use of hydrogen as a clean fuel has been mooted for the past few decades. Its combustion will produce no greenhouse gases, no ozone layer–depleting chemicals, little or no acid rain ingredients, no oxygen depletion, and no pollution. Hydrogen is manufactured by various methods such as direct thermal, thermochemical, electrochemical, and biological methods. The biological route of hydrogen production holds great promise since organic wastes can be used as feedstock for its production. Dark-fermentative hydrogen production shows the highest proportion rate amongst the biological routes. In several cases, bench-scale production systems have shown encouraging results. There is a window of opportunity to further recover energy from the spent media, rich in volatile fatty acids, via the biomethanation process. Such two-stage integration has been envisioned under the eponym of *biohythane*.

A book on biohythane production technologies would certainly help in compiling all the necessary information related to it. I congratulate the authors, Debabrata Das and Shantonu Roy, for seeing the need for such a book and producing it. This book entitled *Biohythane: Future for the Fuel* comprehensively covers all aspects of this process. It includes microbiology, biochemistry, various feedstocks, influence of physicochemical parameters on biohythane production, mathematical modeling and simulation of biohydrogen and biomethane production processes, scale-up and energy analysis, and process economics, policy, and environmental impact.

I strongly recommend this excellent book to energy scientists, environmentalists, engineers, and students who are interested in biohythane production from organic wastes, which are considered an inexpensive gaseous clean energy carrier.

T. Nejat Veziroğlu
President, International Association for Hydrogen Energy

Preface

Energy and persistence conquer all things.
—Benjamin Franklin

Although fossil fuels meet the energy demands for most of the countries today, their contribution toward environmental degradation and climate change due to greenhouse emissions has raised serious global concerns over their usage. In addition, the exhaustion of these reserves has necessitated the search for an alternative source of energy. Thus, this has created the importance of renewable and greener technologies to fulfill the growing energy demands. Recently, hydrogen has emerged as a clean, carbon-neutral, and renewable source of energy, which has the highest energy density (143 GJ ton^{-1}), and on combustion, it produces only water as a by-product. Approximately 368 trillion cubic meters of hydrogen is produced commercially for various purposes using processes such as steam methane reforming, oil/naphtha reforming of refinery/chemical industrial off-gases, coal gasification, and water electrolysis. However, these processes rely directly or indirectly on nonrenewable energy sources, consume a lot of energy, and have a high carbon footprint. For the quest of clean, renewable energy solutions, many technologies have been explored, viz., bio-oil production by hydrothermal liquefaction, biomass gasification, pyrolysis of petroleum for methane production, etc. One such concept that has gained importance in recent times is hythane (hydrogen and methane).

The biological process for clean energy gaseous energy generation encompasses biohydrogen and biomethane production. The carbon footprint of biohydrogen and biomethane production processes is still less compared to chemical processes. Biohydrogen can be produced from organic wastes at ambient temperature and atmospheric pressure, thereby generating a sustainable process that subsequently helps in waste stabilization. The major routes for biological hydrogen production are direct and indirect biopholysis photolysis of water by blue-green algae and microalgae, oxidation of organic acids by photofermentation, and dark fermentation (using mesophilic or thermophilic bacteria). Nevertheless, each of the above-mentioned processes is associated with its respective advantages and limitations. Biophotolysis of water and photofermentation yield a very low rate of hydrogen production, and internal lighting requires additional energy input. Scaling-up of these processes is also difficult. Dark fermentation, on the other hand, is independent of light energy, requires moderate process conditions, and is less energy consuming. In addition, biogas generation process is mainly governed by two groups of microflora: acidogens and methanogens. Little information is available to find out the suitability of acidogens on hydrogen production, which may be considered potential microflora in the dark fermentation process. Thus, the dark fermentation process is considered a most promising method for biohydrogen production amongst all other processes. The spent media of the dark fermentation process contains a significant amount of short-chain fatty acids, viz., acetate, butyrate, propionate, etc. These volatile fatty acids are suitable substrates for methanogens. Therefore, integration of the biohydrogen with biomethane processes under the eponym of "biohythane" could help in the improvement of gaseous energy recovery. The integration of the biohydrogen and biomethanation processes is challenging, and an immediate emphasis is required to develop human resource, expertise, and infrastructure related to it.

The present book encompasses all the fundamentals of the state-of-the-art biohythane production technology, which will be helpful for the research community, entrepreneurs, academicians, and industrialists. It is a comprehensive collection of chapters related to microbiology, biochemistry, feedstock requirements, and scale-up studies of biohythane production processes.

Moreover, the book casts a positive eye on the policies required for applicability and socioeconomic concerns related to biofuels. A comprehensive cost of energy analysis is also a unique feature of this book. This book can be a perfect handbook for young researchers involved in bioenergy, scientists of process industries, policymakers, research faculty, and others who wish to know the fundamentals of the biohythane production technology. Each chapter begins with a fundamental explanation for general readers and ends with in-depth scientific details suitable for expert readers. Various bioengineering and bioenergy laboratories may find this book a ready reference for their routine use.

We hope this book will be useful to our readers!

Debabrata Das

Shantonu Roy

Chapter 1

Introduction

1.1 Background

Sustainable development of human civilization has coevolved with different energy resources it has used. The world population is increasing exponentially. The global energy consumption is likely to grow on the same scale. Recently, the energy consumption, which has surpassed the 11,295 million ton oil equivalent mark and a 53% increase in global energy consumption, is expected by 2030 (European Hydrogen and Fuel Cell Technology Platform, 2005). The major sources of energy are based on fossil fuels, which contribute 88% of global energy need. Under fossil fuels, the contribution of crude oil is 34.8%, followed by coal (29.2%) and natural gas (24.1%) (Ong et al., 2011). The contribution of nuclear energy and hydroelectricity toward energy generation is 5.5% and 6.4%, respectively, which is still much less when compared to conventional fossil fuels. As the demand of fossil fuels is increasing, the well-proven oil and gas reserve of the world would last for 60–70 more years. Furthermore, the emission of greenhouse gases (GHGs) has led to a global phenomenon where atmospheric temperatures throughout the globe started increasing. This phenomenon is termed "global warming." Thus, the need of green renewable energy was realized throughout the world. At present, the renewable energy contributes

Biohythane: Fuel for the Future
Debabrata Das and Shantonu Roy

ISBN 978-981-4745-29-1 (Hardcover), 978-981-4745-30-7 (eBook)
www.panstanford.com

only 11% of the total global energy used. The main objectives aimed toward a sustainable energy vision are:

- Reduction in global anthropogenic carbon dioxide (CO_2) emissions
- Improvement of local (urban) air quality
- Consistent energy supply

Establishment of infrastructure for renewable energy generation is crucial for our economic prosperity.

1.2 Necessity of CO_2-Neutral Energy Sources

Fossil fuels such as oil, coal, and natural gas are the major sources of energy which are required to keep the pace of development of human civilization. These fuels are rich in carbon and have been formed during many millions of years from plant biomass. Combustion of these fuels generates carbon dioxide (CO_2). The presence of carbon dioxide in the atmosphere has been considered as a necessary evil for the environment. CO_2 is considered as a GHG. The presence of CO_2 helps in entrapping the heat radiated by the sun, thereby keeping the temperature of the earth suitable for living. But excess carbon dioxide accumulation in the atmosphere due to nonjudicious use of fossil fuels has led to an increment in global temperatures. This has resulted in melting of polar ice caps, which eventually would lead to an increase in the sea level. This has posed a great threat to low-laying islands and continents. This phenomenon is known as global warming, which has impacted global weather patterns and climate changes (Fig. 1.1).

Global warming has had an irreversible impact on the environment. The necessities of energy for anthropogenic activity and overdependence on fossil fuels have given rise to a situation of energy scarcity in the long term. The Intergovernmental Panel on Climate Change (IPCC) recommends a reduction of global CO_2 emissions by more than 50% in order to stabilize the CO_2 level in the atmosphere at 550 parts per million volume (ppmv) to curb the negative climate effects (Reith et al., 2003).

Under the United Nations Framework Convention on Climate Change, 106 signatory counties joined the framework of the Kyoto

Protocol. According to this agreement, many industrialized nations have committed to curtail CO_2 emission by 5% in 2010, relative to the level in 1990 (UNEP 2001). The trend of CO_2 emission is still increasing. Drastic and comprehensive measures are required to reduce CO_2 emissions. Exploring alternative energy sources based on renewable resources such as sunlight, wind, hydropower, and biomass is the need of the hour. All the alternative energy sources are either directly or indirectly based on solar energy and considered CO_2 neutral since no CO_2 is released to the atmosphere. Going through the current trends, approximately 85% of the world's energy requirement is provided by fossil sources, 7% by nuclear energy, and 8% by renewable sources, primarily through the use of wood as a fuel and hydropower (Shell International, 2001).

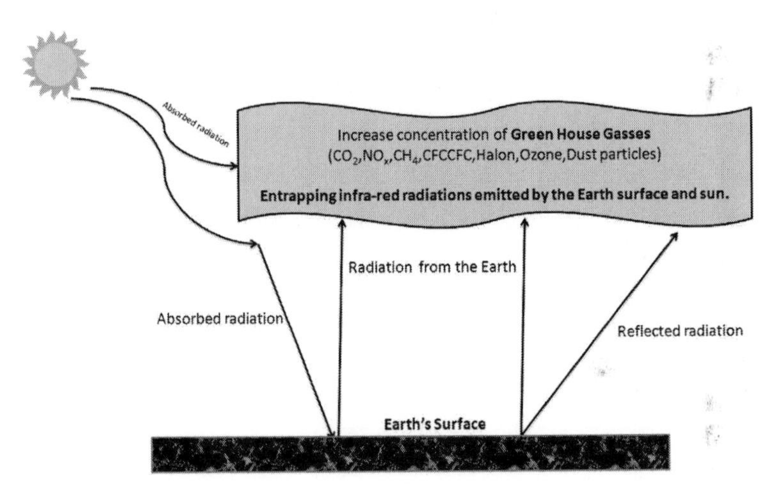

Figure 1.1 Greenhouse gases and their role in global warming.

The present costs of production of electricity based on various renewable sources in the European Union are (Hass et al., 2001):

- 0.30–0.80 euros/kWh for photovoltaic (PV) solar cells
- 0.04–0.25 euros/kWh for hydropower
- 0.07–0.19 euros/kWh for biomass
- 0.04–0.08 euros/kWh for wind turbines

These costs can be compared with current contract prices for electricity from fossil fuels that range from 0.03 to 0.05 euros/kWh.

Likewise, the costs of renewable transport fuels are at present much higher than their fossil counterparts. A major shift toward a renewable energy–based economy is not feasible in the present scenario. Renewable energy production technologies are not competitive with fossil fuel–based energy generation. The socioeconomical impact, environmental impact assessment, and life cycle assessment of implementation of renewable energy–based economy are still scarcely explored. To see light at the end of the tunnel for implementation of a renewable energy–based economy, a breakthrough in renewable energy technologies, exploration of new energy-harnessing techniques, and improved energy storage methods would be required.

1.3 Energizing the Future with Alternative Nonconventional/Renewable Energy Sources

The following are the renewable sources of energy for which technologies are available on a commercial scale (Fig. 1.2):

- Solar energy
- Tidal energy
- Geothermal energy
- Biogas/biohydrogen

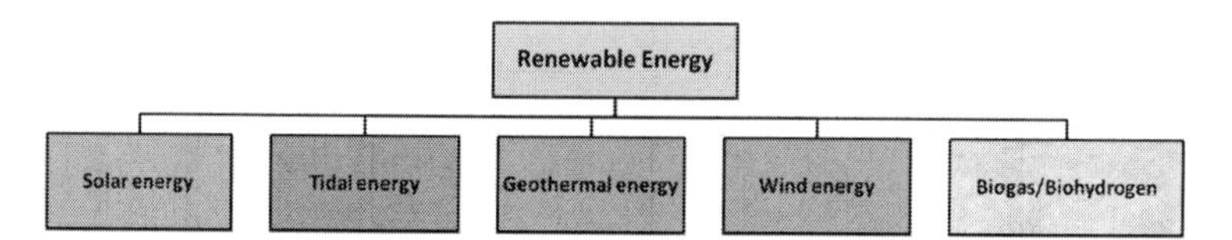

Figure 1.2 Different renewable energy sources.

The energy resources are evenly distributed throughout the world in contrast to fossil fuels whose reserves are present in discrete patches. Similarly, nuclear energy reserves are also prominent in few countries.

Another important characteristic of renewable resources is the magnitude of energy flow, which is three times higher than the present global energy need. In the coming century, significant contribution toward the world's future energy demand would be supplied by renewable energy sources such as biomass, hydropower, wind energy, solar energy (both thermal and PV), geothermal energy, and tidal energy. Major advantages of the use of renewable fuels are given below:

- It is freely available in the form of sunlight, wind, tidal waves, biomass, etc. The dependency on such sources of energy can significantly reduce the fiscal burden of the governments, thereby promoting economic prosperity.
- The use of renewable energy entails considerably very low GHG emission.
- The installation and operation of renewable energy–based power plants are economically cheap. Moreover, the operational costs are restricted for such power plants as they hardly consume conventional fossil fuels.
- Renewable energy can be harnessed at remote areas and can be used locally, thereby saving on high transmission costs.
- Initial capital investment for installation to harness renewable energy resources is high but requires low maintenance.
- It can promote local employment by utilizing available human resources for designing, construction, and operation.
- Encouraging policies implemented by the government in terms of subsidies and initial capital investment could promote people to switch over to renewable energy sources.

On considering the above-mentioned advantages, it can be envisaged that in the near future, through scientific and technological advancements, a boom in the renewable energy sector is imperative. It would not only help in fulfilling the energy demands but also keep the environment healthy and prosperous. The various alternative sources of energy and their advantages and disadvantages are discussed in Table 1.1 (Gross et al., 2003).

1.3.1 Solar Energy

The earth uses about approximately 400 EJ equivalent energy per year from solar energy (Johansson et al., 1993). It accounts for 0.01% of the total energy supplied by solar energy. The yearly supply of solar energy significantly surpasses the actual energy needs, but efficient utilization of it poses technological and economical constraints. The solar energy is harnessed via two broad techniques, viz., PV cells and solar thermal concentrators.

Table 1.1 Current status of renewable energy sources and their bottlenecks

Renewable energy sources		Availability	Current status of installation (MW*)	Bottlenecks toward commercialization
Solar energy	Photovoltaic cells	Abundant	1200	Low power generation
	Solar thermal power plants	Abundant	350	High cost of installation, requirement of a large surface area
Wind energy		Localized	20,000	Low power yield
Tidal energy		Abundant		Inconsistent availability
Geothermal energy		Localized	8200	Inconsistent availability
Biomass		Abundant	35,000	High carbon footprint, char formation, expensive pretreatment processes

**Sources*: UNDP/WEC (2000) and *Wind Power Monthly* (2001)

1.3.1.1 Exploiting solar energy via photovoltaic cells

During the Cold War era, the US space program gave rise to PV cells to power the satellites. PV cells still power modern satellites. Another impetus toward harnessing solar energy came from market expansion related to electronics, which requires very small amounts of power, such as calculators and remote small-scale applications such as telecommunications. In recent years, supportive policies related to use of PV cells for both grid and off-grid electricity have propelled the markets.

The application of solar energy is marred by many factors such as:

- High cost of installation
- Low energy density (approximately 1 kW/m^2)
- Time dependency for electricity generation

However, in the near future, development of cheap solar panels, organic solar cells, and better charge storage could propel the PV industry.

1.3.1.2 Exploiting solar energy via thermal heating

Thermal power generation using solar radiation is also known as concentrating solar power (CSP). Innovations in this technology are reported as far back as 212 BC, when Archimedes used mirrors for the first time to concentrate the sun's rays (European Union [EU], 2004). In 1615, Salomon De Caux had devised a small solar-powered motor consisting of glass lenses and an airtight metal vessel containing water and air.

Solar thermal power consists of a solar collector that concentrates solar radiation and channelizes it to a receiver, which absorbs the radiating energy (Behar et al., 2013) (Fig. 1.3). This radiated energy is then transferred to a heat transfer fluid (HTF). This HTF is then fed into a system to generate high-temperature steam, which is further used to generate electricity.

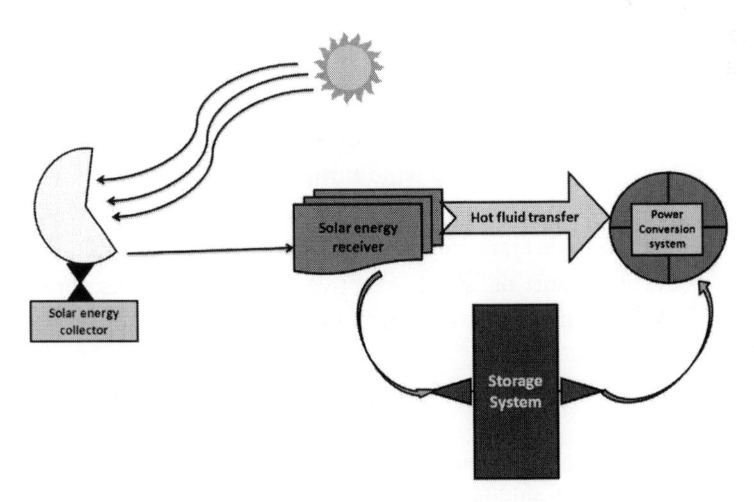

Figure 1.3 Solar thermal power generation.

Harnessing solar intensities by using a mirror collector area and chimney height has gained interest in recent years. In 2013, 2.136 GW of electricity was generated using solar thermal power plants. The United States and Spain are leading in commercializing this technology, followed by China.

1.3.2 Wind Energy

Wind turbine–based energy generation is technologically unique compared to other renewable energies. It has certain discerning demands in terms of the methods used for design, place of installment, etc. Development of modern technological know-how has led to remarkable advances in wind power design. Improvements regarding advance knowledge of aerodynamics, structural dynamics, and micrometeorology have contributed to a 5% annual increase in wind-based energy generation (Herbert et al., 2007). The following are the recent states of wind turbine technologies:

- Development of probabilistic models encompasses factors such as selection of installation location, height, choice of wind generators, wind velocity, wind power potential, etc., for determination of energy output of the wind turbine system.
- Prediction of wind speed data of the site of installation could be suitably explained by Weibull, Rayleigh distribution, and the Markov chain model.
- Installation of a wind generator also requires meteorological data.
- Analysis of vibration problems of wind turbines is now better understood using experimental and theoretical methods.
- Lifetime predictions of wind turbine blades are based on rain flow counting, a linear Goodman fit, and Miner summation.
- Aeroacoustic tests are used to find noise in the aerofoil.
- Development of static reactive power compensators is considered for the improvement of stability of large wind farms.
- Recent advancement includes wind field modeling of wind turbines.
- A computerized control system for monitoring wind power generation could help in decreasing manpower requirements.

As a renewable source of energy, wind power is the fastest-growing sector. But installations of wind turbines are marred by many constraints such as:

- Availability of a constant flow of wind with less variation in flow and direction
- Large area required for installation
- Interference in radar signals, thereby creating difficulties in civil and military aviation
- High cost of installation

1.3.3 Tidal Energy

Tidal waves are the rhythmic rising and lowering of ocean waters. These waves are generated because of gravitational forces between the moon, the sun, and the earth. Lunar gravity imparts more than twice as great a force on the tides as the sun because of the moon's much closer position to the earth. In the open ocean, the amplitude or height of the tidal waves is very small. They are distributed over hundreds of kilometers in the vast ocean. However, the amplitude of the tides can increase dramatically at continental shelves. The potential and kinetic energies of the tidal waves could be harnessed for power generation. The amount of work done in lifting the mass of water above the ocean surface creates potential energy. This energy can be calculated as follows (Gorlov, 2001):

$$E = g\rho A\int zdz = 0.5g\rho Ah^2 \tag{1.1}$$

where E is the energy, g is the acceleration due to gravity, ρ is the seawater density, A is the sea area under consideration, z is a vertical coordinate of the ocean surface, and h is the tide amplitude.

The average potential energy provided by the ocean is $E = 5.04h^2$ kJ. The kinetic energy E_T of the oceanic water could be calculated as a function of mass m and velocity V:

$$E_T = 0.5\ mV^2 \tag{1.2}$$

The total tidal energy equals the sum of its potential and kinetic energy components. Development of new, efficient, low-cost, and environmentally friendly hydraulic energy converters such as triple-helix turbines could help in harnessing tidal energy. Such power stations can provide clean energy to small communities or even individual households located near continental shorelines, near straits, or on remote islands with strong tidal currents.

1.3.4 Geothermal Energy

The energy entrapped as heat in the earth's interior is called geothermal energy (Lund, 2005). Direct use of geothermal energy is one of the oldest, most versatile, and also the most common form of utilization of this energy (Dickson and Fanelli, 2003). The earth's heat flow and geothermal gradient are well established and the presence of different types of geothermal fields documented; the geologic environment of geothermal energy and the methods for geothermal exploration are encouraging different researchers toward the goal of energy generation.

For generation of electricity, geothermal energy has been extensively exploited for decades: 0.4% of the world's total electrical energy has been produced by geothermal sources. An installed electricity setup of more than 7173.5 MW has been achieved globally. At present, the global geothermal electrical capacity installed would be probably comparable to that of biomass (Barbier, 1997). The major disadvantage of harnessing geothermal energy is its nonuniform availability and environmental challenges.

1.3.5 Biohydrogen/Biogas Production

Among all the renewable energy sources, biohydrogen/biogas production holds promise in terms of ease of operation, low cost input, and decentralized energy production. On the basis of the following points, hydrogen could be considered a future fuel (Bockris et al., 1981):

- The most lucrative feature of considering hydrogen as a fuel is the end products formed upon its combustion. The end products are water and energy. So there is no GHG emission. Thus it poses no threat to the environment.
- It can be easily converted to electricity using fuel cells. Fuel cells have a high conversion efficiency as they don't follow Carnot engine theory.

- It can be produced from renewable sources such as biomass, wastewater, and industrial wastes.
- Storage technologies such as gaseous state storage, liquid form storage, and metal hydride storage are available and evolving as time passes. Hydrogen can be transported over large distances using existing gaseous pipelines.

Hydrogen is an important molecule with extensive applications and uses. It can be used instead of fossil fuels for virtually all purposes, ranging from surface and air transportation to heat and electricity generation. The industrial applications of hydrogen include ammonia synthesis, petroleum processing, petrochemical production, oil and fat hydrogenation, metallurgical application, glass and optical fiber manufacturing, and application in electronic industries. At present, hydrogen is produced mainly from fossil fuels, hydrocarbons, and water. Table 1.2 shows the technologies available for hydrogen production and their status in terms of commercial application. All the processes have certain advantages and disadvantages, but biohydrogen production still looks promising in terms of usage of organic wastes as feedstock. Thus biohydrogen can help in realizing the concept of *waste to energy*.

Table 1.2 Economics of different energy generation processes (Nayak et al., 2013)

Energy sources	Source and process (large-scale technology)	Cost of production
Hydrogen production	Natural gas (via steam reforming)	\$4–\$5/kg
	Wind (via electrolysis)	\$8–\$10/kg
	Nuclear (via electrolysis)	\$7.50–\$9.50/kg
	Nuclear (via thermochemical cycles)	\$6.50–\$8.50/kg
	Solar (via electrolysis)	\$10–\$12/kg
	Solar (thermochemical cycles)	\$7.50–\$9.50/kg
	Organic waste (dark fermentation)	\$1.3/MBTU
	Gasoline	\$23.5/MBTU
	Natural gas	\$2–\$7/MBTU

(Continued)

Table 1.2 (*Continued*)

Energy sources	Source and process (large-scale technology)	Cost of production
Ethanol	Synthetic	$0.15/kg
	Fermentation	$0.4/kg
Syngas	Pyrolysis of coal	$24.47/TCM (thousand cubic meters)
	Pyrolysis of organic waste	$10.62/TCM (thousand cubic meters)
Incineration	Cofiring	$ 0.17/kwh
Anaerobic digestion (biomethane)	Biomethane-based generator	$ 0.4/kwh

1.3.5.1 Hydrogen economy and fuel cells

The main hurdle toward commercialization of hydrogen was the challenges associated with the storage of this gas. This has limited its application in civilian areas compared to storage of liquid fossil fuels. It is possible to use hydrogen in internal combustion (IC) engines, either directly or mixed with diesel and compressed natural gas (CNG). It can also be used directly as a fuel in fuel cells to produce electricity. Hydrogen energy is often mentioned as a potential solution for several challenges that the global energy system is facing. The advent of fuel cell technologies has partially addressed the need for hydrogen storage. Hydrogen produced in any system can be stored transiently and can be channelized to fuel cells to convert it to electricity. With low CO_2 emissions, fuel cells have the potential to become major factors in catalyzing the transition to a future sustainable energy system. A large number of countries are now implementing roadmaps for the advancement of fuel cell and hydrogen technologies. The EU has set a roadmap which is a potent example of the development and deployment of hydrogen and fuel cell technologies. It set a target of 1 GW of distributed power generation capacity from fuel cells by 2015 and 0.4–1.8

million hydrogen vehicles sold per year by 2020. The three major technological barriers that must be prevailing over for a transition from a carbon-based (fossil fuel) energy system to a hydrogen-based economy are as follows:

- Cheap and sustainable hydrogen production and delivery must be significantly reduced.
- New generations of hydrogen storage systems are needed that would be cheap and easy to implement for both vehicular and stationary applications.
- The cost of fuel cells and other hydrogen-based systems must be reduced.

The combination of large and small fuel cells running on local hydrogen supply networks for domestic and decentralized heat and electricity power generation would help in realizing the vision of integrated energy systems. Using hydrogen as a fuel has certain safety issues. This encompasses scientific and technological aspects of development of safety protocol and strategies. The psychological and societal acceptance of using hydrogen as fuel is also critical. Even though hydrogen has been maligned for its safety issues, it has an exemplary safety record when it comes to industrial use. As hydrogen is the lightest element on earth, it has a tendency to escape from small pores and channels, and on mixing with atmospheric oxygen it becomes inflammable.

There are many technologies available for hydrogen production. It can be produced from fossil fuels such as natural gas, coal, hydrocarbons, or it can be produced from gasification of biomass, municipal waste, and industrial wastewater. The majority of hydrogen is produced by electrolysis of water. Steam reforming of methane is also a commercial way of producing hydrogen. This process leads to CO_2 emission, but the amount of CO_2 released is far less compared to the combustion of methane. High-temperature pyrolysis was used to convert hydrocarbons, biomass, and municipal solid waste (MSW) into hydrogen. The by-products of this process are carbon-rich charcoal, CO, and CO_2. At present, the cost of this process is higher than that of steam reforming of natural gas. The production of hydrogen by electrolysis of water has a very high efficiency of 70%–75%, but the cost of production is several times higher than that produced from fossil fuels. The advent of technology such as fuel cells that convert hydrogen to electricity has infused new life in the implementation of a hydrogen-based economy. A fuel cell is a device that is similar to a continuously recharging battery which generates electricity by the low-temperature electrochemical reaction of hydrogen and oxygen. The contrasting difference with batteries storing energy is that a fuel cell can produce electricity continuously as long as hydrogen and oxygen are supplied to it. Hydrogen-powered fuel cells produce water as a by-product and produce virtually no pollutant. Fuel cells operate at temperatures much below the IC engine. Fuel cells are not bound by the limitations of the Carnot cycle; thus these can efficiently convert fuel into electricity compared to IC engines. The operating temperature, the type of fuel, and a range of applications of fuel cells depend on the electrolyte the cells use. The electrolyte can be an acid, a base, a salt, or a solid ceramic or polymeric membrane that conducts ions.

1.4 Role of Biomass and Biotechnological Processes for Renewable Energy Production

Organic wastes, comprising mainly cellulose and hemicelluloses, are the most abundant resources of energy which can be renewed indefinitely through trees, plants, and associated residues; poultry, litter, and other animal waste; and industrial waste and paper components of MSW.

As a result of the rising energy demand, the high cost of fossil fuels, dwindling fossil fuel reserves, and the greenhouse effect, biomass is getting recognition globally as a resource for producing an alternate and clean source of energy. Organic wastes have a low-energy-density material with a low bulk density; these degrade on storage and also are difficult and costly to transport, store, and use. These wastes can be converted into fuels using different technologies (Fig. 1.4). At present, about 500 million metric tons per year of organic wastes is generated in India. This biomass has a potential of producing about 18,000 MW energy. India has around 580 odd sugar mills, which generate bagasse. This bagasse has the potential of generating about 5000 MW by the cogeneration technique.

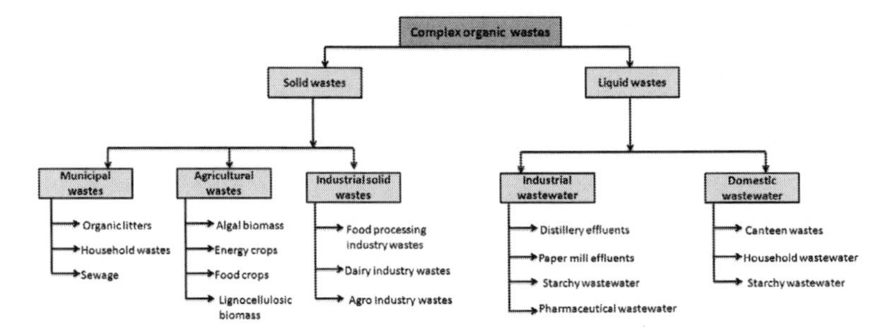

Figure 1.4 Different types of organic wastes.

1.4.1 Availability of Different Organic Wastes and Their Characteristics for Energy Generation

1.4.1.1 Lignocelluloses and agroforestry-based biomass

In plants, starch and cellulose are the two abundant polymer molecules whose basic unit is glucose. Plants store energy in the form of starch, whereas cellulose forms the structural skeleton of cell walls for leaves, stems, stalks, and woody portions. Moreover, plant cell walls and some algae also contain xylans, which are a polymeric sugar having xylose (a pentose sugar) as the basic unit. Similar to cellulose, xylans also have predominantly β-glycosidic bonds between xylose subunits. Thus, due to the large abundance of lignocellulosic raw materials, they might be considered suitable feedstock for future biofuel industries. Lignocelluloses are biopolymers consisting of tightly bound lignin, cellulose, and hemicellulose. Sugar-rich plants such as *Miscanthus*, corn, and beetroot could be considered a source of fermentable sugars. Cultivation of such crops for biofuel purposes has evoked the debate on food versus fuel issues. Use of switch grass, rice husk, fodder straw, etc., for biofuel production has gained importance in recent years. Lignocellulosic crops like these need a systematic pretreatment step to remove the lignin content. After lignin removal, the crystalline cellulose is still not accessible to microbes. Further saccharification of this crystalline cellulose yields simple sugars that could be used for biohydrogen production. Requirement of pretreatment and saccharification processes increases the operational cost of the process. Moreover, many growth inhibitors such as furfurals are produced during the pretreatment and saccharification processes. Lignocellulosic biomass might be in high abundance, but requirement of harsh pretreatment and saccharification processes limits its use as feedstock. In recent times, algal biomass has gained importance as feedstock for fermentation. Few reports are available on usage of algal biomass as feedstock for hydrogen production. These biomasses are devoid of recalcitrant polymers such as lignin. The main food reserve in algae is either starch or semicrystalline cellulose.

1.4.1.2 Food industry wastes

The food processing industry is one of the booming businesses that contribute a chunk toward India's gross domestic product (GDP). The processed-food industry generates a lot of organic-rich wastewater. Disposal of food wastes possesses a great environmental threat. Food wastes contain about 90% volatile suspended solids and a high organic content that makes them suitable feedstock for microbial fermentation.

1.4.1.3 Dairy industry wastewater

Dairy has been an essential part of India's rural economy. The milk production capacity of dairy industries is about 135 million tons/year. Industries based on milk as a starting material generate a lot of organic-rich wastewater. These effluents have a high biochemical oxygen demand (BOD) and chemical oxygen demand (COD), which makes them hazardous for the environment, if discharged untreated. The organic content of the wastewater makes it an ideal contender as feedstock for fermentative bacteria. Usage of such wastewater has been successful in the case of biogas production. Hydrothermal pyrolysis could also be used to produce a variety of liquid fuels.

1.4.1.4 Alcohol industry wastes

India produces 1680 million liters of distillery effluents per year. For production of 1 L of ethanol, approximately 12 L of spent media gets generated. Distillery or alcoholic beverage industry wastewaters are rich in biodegradable organic material, such as sugars, hemicelluloses, dextrin, resins, and organic acids. These wastewaters have a high COD (80–160 g L^{-1}). These wastewaters were also considered for biogas production. Other technologies of energy generation include incineration of distillery silage.

1.4.1.5 Municipal wastes

India produces more than 38,254 million liters of sewage every day. Migration of rural populations to cities due to industrialization and population explosion are the major contributors toward MSW generation. The figure could further increase in the near future by virtue of development of the country's economy. Organic fractions of MSW are widely available renewable resources, which can be explored as feedstock for biofuel production. MSW is rich in polysaccharides and proteins. The dried sewage sludge is generally incinerated to generate heat energy. Anaerobic digestion of these wastes generates a good amount of methane, which could be used to produce electricity.

1.5 Biofuels from Organic Wastes

Different energy generation technologies have been explored for production of fuel from organic wastes (Fig. 1.5). Biological routes of fuel production are based on fermentation, such as biomethane, biohydrogen, bioethanol, biobutanol, etc. On the other hand, biodiesel is produced under autotrophic/mixotrophic conditions by microalgae. Nonbiological routes of fuel production using organic wastes involve various techniques such as incineration technology, hydrothermal liquefaction, syngas production, and pyrolysis. These methods are energy intensive and marred by many disadvantages such as char formation, high carbon footprint, and scaling-up challenges.

1.5.1 Bioethanol

Use of bioethanol for fuel has gained importance in recent years. Bioethanol production requires carbohydrate-rich biomass such as cellulosic biomass, agricultural waste, and wood waste as feedstock. Moreover it is biodegradable, is environment friendly, and has a lesser carbon footprint compared to petroleum fuels. Simple sugars generated from saccharification of complex sugars are converted into ethanol via a defined metabolic pathway. The microbes capable of performing ethanol fermentation belong to groups of solventogenic microbes (Eq. 1.3):

$$C_nH_{2n}O_n \longrightarrow \frac{n}{3}CH_3CH_2OH + \frac{n}{3}CO_2 + 227.0\ \text{kJ/mol} \tag{1.3}$$

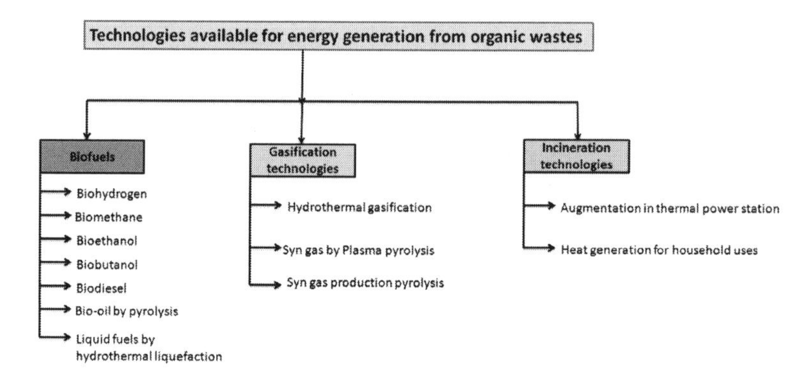

Figure 1.5 Possibilities of energy generation from different organic wastes.

The ethanol thus produced can be recovered by fractional distillation. Ethanol on combustion gives 1370.7 kJ/mol (Eq. 1.4) of energy that can be harnessed for cooking and automobile combustion engines.

$$CH_3CH_2OH + 3O_2 \xrightarrow{\text{Combustion}} 2CO_2 + 2H_2O + 1370.7\ \text{kJ/mol} \tag{1.4}$$

A few characteristics of bioethanol are distinct from other energy carriers:

- Already existing infrastructure for storage and distribution can be used for it.
- Blending with petroleum fuel is easy and causes no harm to an IC engine. There is lower emission of harmful, unburned hydrocarbon and carbon monoxide.

Very few countries possess the technologies and skills of producing ethanol which is used as a fuel. The acceptance regarding usage of ethanol as a blend in fuel has gained momentum globally. The dependency on starch- or sugar-rich crops for production of ethanol has mooted a debate on its sustainable production and as a replacement for fossil fuels. This has led to a debate related to "food vs. fuel." This conflict of interest between food and fuel for arable land can potentially lead to an increase in food prices. In recent times, lignocellulosic biomass has gained importance as feedstock for production of ethanol. Moreover, higher availability of lignocellulosic biomass in the form of an agricultural by-product negates the debate of "food vs. fuel." Extensively used lignocellulosic biomasses for ethanol production are sugarcane bagasse, wheat straw, rice husk, rice straw, corn straw, etc. Various pretreatment and sacchharification processes have been studied to facilitate the availability of simple sugars that are entrapped in complex polymeric forms. The pretreatment techniques are physical, chemical, or biological in nature. These techniques are required to break the rigid and complex plant cell wall. The plant cell wall is made up of lignin and cellulose hemicelluloses, which are polymeric form of simple sugars. The polymeric sugars present in naive lignocellulosic biomass would not be accessible to microbes for fermentation.

1.5.2 Biobutanol

The importance of liquid fuels is due to their ease in transportation. This gives the necessary impetus toward research on liquid biofuels, such as alcohol from organic matters. At present, bioethanol and biodiesel production solely cannot suffice the demand of biofuel. This has led to research focusing on biofuel with higher calorific values. Biobutanol is considered superior to bioethanol by virtue of its energy density and nonhygroscopic properties. The acetone–butanol–ethanol (ABE) fermentation is well established and one of the oldest fermentation technologies. There are many hindrances toward commercialization of biobutanol production. Hindrances include availability of cheap feedstock, choice of feedstock, low product yield, intolerance of higher concentration of butanol by the production strain, undesired end products, and separation of butanol in downstream processing. Biobutanol is produced exclusively by obligatory anaerobes such as *Clostridium* species. Creation of complete anaerobicity inside a large reactor is a cumbersome and expensive process.

1.5.3 Biodiesel

Biodiesel consists of monoalkyl esters that are derived from different oil-rich sources, for example, organic oils extracted from plants, animal, algae, etc., via the transesterification process. The chemical reaction for biodiesel transesterification involves hydrolysis of the ester bond between glycerol and the fatty acid chain and then further esterification with methanol, as shown in Eq. 1.5.

$$\text{Triglyceride} + 3\ \text{Methanol} \xrightleftharpoons{\text{Catalyst}} \text{Glycerine} + 3\ \text{Methyl Esters (Biodiesel)} \quad (1.5)$$

The presence of a catalyst and alkali such as potassium hydroxide enhances transesterification. Since it's a reversible reaction, an excess of methanol could be used to force the reaction in the forward direction. Solvent recovery could be done by the soxhlet technique and the solvent can be reused. Elevating the temperature to 60°C could increase the kinetics of the reaction, and the process could be completed in 90 min.

Edible oils such as soybean, rapeseed, canola, sunflower, palm, coconut, and also corn are the most common feedstock for biodiesel production. The dependency on edible oil has given strength to the "food vs. fuel" debate. India imports most of its edible oil. So establishment of biodiesel technologies based on edible oil is not feasible. Among different oleaginous seeds, biodiesel production using *Jatropha curcas* seeds was found to be more promising in terms of yield and productivity. Jatropha oil is nonedible and the per hectare yield is high compared to other oil seeds. Jatropha oil has low acidity, good stability, and low viscosity. Beside this, it has a higher cetane number compared to diesel. This makes it a good alternative fuel with no modifications required in the engine.

Many reports suggest that on subjecting stressed conditions, algae tend to accumulate more lipids compared to favorable growth conditions. Under favorable growth conditions, glycerol-based membrane lipids are formed from the synthesized fatty acids via esterification, which constitutes 5%–20% w/w of the dry cell weight (DCW). The fatty acid compositions produced under favorable growth conditions are medium chain (C10–C14), long chain (C16–C18), and very long chain (C20) and fatty acid derivatives. The fatty acid composition changes under unfavorable or stress conditions. Under stress conditions, the biosynthesis of lipids is redirected toward production and storage of neutral lipids (20%–50% w/w DCW). Triacylglycerols (TAGs) are the bulk constituents of neutral lipids. TAGs serve as energy stores and are not involved in the formation of structural phospholipids in cell membranes (Hu et al., 2008). The TAGs are stored as densely packed lipid bodies in the cytoplasm of the algal cell and can be visualized by stains such as Sudan black or Nile Red. Lipid accumulation also takes place in the interthylakoid space of the chloroplast in certain green algae. However, maintaining operational conditions such as low temperature, low light intensity, and nitrogen deficiency leads to accumulation of high-grade oil in microalgae, posing a major challenge in large-scale photobioreactors. In the present scenario, production of algal biomass with 20% lipid content in bulk for a sustained period of time is difficult and economically not viable.

1.5.4 Bio-Oil

Biomass is converted to bio-oil through pyrolysis in the absence of air at atmospheric pressure, a low temperature (450°C–550°C), high heating rate (103–104 K/s), and short gas residence time. In fast pyrolysis, high yields were observed along with a high fuel-to-feed ratio, making this process a promising technology for commercial biofuel technology. The liquid product from biomass pyrolysis is known as biomass pyrolysis oil—bio-oil, pyrolysis oil, or bio-crude for short. The chemical composition of bio-oil tends to change toward thermodynamic equilibrium during storage. Under high temperatures, complex polymers get depolymerized, along with the production of fragments of cellulose, lignin, hemicellulose, etc. Bio-oil is a complex mixture of acids, alcohols, aldehydes, esters, ketones, sugars, phenols, guaiacols, syringols, furans, lignin-derived phenols, and extractible terpenes with multifunctional groups.

1.5.5 Biohydrogen

Considering the pros and cons of many fuels, hydrogen could be projected as a promising future fuel because it is considered as a clean and renewable source of energy. The replacement of conventional fossil fuels could be conceptually done by hydrogen. Thus it could decrease our dependency on hydrocarbon-based fuels. It has the highest energy density (143 GJ ton^{-1}) by virtue of its high calorific value and low molecular weight. On combustion it produces water and energy as products. So hydrogen can be considered as the only carbon-free fuel. Thus on burning hydrogen, there is no GHG emission, no toxic gas emission, the least propensity of acid rain, and no ozone layer depletion. Even though hydrogen is the most abundant element in the universe, it must be produced from other hydrogen-containing compounds such as biomass, organic wastewater, or water. In recent times, the idea of a hydrogen-based economy has gained importance. India imports 80% of its crude oil needs. This creates immense pressure on the government's fiscal deficit. The convention techniques involved in hydrogen production were steam methane reforming, oil/naphtha reforming of refinery/chemical industrial off-gases, coal gasification, and water electrolysis. It is interesting to observe that for production of hydrogen, fossil fuels are mostly required. This makes the process energy intensive. It also has a high carbon footprint. If hydrogen is produced by using renewable feedstock, then it can truly be considered as a renewable energy source. Biohydrogen is produced from organic wastes at ambient temperature and atmospheric pressure. Thus energy generation could be coupled with waste management. Biological hydrogen is mainly produced by the following processes: photolysis of water (direct and indirect biopholysis) by blue-green algae and microalgae, oxidation of organic acids by photofermentation, and dark fermentation (using mesophilic or thermophilic bacteria). However, the above-mentioned processes have their own advantages and disadvantages. Processes such as biophotolysis of water and photofermentation are marred by a very low rate of hydrogen production and require light as an additional energy source. They also face difficulty on scaling up. Amongst the various other processes, dark fermentation appears to be more promising. It is independent of light energy, requires moderate process conditions, and consumes less energy. Compared to other biological hydrogen production processes, dark fermentation has relatively a higher hydrogen production rate and yield.

In dark-fermentative hydrogen production, glucose is considered the principle substrate. Complex polymeric organic substrates are hydrolyzed to simple sugars like glucose. Glucose is the simplest sugar, which is preferred by most of the microbes. It is further metabolized via the glycolytic pathway to produce pyruvate. Subsequently under anaerobic conditions, the fate of pyruvate is to get converted to acetic acid and butyric acid. On doing so, hydrogen is produced as a by-product. The pyruvate ferredoxin oxidoreductase (PFOR) enzyme oxidizes pyruvate to acetyl coenzyme A (acetyl-CoA). This pyruvate oxidation step requires ferredoxin (Fd) reduction, which in its reduced form is oxidized by [FeFe] hydrogenase and catalyzes the formation of hydrogen. The overall reaction is shown in Eqs. 1.6 and 1.7.

$$\text{Pyruvate} + \text{CoA} + 2\ \text{Fd (ox)} \rightarrow \text{Acetyl-CoA} + 2\ \text{Fd (red)} + CO_2 \quad (1.6)$$

$$2H^+ + \text{Fd(red)} \rightarrow H_2 + \text{Fd(ox)} \quad (1.7)$$

Stoichiometry shows that 4 mol of hydrogen can be produced per mol of glucose if acetate is the sole end product of pyruvate oxidation, whereas if butyrate is the sole end product then only 2 mol of H_2 could be produced per mole of glucose. The overall biochemical reaction with acetic acid and butyric acid as the metabolic end products is shown in Eqs. 1.8 and 1.9, respectively.

$$C_6H_{12}O_6 + 2H_2O \rightarrow 2CH_3COOH + 2CO_2 + 4H_2 \quad (1.8)$$

$$C_6H_{12}O_6 \rightarrow CH_3CH_2CH_2COOH + 2CO_2 + 2H_2 \quad (1.9)$$

The facultative anaerobic bacteria such as *Escherichia coli* and *Enterobacter* sp. follow a different pathway for hydrogen production. It involves the formation of acetyl-CoA and formate from oxidation of pyruvate. This reaction is catalyzed by pyruvate formate lyase (PFL) (Eq. 1.10):

$$\text{Pyruvate} + \text{CoA} \rightarrow \text{Acetyl-CoA} + \text{Formate} \quad (1.10)$$

The formate is then further cleaved to produce carbon dioxide and hydrogen. This reaction is catalyzed by the formate hydrogen lyase (FHL) enzyme (Eq. 1.11).

$$HCOOH \rightarrow CO_2 + H_2 \quad (1.11)$$

For commercial production of biohydrogen, cheap feedstock/raw material should be used. Most of the studies on biohydrogen are based on utilization of simple sugars such as glucose, sucrose, maltose, and lactose. These simple sugars are expensive, and usages of such raw material are not economically viable. To address this issue, production of biological hydrogen using different organic waste resources as substrate is a cheap and promising approach. There is a relatively high abundance of complex sugars (polysaccharides) in nature. Most of these polymeric sugars (cellulose, hemicellulose, amylase, etc.) are inaccessible to microorganisms. To tap the energy bound in these polymeric sugars, detailed research is required targeting the pretreatment and saccharification techniques. As mentioned earlier, biohydrogen could be considered renewable and cheap when its production is based on low-value and renewable resources.

In terms of large-scale or the precommercial stage of biohydrogen production, few reports are available. The biggest scale of bioreactors reported for hydrogen production is of 100 m^3 volume (Vatsala et al., 2008). It showed the feasibility of using a distillery effluent in dark-fermentative hydrogen production on a large scale. The maximum hydrogen yield of 2.76 mol per mol of glucose was observed, which is still less than the theoretical maximum. About 800 L mesophilic continuous hydrogen production in a packed bed reactor was demonstrated at the Indian Institute of Technology, Kharagpur, India (Fig. 1.6). A decentralized energy solution could be possible if the technological challenges related to scaling-up process and storage are overcome in the near future.

Figure 1.6 Commissioned 800 L bioreactor for biohydrogen production at the Indian Institute of Technology, Kharagpur, India.

1.5.6 Biomethane

Biomethane production using anaerobic digestion of solid organic wastes such as biowaste, sludge, cattle manure, energy crops, and other biomass is a well-established technology. The advantage of this technology is that it can utilize a wide range of feedstock. Methane formation from water-saturated decaying organic plant materials was first time reported by Volta. The anaerobic degradation process is a complex system, which is an outcome of a dynamic microbial activity, along with the involvement of biochemical and physicochemical factors. In this process the complex high-molecular-weight carbohydrates, fats, and/or proteins are hydrolyzed by the enzymatic action of microbes. These complex organic fractions of the wastes are fermented by acidogenic bacteria to form volatile fatty acids (VFAs), hydrogen, CO_2, etc. VFAs are suitable substrates for methanogenic bacteria. They convert these metabolites to CO_2 and CH_4. The synergy between acidogenic and methanogenic microbes plays a critical role toward the stability of the process.

1.5.7 Nonbiological Gasification Technologies for Organic Waste

1.5.7.1 Hydrothermal gasification

By definition, a hydrothermal process involves an aqueous system which is subjected to high temperature and pressure. Since the pressure remains above the saturation pressure at the respective temperature, the liquid water phase remains predominant. Hydrothermal gasification of biomass and other organic matter has many advantages over steam reforming. On manipulating the process parameters, methane or hydrogen-rich gas can be produced in this process. Another big advantage is the efficient removal of tar from the system. Tar was solubilized in water under high temperature and pressure as at this condition water behaves as a nonpolar solvent. Hydrothermal gasification can be applied for gasifying a variety of wet biomass such as manure and sewage sludge (biosolids). In contrast to anaerobic digestion, higher biomass utilization has been reported. Recalcitrants such as lignin are also gasified under hydrothermal conditions. This is an emerging field of biofuel generation, and it would need much more effort to commercialize the process.

1.5.7.2 Syngas by plasma pyrolysis

In plasma gasification, an ionized gas at high temperature can be used to catalyze the conversion of organic biomass to synthetic gas and solid waste (slag). It is used commercially as a form of waste treatment and has been tested for the gasification of biomass and solid hydrocarbons, such as coal, oil sands, and oil shale. At high temperature, the organic wastes get vaporized to the gaseous phase. The feedstock for plasma pyrolysis is MSW, organic wastes, biomedical waste, and hazardous materials (hazmat). The presence of inorganic material such as metal and construction waste leads to slag formation. This decreases syngas production. Slag is a chemically inert material. It can be used for recovery of metals. The smaller particle size improves the efficiency of plasma pyrolysis. The chemical compositions of syngas are carbon monoxide (CO), hydrogen, and CH_4. The gasification efficiency of plasma pyrolysis is around 99%. Moreover, this process is ecologically clean. As there is incomplete combustion during pyrolysis, many toxic materials are not produced. Toxic compounds such as furans, dioxins, nitrogen oxides, or sulfur dioxide are generally formed during proper combustion of organic wastes. The major disadvantages of plasma technologies are the requirement of large initial investment costs when compared with landfill reduction in sampler orifice diameter during prolonged usage of plasma flame and requirement of regular maintenance.

1.5.7.3 Energy generation by incineration

The most commonly used waste disposal method is combustion/incineration. The heat generated during this process can be tapped for electricity generation. The technology involved in harnessing biomass-based thermal power is that most of the components of the plant have commonality with conventional coal-based thermal power plant. The only difference is the boiler. The overall efficiency of the process is about 23%–25%. The steam turbine thus generated could be used to produce power or could be used partly or fully for another useful heating activity (cogeneration). India has as many as 288 biomass power and cogeneration projects. The total power generation of 2665 MW capacity has already been installed in the country.

1.6 Comparative Studies with the Conventional Process

Among the above-mentioned processes, biofuel generation from organic waste is a less energy-intensive process. Particularly biological hydrogen production showed better potential of commercialization compared to other biofuels. Table 1.2 shows the comparison in terms of cost of production of fuel on using different technologies. Need of pretreatment and saccharification increases the production cost of biofuel from lignocellulosic wastes. Moreover, substrate conversion efficiency and rate of hydrogen production are quite low compared to the conventional process. The cost of electricity generated from a biomass-based thermal power plant is still not comparable with that from a coal-based thermal power plant. Disposal of ash and emission of GHGs are issues that are concomitant with such technologies.

1.7 Biohythane: Potential as a Future Fuel

1.7.1 History of Hythane as a Fuel

Hythane® (a combination of hydrogen and methane) was developed in the laboratory of a company called Hydrogen Components, Inc. (HCI). HCI was founded by lead developer Frank Lynch, who focused on the development of an advanced technology for operation of hydrogen engines and also ventured in hydrogen storage projects by metal hydride. Apprehending the scope of using hydrogen as a future fuel and to bring it to the common people, HCI worked to create a blend of hydrogen and natural gas that would yield higher combustion efficiency and drastic reduction in emission of GHGs, while keeping such blending cost-effective. Thus, the term "Hythane®" was conceptualized.

Hythane® is a mixture of natural gas and hydrogen, usually 5%–7% hydrogen by energy. Natural gas is generally about 90% methane, along with small amounts of ethane, propane, higher hydrocarbons, and "inert" gases like carbon dioxide or nitrogen.

The combination of hydrogen and methane as a vehicular fuel proposes many advantages (Moreno et al., 2012):

- Lower flammability of methane limits its fuel efficiency. Addition of hydrogen could improve the lean flammability range significantly.
- In lean air/fuel mixtures, the flame speed of methane is low, whereas hydrogen has eightfold more flame speed.
- Hydrogen is a powerful combustion stimulant for accelerating methane combustion within an engine, and hydrogen is also a powerful reducing agent for efficient catalysis at lower exhaust temperatures.

Hydrogen production through dark fermentation has certain limitations. Gaseous energy recovery in terms of only hydrogen might not be sufficient to make this process commercially viable. Only 20%–30% of total energy can be recovered through hydrogen production. Despite integration with photofermentation, theoretically, 12 mol hydrogen can be recovered per mol of glucose, but due to the scaling-up problem of photofermentation such a two-stage process cannot be commercialized. To make dark-fermentative hydrogen production worthy of commercialization, it is necessary to integrate it with the biomethanation process. The spent media of the dark fermentation is rich in VFAs that would be an ideal substrate for acidogenic methanogens. Biomethanation technologies are well established and are easy to scale up. The integrated process leads to 50%–60% gaseous energy recovery (Fig. 1.7). The simplicity of the reactor design would lead to decrease in the operational cost of the entire process. Biohythane production could be envisioned as a renewable source of energy only when it would be produced from renewable sources. Any organic compound which is rich in carbohydrates, fats, and proteins could be considered as a raw material for biohythane production.

The advent of fuel cell technology, which converts hydrogen to electricity, has infused new life in the implementation of a hydrogen-based economy. The path of a hydrogen economy would be realized through the implementation of a fuel cell system with biohydrogen production systems. Till now few steps have been taken on the demonstration of integration of biohydrogen production with fuel cells. It would be interesting to see the performance of continuous biohydrogen production when connected to fuel cells. The biohydrogen setup should be put strategically near to those places where supply of feedstock is cheap and easily available. The electricity generated by such a process could be helpful for rural electrification. The development of such a process would lead to decentralized use of hydrogen. Resourceful tapping of energy from feedstock can be achieved by the following if a high-energy carrier is produced from it, followed by electricity production (feedstock → energy carrier → electricity). Considering the above-mentioned fact, biohydrogen production proposes to be more energy efficient compared to biomethane production. Renewable biomethane in the second stage, however, would be advantageous as it could be channelized to the existing infrastructure of CNG. Under the biohythane concept second-stage biomethane could be used separately as a fuel or could be mixed with biohydrogen in a certain ratio to make it suitable for IC engines. The production of methane or any other hydrocarbon as a major catabolic product is unique to this group of microbes, which share many other characteristics that are not common among other microbes. Biomethane is produced by a wide range of microorganisms. This includes obligate anaerobic microbes grouped according to their choice of substrate for methane production, as mentioned below:

- Hydrogen-consuming capnophilic methanogens
- Methylotrophic methanogens
- Acetoclastic methanogens

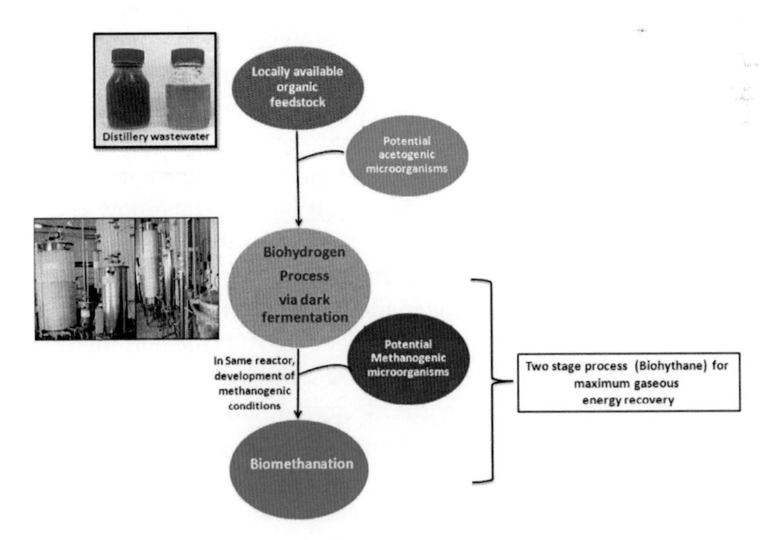

Figure 1.7 Biohythane concept for maximum gaseous energy recovery.

Methylotrophic methanogens are those microorganisms which would consume methanol to produce methane. The prospect of having methylotrophic methanogens in the second stage of biohythane production would be the least. This is because during dark-fermentative hydrogen production, methanol production is rarely reported.

1.7.2 Integration Challenges and Future

Fermentative hydrogen production and its feasibility as a clean fuel have been explored largely at the laboratory scale. In most of the cases continuous stirred tank reactors (CSTRs) are used (Hawkes et al., 2007). A variety of substrates have been used as feedstock for continuous hydrogen production in CSTRs. This feedstock includes domestic wastes, agricultural wastes, food wastes, animal wastes, pig slurry, and swine manure (Shin and Youn 2005; Kotsopoulos 2009; Zhu et al., 2009).

The advent of different bioreactor configurations created a new hope toward efficient biohydrogen production (Fig. 1.8). Use of an anaerobic sludge blanket reactor (ASBR) has shown promising results compared to CSTRs. In both cases, food waste was used as feedstock (Kim et al., 2008). When it comes to the use of agricultural wastes, which have high solid content, bioreactor configurations such as CSTRs and ASBRs don't seem to perform properly (Li and Fang 2007). The instability of the above-mentioned systems toward highly variable feedstock composition and metabolic shift due to change in the microbial population dynamics are some of the contrasting problems. A few novel ideas regarding reactor designing have enabled researchers to overcome the above-mentioned problems. One such report is where an inclined plug-flow reactor which is cylindrical in shape has been kept at a 20° angle to the base. This facilitates infusion of the waste inside the reactor. A specialized screw-type arrangement was made inside the reactor. This design helped in infusing the feed from the feed tank (at the bottom) to the outlet (at top). These screw-type arrangements also had 14 leads that helped in maintaining the desired retention time (7 days) (Jayalakshmi et al., 2009). Many researchers are proponents of batch fermentation in the start-up phase so as to enhance stable granule formation and consequently enhanced seed source activity (Chou et al., 2008).

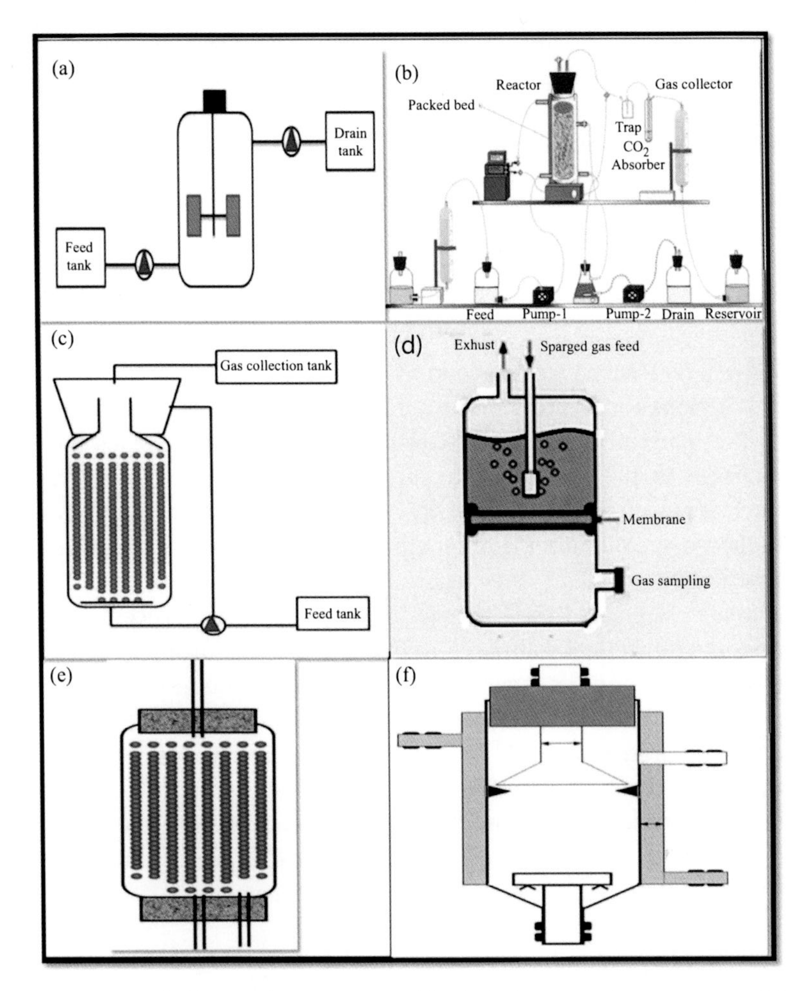

Figure 1.8 Different bioreactors used in dark fermentation processes: (a) continuous stirred tank reactor (CSTR), (b) packed-bed reactor, (c) fluidized-bed reactor, (d) membrane bioreactor, (e) trickling filter-based reactor, and (f) upflow anaerobic sludge blanket (USAB) reactor.

For improved gaseous energy extraction, second-stage biomethanation has been advocated for a long time (Venetsaneas et al., 2009). Thus integration of two different types of metabolically different microbial systems poses many challenges. Such a two-phase anaerobic digestion system was first proposed by Pohland and Ghosh in 1971 (Pohland et al., 1971). As acidogens are fast-growing microbes, they require generally short hydraulic retention times

(HRTs), acidic pH, and reduced partial pressure systems, whereas methanogens require alkaline pH (7–7.8) and long HRTs for the biomethanation process. Few reports are available where successful integration of reactors for hydrogen and methane production has been studied where food waste has been used as feedstock (Chu et al., 2008).

1.8 Conclusion

A fossil fuel–based world economy has come to a flash point where it is looking for a carbon-neutral fuel to power future civilizations. In the quest of harnessing clean energy, many technologies have emerged in the recent times, such as wind energy, solar power, nuclear power, geothermal energy, small hydroelectric plants, biogas, biohydrogen, biodiesel, and bioethanol. Each of these technologies has unique pros and cons regarding commercial operations. This chapter enunciated the prospect of the biohythane process for a possible future fuel compared to other biofuels. It also highlighted the challenges and prospects of using biohythane under the already existing CNG infrastructure. Moreover, biomethane could be used as a fuel separately or could be mixed with biohydrogen in a certain ratio to make it suitable for IC engines. The intricate relationship between acidogens and methanogens plays a crucial role in the operation of such two-stage systems. A detailed understanding of the same has been provided in subsequent chapters.

References

Barbier E (1997). Nature and technology of geothermal energy: a review, *Renew Sust Energy Rev*, **1**(1), 1–69.

Behar O, Khellaf A, Mohammedi K (2013). A review of studies on central receiver solar thermal power plants, *Renew Sust Energ Rev*, **23**, 12–39.

Chou C, Wang C, Huang C, Lay J (2008). Pilot study of the influence of stirring and pH on anaerobes converting high-solid organic wastes to hydrogen, *Int J Hydrogen Energy*, **33**(5), 1550–1558.

Chu CF, Li YY, Xu KQ, Ebie Y, Inamori Y, Kong HN (2008). A pH-and temperature-phased two-stage process for hydrogen and methane production from food waste, *Int J Hydrogen Energy*, **33**(18), 4739–4746.

Dickson MH, Fanelli M (eds) (2003). *Geothermal Energy: Utilization and Technology*, UNESCO renewable energy series, Earthscan, London, p. 205.

European Hydrogen and Fuel Cell Technology Platform (2005). *Deployment Strategy*. https://www.hfpeurope.org/hfp/keydocs.

European research on concentrated solar thermal energy (2004). Directorate-general for research sustainable energy systems, European Union (EU).

Gorlov AM (2001). *Tidal Energy*, Academic Press, Dan Diego, USA, 2955–2960.

Gross R, Leach M, Bauen A (2003). Progress in renewable energy, *Environ Int*, **29**(1), 105–122.

Haas, R, et al. (2001). *Review Report on Promotion Strategies for Electricity from Renewable Energy Sources in EU Countries.*

Hawkes FR, Hussy I, Kyazze G, Dinsdale R, Hawkes DL (2007). Continuous dark fermentative hydrogen production by mesophilic microflora: principles and progress, *Int J Hydrogen Energy*, **32**(2), 172–184.

Herbert GJ, Iniyan S, Sreevalsan E, Rajapandian S (2007). A review of wind energy technologies, *Renew Sust Energy Rev*, **11**(6), 1117–1145.

Jayalakshmi S, Joseph K, Sukumaran V (2009). Biohydrogen generation from kitchen waste in an inclined plug flow reactor, *Int J Hydrogen Energy*, **34**(21), 8854–8858.

Johansson TB, Kelly H, Reddy AKN, Williams RH (eds) (1993). *Renewable Energy, Sources for Fuels and Electricity*, Earthscan/Island Press, London.

Kim JK, Han GH, Oh BR, Chun YN, Eom CY, Kim SW (2008). Volumetric scale-up of a three stage fermentation system for food waste treatment, *Bioresour Technol*, **99**(10), 4394–4399.

Kotsopoulos TA (2009). Biohydrogen production from pig slurry in a CSTR reactor system with mixed cultures under hyper-thermophilic temperature (700°C), *Biomass Bioenergy*, **33**(9), 1168–1174.

Li C, Fang HHP (2007). Fermentative hydrogen production from wastewater and solid wastes by mixed cultures, *Crit Rev Environ Sci Technol*, **37**(1), 1–39.

Lund JW, Freeston DH, Boyd TL (2005). Direct application of geothermal energy worldwide review, *Geothermics*, **34**(6), 691–727.

McKay, D (2002). *Energy Needs Choices and Possibilities-Scenarios to 2050*, Shell International, *VDI Berichte*, **1734**, 1–20.

Moreno F, Muñoz M, Arroyo J, Magén O, Monné C, Suelves I (2012). Efficiency and emissions in a vehicle spark ignition engine fueled with hydrogen and methane blends, *Int J Hydrogen Energy*, **37**, 11495–11503.

Nayak BK, Pandit S, Das D (2013). Biohydrogen. In: Kennes C, Veigaría C (eds), *Air Pollution Prevention and Control*, John Wiley & Sons, Ltd., 345–381.

Ong HC, Mahlia TMI, Masjuki HH (2011). A review on energy scenario and sustainable energy in Malaysia, *Renew Sust Energy Rev*, **15**(1), 639–647.

Pohland FG, Ghosh S (1971). Developments in anaerobic stabilization of organic wastes-the two-phase concept, *Environ Lett*, **1**(4), 255–266.

Reith JH, Wijffels RH, Barten H (2003). *Bio-Methane and Bio-Hydrogen: Status and Perspectives of Biological Methane and Hydrogen Production*, Dutch Biological Hydrogen Foundation, The Hague, The Netherlands.

Shell International (2001). *Energy Needs, Choices and Possibilities, Scenarios to 2050*, Shell International Ltd., London.

Shin H, Youn J (2005). Conversion of food waste into hydrogen by thermophilic acidogenesis, *Biodegradation*, **16**(1), 33–44.

UNDP/WEC (2000). World Energy Assessment: Energy and the Challenge of Sustainability, New York: UNDP.

UNEP (2001). *UNEP Figures Confirm 2010 Kyoto Targets Will Not Be Met.* http://www.edie.net/.

Vatsala TM, Mohan Raj S, Manimaran A (2008). A pilot-scale study of biohydrogen production from distillery effluent using defined bacterial co-culture, *Int J Hydrogen Energy*, **33**, 5404–5415.

Venetsaneas N, Antonopoulou G, Stamatelatou K, Kornaros M, Lyberatos G (2009). Using cheese whey for hydrogen and methane generation in a two-stage continuous process with alternative pH controlling approaches, *Bioresour Technol*, **100**(15), 3713–3717.

Wind Power Monthly (2001). *The Windicator. Various Issues.* http:// www.windpowermonthly.com/.

Zhu J, Miller C, Li YC, Wu X (2009). Swine manure fermentation to produce biohydrogen, *Bioresour Technol*, **100**(22), 5472–5477.

Chapter 2

Microbiology of the Biohythane Production Process

2.1 Introduction

The biological route of hydrogen production contributes to approximately 1% of total hydrogen generation processes. It offers potentially lucrative research toward the generation of a carbon-neutral, sustainable energy system, which is virtually inexhaustible (Das, 2001). Biological hydrogen production requires ambient temperature and atmospheric pressure. Certain microorganisms control this. Moreover, a decentralized smaller-scale system could be developed for biological hydrogen production where suitable and cheap substrates are available so that transportation cost can be eliminated as well as unnecessary energy expenses can be reduced. Biological routes of hydrogen production are water biophotolysis, photofermentation, dark fermentation, and a combination of dark fermentation and photofermentation in a hybridized metabolic system. Photocatalytic splitting of water molecules is the primary method of hydrogen production in photosynthetic algal species and cyanobacteria. Biophotolysis involves transfer of electrons from water to hydrogenase through the photosystems and is mediated by the electron carrier ferredoxin (Fd). Oxygen generated in the process inhibits the activity of hydrogenases greatly (Benemann et al., 1973) which affect hydrogen production.

Biohythane: Fuel for the Future
Debabrata Das and Shantonu Roy

ISBN 978-981-4745-29-1 (Hardcover), 978-981-4745-30-7 (eBook)
www.panstanford.com

Various genera of green algae are known to produce hydrogen via indirect photolysis of water under anaerobic conditions. In addition to hydrogenase, biophotolysis is also carried out by another hydrogen-evolving enzyme in cyanobacteria and blue-green algae (Smith et al., 1992). The rate of hydrogen production by algae is extremely low. In the case of *Chlamydomonas reinhardtii*, the maximum cumulative hydrogen production of 3.1 ± 0.3 mL L^{-1} was observed (Tamburic et al., 2011). A low hydrogen production rate is one of the major bottlenecks for large-scale hydrogen production from photolysis of water. Photofermentation by phototrophic bacteria has shown great potential for the biohydrogen production system compared to algal biophotolysis of water (Fascetti, 1998). The purple nonsulfur bacteria use organic acids as electron donors and produce hydrogen by absorbing light (Tsygankov, 2001). The advantages of photofermentation over biophotolysis are high theoretical conversion yields and no oxygen evolution. Therefore, no hydrogenase or nitrogenase deactivation takes place. The major bottleneck for photofermentation is the shedding effect of the pigments produced by the purple nonsulfur bacteria during photofermentation as accumulation of pigments leads to poor light penetration (Gilbert et al., 2011). The emergence of the dark-fermentative hydrogen production process has led the research toward making the biohydrogen production process commercially viable. There are many advantages of dark-fermentative hydrogen production, viz., independent of light energy, high rate of hydrogen production compared to other biological routes, and less energy intensive. The volatile fatty acids produced during dark fermentation could be an ideal feedstock for acetoclastic methanogens. Integrating biomethanation with dark-fermentative hydrogen production might help in the improvement of total gaseous energy recovery.

2.2 Hydrogen Production by Dark Fermentation

Several microbial strains are known to help produce hydrogen through anaerobic fermentation of different carbohydrates. Substrates other than carbohydrates are less promising for fermentative hydrogen production. Some amino acids can only produce hydrogen through fermentation, while at very low hydrogen partial pressure, lipids could be possibly used for hydrogen production. In dark-fermentative studies either a pure substrate (commonly glucose) or carbohydrate-rich waste has been used for hydrogen production (Abo-Hashesh et al., 2011). Wastes are always the preferred substrate for biohydrogen production, but these require presence of omnivorous heterotrophic organisms or a consortium of microorganisms with a wide range of catabolic activities.

In a glycolytic pathway, ATP and NADH are synthesized by metabolic oxidation of carbohydrates to pyruvate. Further metabolism of pyruvate depends on the microorganisms and environmental conditions in which the microorganisms are surviving. Under anaerobic conditions, different arrays of enzymes participate in conversion of pyruvate to hydrogen. Theoretically, 1 mol of hydrogen can be produced per mol of pyruvate. In facultative anaerobes, pyruvate is converted to formate and acetyl coenzyme A (acetyl-CoA). The formate can be further metabolized to yield hydrogen and CO_2 through an orchestrated interplay of different membrane-associated hydrogenases (Hallenbeck et al., 2012). Alternatively, in obligate anaerobes, pyruvate immediately oxidizes into acetyl-CoA, giving CO_2 and reduced ferredoxin. The reduced ferredoxin can drive hydrogen production by channelizing electrons to Fe-Fe hydrogenases. In both cases, a variety of metabolic end products such as ethanol, acetate, butanol, butyrate, and acetone are produced from acetyl-CoA. The redox state of the substrate to regenerate NAD^+ is essential for cellular metabolism that exhausted during the glycolytic pathway.

The amount of NADH generated during glycolysis is a function of the oxidation state of the substrate. As a single molecule of glucose catabolizes to yield 2 mol of pyruvate, theoretically 2 mol of hydrogen per mol of glucose can be produced. Moreover, NADH formed during oxidation of glyceraldehyde-3-phosphate (G-3-P) can also contribute toward 2 mol of hydrogen per mol of glucose consumed. Since the equilibrium midpoint potential ($E^{o\prime}$) of the NAD^+/NADH couple is −320 mV but that of the H^+/H_2 couple is −420 mV, the generation of hydrogen from NADH is a thermodynamically unfavorable reaction. Therefore, under standard equilibrium conditions (1 atm. H_2), energy must be provided into the system to make the hydrogen production process thermodynamically feasible ($\Delta G^{o\prime} = -nF\Delta E^{o\prime} = +19.3$ kJ/mol). Varieties of hydrogenases are present in microorganisms that are dependent on NADH for hydrogen production, for example, fermicutes, thermophiles, and extremophiles (Fig. 2.1). In general, facultative anaerobes like *Escherichia coli, Klebsiella pneumonia, Citrobacter, Enterobacter aerogenes,* and *Bacillus couagulans* do not possess such type of NADH-dependent hydrogenase. This restricts their hydrogen production potential to a maximum of 2 mol of hydrogen per mol of glucose. If all the excess NADH is converted to hydrogen, the organisms that contain the relevant pathways would be able to produce 4 mol of hydrogen per mol of glucose. However, as mentioned earlier, this is unrealistic thermodynamically. There are two hypotheses of electron transfer from NADH to protons: (a) NADH directly reducing a specific hydrogenase and (b) NADH might shuttle electrons to ferredoxin via NADH ferredoxin oxidoreducatse (Schut and Adams, 2009). The molecular details of this unique energy coupling mechanism are not presently known, and the actual energetic will be determined by the prevailing hydrogen partial pressures and cellular concentrations of NAD^+, NADH, Fd_{ox}, and Fd_{red}.

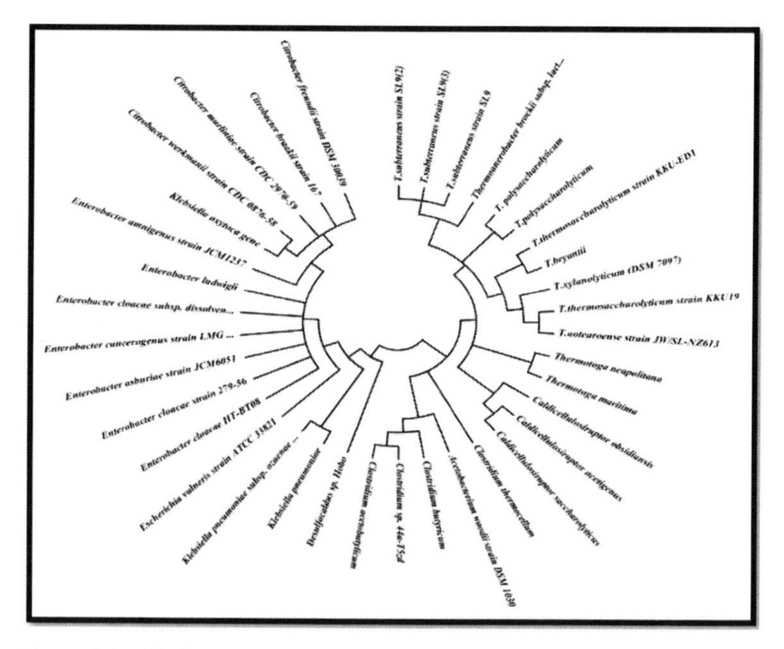

Figure 2.1 Phylogenetic representation of various hydrogen-producing microorganisms. (*Desulfocaldus* sp. Hobo was taken as an outgroup.)

Recently, much advancement has been made in dark-fermentative hydrogen production. Various types of immobilized systems were developed to achieve high volumetric rates of hydrogen production. Other advances emerged through application of metabolic engineering so that microbes can be manipulated to maximize the yield predicted from the metabolic path.

2.2.1 Hydrogen-Producing Microorganisms

Very few microorganisms are known for their ability of converting biomass to hydrogen. Hydrogen-producing microbes belong to different domains such as facultative anaerobes, obligate anaerobes, methylotrophs, and photosynthetic bacteria (Nandi and Sengupta, 2008). However, fermentation processes contributes toward hydrogen which is present in the biosphere. On the basis of the hydrogen-producing mechanism, microbes are broadly categorized as either dark-fermentative or photofermentative microorganisms.

Dark-fermentative hydrogen production is more advantageous due to its relatively low cost, low energy demands, moderate operative conditions, and minimal pollution generation (Angenent et al., 2004). Other advantages of dark fermentation over photofermentation are rapid bacterial growth rates, relatively high hydrogen production capacities, no dependency on light sources, no oxygen limitation problems, and low capital costs (Das, 2001). Therefore, dark-fermentative hydrogen production has attracted attention of researches. Among a large number of microbial species, strict anaerobes and facultative anaerobic chemoheterotrophs are efficient hydrogen producers. The reduction of protons to hydrogen in microorganisms has evolved to dissipate excess electrons within the cells and to allow additional energy steps in metabolism.

2.2.1.1 *Clostridium*

The genus *Clostridium* belongs to the low G+C gram-positive group of bacteria. They are rod-shaped, fermentative, spore-forming, obligate anaerobes. Their size varies from 0.3–2.0 × 1.5–20 μm. They have a lower doubling time and can persevere in unfavorable conditions, where the other anaerobic bacteria fail to survive. This makes the organisms potent for industrial applications. *Clostridium* sp. came into picture the first time during the First World War when *Clostridium* was used in fermentation for solvent and alcohol production (Weizmann and Rosenfeld, 1937). Hydrogen is a common by-product in such fermentation processes.

Obligate anaerobic clostridia are potential hydrogen producers and well known for higher hydrogen yield (Kamalaskar et al., 2010; Valdez-Vazquez and Poggi-Varaldo, 2009). *C. butyricum, C. welchii, C. pasteurianum,* and *C. beijerinckii* are newly isolated *Clostridium* sp. that were used individually as well as in synthetic-mixed consortia for hydrogen production. The *C. beijerinckii* AM21B was isolated from the termite gut and showed the highest hydrogen yield of 1.8 to 2.0 mol per mol of glucose (Taguchi et al., 1996). This strain is capable of utilizing a wide range of other carbohydrates, such as xylose, arabinose, galactose, cellobiose, sucrose, and fructose. Another *Clostridium* sp. (strain no. 2), also isolated from termites, produces hydrogen more efficiently from xylose and arabinose (13.7 and 14.6 mmol/g or 2.l and 2.2 mol/mol) compared to glucose (11.1 mmol/g or 2.0 mol/mo1) (Taguchi et al., 2011). These results suggest that both *Clostridium* sp. can be used for hydrogen production using cellulose and hemicellulose as substrates present in plant biomass. Hydrolysis of biomass for the production of a fermentable substrate can be done along with fermentation, where saccharification and fermentation occur simultaneously or in a separate saccharification process preceded by fermentation. Thus, the ability of clostridia to produce hydrogen looks very promising.

2.2.1.2 *Enterobacter*

The *Enterobacter* genus belongs to the class Gamma Proteobacteria. They are generally gram-negative, rod-shaped, motile (peritrichous flagellated) or nonmotile, facultative anaerobes. Their size varies from 0.3–1.0 × 1–6 μm. They have high growth rates, are capable of utilization of a wide range of carbon sources, and are resistant to lower traces of dissolved oxygen, and hydrogen production is not inhibited by high hydrogen pressures (Tanisho et al., 1987). However, the yield of hydrogen was lesser in *Enterobacter* sp. compared to *Clostridium* sp. when glucose was used as a substrate. Under batch fermentation, a hydrogen yield of l.0 mol per mol of glucose and a production rate of 21 mmol L^{-1} h^{-1} was observed (Tanisho, 1998). In a continuous hydrogen production process using a continuous stirred tank reactor

(CSTR), hydrogen production was monitored for 42 days using the same strain and considering molasses as the substrate. The highest hydrogen production rate and yield of 17 mmol L^{-1} h^{-1} and 1.5 mol/mol was observed, respectively. In contrast to batch fermentation, the metabolic end product showed prominence of lactate, whereas butyrate and acetic acid were produced in lower amounts (Tanisho and Ishiwata, 1994). It was also observed that flushing the culture medium with argon enhanced the hydrogen yield to 1.6 mol per mol of glucose. To enhance hydrogen production rates, mutants of *E. aerogenes* and *E. cloacae* were developed. In these mutants, the production of metabolites such as alcohols and organic acids was blocked. Production of these metabolites competes for reductants (NADH) which are also required for hydrogen production. A double-mutant strain of *E. aerogenes* was developed, which showed lower production of ethanol and butanediol. This lead to twofold improvement of hydrogen yield compared to the wild type (Rachman et al., 1998).

The *E. cloacae* IIT-BT 08 (presently known as *Klebsiella pneumonia* IIT-BT 08) strain was isolated from a leaf extract. These bacteria produce hydrogen using a wide ranges of substrates. In batch fermentation, the maximum hydrogen yield of 2.2 mol per mol of glucose was observed (Kumar and Das, 2000). The maximum hydrogen production rate measured was 35 mmol L^{-1} h^{-1} using sucrose as a substrate. A double-mutant strain of *E. cloacae* IIT-BT 08 was also developed for improvement of hydrogen production (Kumar et al., 2001). In batch fermentation, it showed 1.5 times increased hydrogen yield on glucose, that is, 3.4 mol of hydrogen per mol of glucose.

In continuous hydrogen production, the *E. aerogenes* double mutant (Rachman et al., 1998) was reported to give the maximum hydrogen production rate of 58 mmol L^{-1} h^{-1} at a dilution rate of 0.67 h^{-1}, which was nearly two times higher compared to the wild type. The whole-cell immobilized system was further investigated in a packed-bed reactor. The column packed with spongy material having immobilized whole cells (*E. aerogenes*) produced hydrogen on a starch hydrolysate. A maximum hydrogen yield of 1.5 mol per mol of glucose was observed at a dilution rate of 0.1 h^{-1} (Palazzi et al., 2000).

2.2.1.3 *Escherichia coli*

Escherichia coli also belongs to the class Gammaproteobacteria. *E. coli* are generally gram-negative, rod shaped, and motile. These strains are capable of producing hydrogen and CO_2 from formate in the absence of oxygen (Stickland, 1929). The principal enzyme complex responsible for hydrogen production is the formate lyase (FHL) enzyme system. FHL is a membrane-bound multienzyme complex. It consists of two subunits, viz., a formate dehydrogenase and a hydrogenase. However, inconsistency exists on the pathway leading to hydrogen production, either via formate or without formate as an intermediate. Production of hydrogen from formate and glucose by immobilized *E. coli* showed 100% and 60% efficiency, respectively. The hydrogen yield using *E. coli* was 0.9–1.5 mol per mol of glucose (Blackwood et al., 1956). On using immobilized whole cells, higher hydrogen yields of 1.2–1.5 mol of hydrogen per mol of glucose were observed.

2.2.1.4 *Citrobacter*

Citrobacter are known to produce hydrogen both chemoliho-trophically and organotrophically. They also belong to the gamma proteobacteria group. They are generally facultative anaerobic, gram-negative bacilli of the *Enterobacteriaceae* family. The *Citrobacter* sp.Y19 isolated from sludge digesters could produce hydrogen from CO and H_2O by the water-gas shift reaction under anaerobic conditions (Jung, 2002). Hydrogen production studies were carried out in serum bottles, and the highest hydrogen production rate of 15 mmol L^{-1} h^{-1} was observed. The chemoorganotrophic mode of hydrogen production was reported in *C. freundii*. It used cane molasses as a substrate to produce hydrogen in the conventional fermentative pathway (Vatsala, 1992).

2.2.1.5 *Bacillus*

Different strains of the genus *Bacillus* have been identified as potential hydrogen formers. They are generally g, facultative, motile, and mesophilic bacteria. The optimal temperature for their growth is around 30°C. However, many strains can withstand exposure to much higher temperatures. Strains of *Bacillus* are known to secrete useful enzymes. Under harsh conditions, these can form spores.

B. licheniformis isolated from cattle dung was reported to produce hydrogen (Kalia and Purohit, 2008). Its hydrogen yield was, though, inferior (0.5 mol per mol of glucose) compared to *Clostridium* sp. It generally follows the lactic acid pathway. The immobilized whole-cell system of *B. licheniformis* showed a better hydrogen yield of 1.5 mol per mol of glucose (Kalia et al., 1994). Another species of a potential hydrogen-forming organism belonging to this genus was identified as *B. coagulans* (isolated from sewage sludge) and showed a higher hydrogen yield (2.2 mol per mol of glucose) compared to *B. licheniformis* (Kotay and Das, 2007). From the industrial point of view, handling of facultative anaerobic hydrogen formers is relatively easy compared to strict anaerobes such as clostridia and methanogens.

2.2.2 Thermophilic Dark Fermentation

Dark fermentation at thermophilic temperatures showed favorable kinetics and stoichiometry of hydrogen production compared to the mesophilic system. It also reduces the risk of methanogenic and pathogenic contaminations. Metabolism at higher temperatures becomes thermodynamically more favorable and less affected by the partial pressure of hydrogen (pH_2) in the liquid phase. Many industrial effluents are discharged at high temperature, which have an inherent high organic content. Effluents from the distillery industry, sugar industry, food processing, etc., are often discharged at higher temperatures. These discharges need to be cooled down to ambient temperatures before they can be treated via anaerobic digestion. This process is not cost effective, and there is always a risk of losing the biological activity while cooling (Jo et al., 2008). Thermophilic hydrogen-producing bacteria can be used directly for the treatment of such effluents. Thermophiles can grow at higher temperatures as their cell membrane is rich in saturated fatty acids and they have vast repertoires of thermostable proteins (Zhang et al., 2003). Thermophilic bacteria can be classified into three groups on the basis of optimum temperature of growth. For example, moderate thermophiles have an optimum temperature of 45°C–55°C for their growth, true thermophiles have an optimum temperature of 55°C–75°C for growth, and extremophiles have an optimum temperature above 75°C. Many microbial species growing at thermophilic conditions, such as *Thermoanaerobacterium*, *Thermoanaerobacter*, clostridia, *Thermotoga*, and *Caldicellulosiruptor* sp., have been reported as hydrogen producers (Zeidan and Van Niel, 2009).

Moderate thermophiles were isolated from Iceland hot springs. These were grown at a temperature range of 50°C–60°C. Enrichment with hot spring sediments led to the domination of a few genera like *Thermoanaerobacter*, *Thermoanaerobacterium*, and *Clostridium* sp. (Brynjarsdottir et al., 2013). These genera were reported to produce hydrogen. Another group of bacteria was identified that were growing at extreme thermophilic temperature ranges such as 70°C and 75°C. Most of them belong to *Caldicellulosiruptor* and *Thermotoga* sp. In these genera, the metabolic end product was directed more toward hydrogen than ethanol (Vanniel, 2002). Some of the genera capable of producing hydrogen are discussed next.

2.2.2.1 Thermoanaerobacterium sp.

Two xylan-degrading bacteria were isolated from Frying Pan Springs in Yellowstone National Park in 1993, which was later identified as *Thermoanaerobacterium* sp. (Lee et al., 1993). Phylogenetically, these microorganisms have interrelationships with *Clostridium* sp. These are low G+C containing bacteria, having a straight rod shape, gram-negative, filamentous, motile, and peritrichous in nature. These form spores during nutritionally adverse conditions. Their metabolic end products are diversified, such as ethanol, acetate, CO_2, hydrogen, and lactate.

2.2.2.2 Thermoanaerobacter sp.

This genus was listed under the irregular, non-spore-forming, gram-positive rods in *Bergy's Manual* of *Systematic Bacteriology* (Wiegel and Ljungdahl, 1981). The genera *Thermoanaerobacter* and *Thermoanaerobium* included the first thermophilic, anaerobic bacteria that produce hydrogen along with ethanol and lactate as sugar fermentation products. These are obligate anaerobes, non-spore-forming (exception *Thermoanaerobacter finnii)* bacteria. These microorganisms can utilize a variety of sugars but cannot degrade cellulose. Hydrogen, ethanol, lactate, acetate, and CO_2 are the major products. No butyrate production was reported in these species. Some species have been reported to produce up to 4 mol per mol of glucose, which is the theoretical maximum hydrogen yield potential under nitrogen-flushed conditions (Soboh et al., 2004).

2.2.2.3 Clostridium sp.

This genus is the most studied one because of its potential biofuel production characteristics. The diverse species among the genus were found by molecular analysis. The genus *Clostridium* belongs to the family Clostridiaceae, order Clostridiales, class Clostridia, and phylum Firmicutes. They are rod-shaped, gram-positive, motile, often spore-forming, and obligate anaerobic organisms. They can degrade cellulose via their cellulase enzymes and can ferment the lignocellulosic biomass to hydrogen. With cellulose, the highest hydrogen yield of 1.6 mol per mol of hexose was reported (Levin et al., 2006). These microbes are particularly important in converting lignocellulosic biomass to biohydrogen.

2.2.2.4 Caldicellulosiruptor sp.

The hydrogen production ability of *Caldicellulosiruptor saccharolyticus* was explored at extreme temperatures. It was characterized within the *Bacillus/Clostridium* subphylum on the basis of their physiological characteristics and phylogenetic position (Rainey et al., 1994). These species are obligatory anaerobic, extremely thermophilic, and non-spore-forming, gram-positive bacteria. The natural habitats for such microbes are hot springs and lake sediments. These microbes are known to have a plethora of hydrolytic enzymes by which they can utilize a wide range of substrates like cellulose, cellobiose, xylan, and xylose. Because of their wide range of hydrolytic enzymes, these species have the potential to use lignocellulosic wastes for hydrogen production. The predominant metabolite formed by this organism is acetate and lactate. With paper pulp it shows the maximal volumetric hydrogen production rate of 5 to 6 mmol L^{-1} h^{-1} (Kádár et al., 2004).

2.2.2.5 Thermotoga sp.

The *Thermotoga* genus was first isolated from geothermal heated sea floors in Italy and Azores (Huber et al., 1986). These extremophilic bacteria are capable of growing and producing hydrogen at 90°C, the highest-reported temperature for hydrogen production. The genus *Thermotoga* derived its name due to the presence of the characteristic outer sheet-like structure called toga. These strains are rod-shaped, gram-positive, obligate anaerobes. The natural habitats of these strains have high temperature, pressure, and sulfur-containing environments. *Thermotoga* strains can use elemental sulfur or thiosulfate or both as their electron source. The end products of metabolism in such strains are mostly acetate, hydrogen, and CO_2, with trace amounts of ethanol. *T. maritima* and *T. neoplanita* were reported as having hydrogen-producing ability (Finkelstein et al., 2002; Huber et al., 1986).

2.2.3 Challenges of Cultivating Obligate Anaerobic Bacteria for Biohydrogen Production

There are many difficulties associated with cultivation of obligate anaerobes. The presence of dissolved oxygen in the fermentation media can cause death of the microorganisms. Even a little oxygen concentration of 10^{-56} mol L^{-1} proves lethal for the obligate anaerobic microbes growing in broth. Thus an anaerobic condition is a prerequisite in fermentation, and it can be maintained either by purging an inert gas like nitrogen or argon or by adding scavengers such as cysteine HCl or thioglycolate (Hungate, 1969). Dissolved oxygen present in wastewater could prove to be a major bottleneck for hydrogen production by obligate anaerobes. The redox potential of the medium also plays a critical role in the growth of the anaerobes. The redox potential of the medium should be in the order of −110 mV or above for cultivation of obligate anaerobes. The resazurin dye can be employed at very low concentrations (0.5 to 1 g L^{-1}) to observe the redox state of the fermentation media.

2.2.4 Co-Cultures and Mixed Cultures

Recently, hydrogen production using a mixed culture gained importance (Venkata Mohan, 2009). Continuous hydrogen production using a co-culture of *Clostridium butyricum* anxd *Enterobacter aerogenes* showed having higher hydrogen yield properties of the strict anaerobe and oxygen consumption by the facultative anaerobe (Yokoi et al., 1998). Such type of bioagumented system reduces the dependency on the need for oxygen scavengers in fermentation media. The presence of *E. aerogenes* rapidly restores anaerobic conditions by naturally scavenging the dissolved oxygen. Moreover, a mixed culture provides other advantages such as production of a "cocktail" of different hydrolytic enzymes required for utilization of complex substrates. A single isolate might not be able to produce all the enzymes needed for solubilization of complex substrates. Mixed microbial consortia can be developed from various sources, such as fermented soybean meal or sludge from anaerobic digesters of municipal sewage or organic waste and sludge from kitchen wastewater (Venkata Mohan et al., 2008). The mixed microflora often contains unwanted bacteria such as methanogens, hydrogen-oxidizing bacteria, etc. Therefore, a systematic enrichment process should be followed to create a selection pressure to selectively enrich hydrogen-forming bacteria. Techniques such as heat treatment, bromoethane sulfonate (BES) treatment, acid treatment, and alkali treatment were commonly used for the enrichment of mixed cultures (Goud and Mohan, 2012; Venkata Mohan et al., 2008). Heat treatment inhibits the activity of the hydrogen consumers, while the spore-forming, hydrogen-producing anaerobic bacteria survive. BES may be used as a methanogen inhibitor. It inactivates coenzyme M in methanogens and inhibits methane production. Additionally, operation at a lower hydraulic retention time (HRT) helps in washing out slow-growing methanogens and retains acidogenic hydrogen-forming microbes. Industrially, the use of mixed cultures for hydrogen production from organic waste might be advantageous over pure cultures. One of the studies showed that enriched anaerobic microflora utilized cellulose as a substrate for hydrogen production with a hydrogen yield of 2.4 mol per mol of hexose in batch experiments at 60°C (Ueno et al., 1995). Furthermore, stable hydrogen production was observed for 190 days using industrial wastewater of a sugar factory (Ueno et al., 1996). The highest hydrogen production rate of 1.4 mmol L^{-1} h^{-1} was observed at an HRT of 3 days. For efficient usage of a mixed culture, retention of hydrogen-producing microorganisms and understanding of cooperative interaction between different strains need to be explored in the near future.

2.3 Microbial Basis of Methanogenesis

Methane formation from water-saturated, decaying organic plant materials was first reported by Volta. Nearly after a century, the relation of methane production with microbes was established. Tappeiner provided more adequate proof of the microbiological origin of methane in 1882. Three identical anaerobic cultures were provided with organic parts of plants as the substrate and the intestinal content of ruminants as the seed culture. The rumen of herbivorous animals harbors cellulose-degrading organisms, but the role of such organisms toward methane production was not clear to animal physiologists. With the understanding of syntrophism among different anaerobic microorganisms, a clear view of the complex process of methane production was obtained.

2.3.1 Diversity and Taxonomy of Methanogens

The characteristic of methanogens is their ability to produce methane and other hydrocarbons, which sets them apart from other microbes. Methanogens belong to *Archaeobacteria*, which are distinguished from *Eubacteria* by virtue of many contrasting characteristics, such as the presence of isoprenoid-rich membrane lipids, which is linked with glycerol, and the absence of a mauramic acid–based peptidoglycan cell wall and distinct ribosomal RNA (Balch et al., 1979; Raskin et al., 1994). On the basis of metabolic characteristics, methanogens can be categorized into three groups: CO_2-reducing, methylotrophic, and aceticlastic pathways. The CO_2-reducing methanogens convert CO_2 or bicarbonate to methane, for which they require two electrons (Rouviere and Wolfe, 1988). Many methanogens use hydrogen as the sole source of electrons. In the natural habitat, the source of hydrogen may be geological eruption or hydrogen produced by other acidogenic microbes. Under anaerobic conditions, hydrogen is rapidly consumed by methanogens, so it never gets accumulated in the system. Thus, hydrogen plays an important role of an extracellular intermediate.

Formate serves as an electron donor for many hydrogenotrophic methanogens that convert CO_2 to CH_4. Similar to hydrogen, formate is also an important intermediate in methane production, even though its concentration remains low under a methanogenic environment

(Boone et al., 1989). Very few methanogens also can reduce CO_2 to CH_4 by oxidizing primary and secondary alcohols (Bleicher et al., 1989). On the other hand, methylotrophic methanogens can reduce CO_2 to CH_4 by oxidizing methyl group containing substrates such as methanol, trimethylamine, and dimethyl sulfide (Hippe et al., 1979; Mathrani and Boone, 1985). The methyl group is transferred to a methyl carrier (ultimately to coenzyme M) and reduced to methane.

2.3.2 Taxonomy of Methanogens

Initially, the species of methanogenic bacteria was classified along with nonmethanogens on the basis of morphological characteristics. The eighth edition of *Bergey's Manual* recognized the physiological unity of methanogens and brought them under a single group (Bryant and Boone, 1987). A detailed study of the diversity of methanogens was done using a ribotyping method, which includes cataloging and sequencing of 16S rRNA. Surprisingly, methanogens showed a phylogenetic relation with some extremehalophiles and extremely thermophilic, sulfur-dependent organisms which belonged to the kingdom *Archaeobacteria.* The methanogens and other *Archaeobacteria* were classified under a new taxonomic-level higher kingdom called urkingdom *Archaea* (Woese et al., 1990). Under the urkingdom *Archaea* the methanogens were included in the *Euryarchaeota.* The *Euryarchaeota* also includes extremehalopliles, *Thermoplasma,* and some nonmethanogenic, thermophilic extremophiles (Fig. 2.2).

2.3.3 Taxonomical Classification of Methanogens

Within the kingdom *Archaeobacteria,* methanogens can be classified into five orders. The orders *Methanobacteriales, Methanococcales,* and *Methanomicrobiales* were the three distinct methanogens described in *Bergey's Manual of Systematic Bacteriology* (Mah and Kuhn, 1984). The methylotrophic and aceticlastic methanogens were further separated from the order *Methanomicrobiales* and reordered under *Methanosarcinales* (Bélaich et al., 1990). Moreover, a new order *Methanopyrales* was discovered, which was phylogenetically distinct from all the known methanogens (Burggraf et al., 1991).

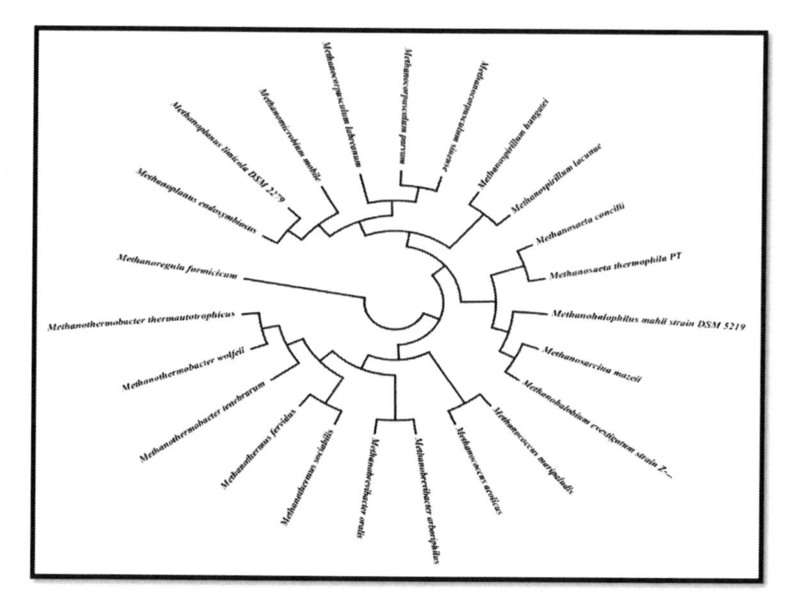

Figure 2.2 Phylogenetic representation of various methane-producing microorganisms.

2.3.3.1 Methanobacteriales

The order *Methanobacteriales* comprises two families, *Methanobacteriaceae* and *Methanothermaceae. Methanobacteriales* are mainly rod-shaped methanogens which use CO_2 as the energy source. The only exception to the above group is *Methanosphaera sp.*, which are cocci in shape and use hydrogen as the energy source to reduce methanol to methane. The *Methanobacteriales* strains are generally gram positive because they have pseudomurein cell walls.

The family *Methanobacteriaceae* is highly diverse. It includes several genera such as *Methanobacterium, Methanothermobacter, Methanobrevibacter,* and *Methanosphaera. Methanobacterium jormicicum* is the oldest described species of *Methanobacterium. M. bryantii* has some physiological and morphological differences from *M. jormicicum,* including the inability of the latter to catabolize formate. *Methanobrevibacter* sp. are commonly found in the gastrointestinal tract or feces of mammals. They are very short rods or cocco-bacilli in shape and have complex organic requirements. They use hydrogen or formate as the energy source to reduce CO_2

to CH_4. Another cocci-shaped member of *Methanobacteriaceae* is *Methanosphaera.* They are nonmotile and gram positive, occur singly or in small groups, and grow by using hydrogen to reduce methanol (CH_3OH) to methane. Their pseudomurien cell wall has serine. Two members of this genus are *Methanosphaera stadmaniae* and *Methanosphaera cuniculi* (Biavati et al., 1988).

Thermophilic *Methanobacteriales* have a separate genus of *Methanothermobacter* sp. These include *Methanothermobacter thermoautotrophicus* and *Methanothermobacter wollfi.* These organisms don't have the ability to utilize formate and depend on hydrogen and CO_2 as the energy source. On the other hand, *Methanobacterium thermoformicicum,* which also belongs to *Methanobacteriales,* is a thermophilic methanogen capable of utilizing formate.

Thermophilic methanogens belonging to order *Methanobacteriales* are classified as *Methanothermaceae.* The genus *Methanothermus* contains extremely thermophilic methanogens that grow at a temperature of 83°C–85°C. These are rod-shaped methanogens that grow on CO_2 and hydrogen (Lauerer et al., 1986).

2.3.3.2 Methanococcales

The order *Methanococcales* consists of coccoid-shaped, marine methanogens that include three thermophilic species (*Methanocaldococcus, Methanothermococcus,* and *Methanoignis*) and one mesophilic species (*Methanococcus*). They are halophilic, chemolithotrophic microorganisms that produce methane by reducing CO_2. They use hydrogen or formate as the energy source. Thermophilic species such as *Methanothermococcus thermolithotrophicus* were isolated from deep-sea hydrothermal vents and grown rapidly at 65°C. These are motile, gram-negative cocci, placed within the family *Methanococcaceae.* No muramic acid and glycoprotein are observed in their cell wall (Huber et al., 1982). *Methanocaldococcus jannaschii* is also a hyperthermophilic marine cocci that grows rapidly at 85°C and is the first methanogenic archaea whose complete genome was sequenced. It is also a barophilic microorganism and requires 200 atm pressure for growth. Another hydrogenotrophic, hyprethermophilic methanogenic microorganism was identified as *Methanoignis igneus.* The presence of di-myoinositol-1,1′-phosphate and β-glutamate in the membrane structure helps it survive in extreme conditions.

2.3.3.3 Methanomicrobiales

The order *Methanomicrobiales* consists of three families, viz., *Methanomicrobiaceae, Methanosarcinaceae,* and *Methanocorpusculaceae. Methanomicrobiales* sp. require acetate as a source of cell carbon. Many other species have additional complex nutritional requirements. This order also includes mesophilic or thermophilic microorganisms that are slightly halophilic. A protein layer (S-layer) is present in the cell wall of these organisms and they have a pleomorphic, irregular, coccoid shape. An external sheath is also present in some species such as *Methanospirillum hungateii.* The presence of the S-layer makes the microorganisms of these groups osmotically sensitive. A mild solution of detergents such as 2% v/v sodium dodecyl sulfate (SDS) or hypertonic solutions are detrimental for such microorganisms. *Methanospirillum hungateii* have a helical spiral shape covered by a sheath. The *Methanocorpusculaceae* are cocci-shaped, hydrogenotrophic methanogens (*Methanocorpusculum*). For reduction of CO_2 to methane various electron donors (such as hydrogen, formate, or alcohols) are utilized by *Methanocorpusculum* sp. *Methanolacinia paynteri* and *Methanomicrobium mobile* are representatives of their group. *Methanoplanus* comprises two species, *Methanoplanus limicola* and *Methanoplanus endosymbiosus.* These microorganisms were isolated from marshy swamps and are weakly motile with a plate-shaped, hexagonal cellular envelop structure. They are gram-negative acetoclastic microorganisms and utilize hydrogen and formate to produce methane (Wildgruber et al., 1982). The genus *Methanoculleus* contains four mesophilic species, viz., *Methanoculleus bourgensis, Methanoculleus olentangyi, Methanoculleus thermophilicus,* and *Methanoculleus marisnigri,* which use hydrogen /formate as methanogenic substrates (Asakawa, 2003).

2.3.3.4 Methanosarcinales

The *Methanosarcinales* genera can be classified as *Methanosarcina, Methanohalobium, Methanococcoides, Methanohalophilus, Methanosaeta* (*Methanothrix*), and *Methanolobus.* The nutritional requirements of the members of *Methanosarcinaceae* are methyl-group-containing compounds such as methanol, methylamines, or methyl sulfides; thus they are called methylotrophic microorganisms. The microorganisms belonging to this group can utilize trimethylamine and dismutate it to ammonia, carbon dioxide, and methane. Similarly, methanol can also catabolize to methane and carbon dioxide. Some of the members of *Methanosarcinaceae* are hydrogenotrophic or acetoclastic and thus can reduce CO_2 to methane or can utilize acetate to methane and carbon dioxide.

Another unique feature of *Methanosarcinaceae* is that they are methylotrophic and none of them can use formate as a catabolic substrate. The family *Methanosaetaceae* contains the aceticlastic genus *Methanosaeta* (*Methanothrix*). They are gram-negative, nonmotile rods (length 2.5 to 6.0 μm) with flat ends. *Methanosaeta* utilizes acetate as the energy source through an aceticlastic reaction. No other substrate supports its growth other than acetate. A sheath-like structure confers a rod-shaped structure to these organisms, and the organisms grow within this sheath. They often form a long chain-like structure.

2.3.3.5 Methanopyrales ord. nov.

Methanopyraceae was classified as a separate group of methanogens. The genus *Methanopyrus* contains a single species, *Methanopyrus kandleri*. It generally has a rod shape, is gram-positive, and grows at a very high temperature. The cell wall consists of a unique type of pseudomurein, which contains ornithin in addition to lysine and no *N*-acetylglucosamine (Kurr et al., 1991). They are hydrogenotrophic bacteria that reduce CO_2 to methane.

2.4 Microbial Interactions

2.4.1 Competition or Methanogenic Substrates: General Considerations

After autogenic hydrogen production, the spent medium is rich in volatile fatty acids such as acetate, butyrate, and ethanol. This is utilized by methanogenic bacteria by adjusting the pH of the spent medium to the alkaline range (pH 7.2 to pH 8). There are three major groups of bacteria that compete with methanogens for the substrate. The methanogenic seed cultures generally have sulfate-reducing bacteria and metal-reducing bacteria as co-contaminats. The sulfate-reducing bacteria belongs to gram-negative proteobacteria. They can use much greater diversity of electron donors than methanogens. The organic acids, alcohols, amino acids, and aromatic compounds are the potential electron donors for sulfate-reducing bacteria. Hydrogenotrophic microorganisms are gram-positive eubacteria and can also use a variety of substrates, including sugars, purines, and methoxyl groups of methoxylated aromatic compounds.

In a habitat where the organic substrate (electron donor) was limiting, a hierarchy for competition for electron donor was observed. Metal-reducing bacteria such as Fe^{3+} reducers can outcompete other organisms if potential electron acceptors are present in the system (Table 2.1). This is followed by a succession of sulfate-reducing bacteria, methanogens, and acetogens.

Table 2.1 Hydrogen and acetate utilization by Fe^{3+}-reducing bacteria, sulfate-reducing bacteria, methanogens, and acetogens

Reactants	Products	$\Delta G^{0'}$ (kJ/rxn)
$4H_2 + 8Fe^{3+}$	$8H^+ + 8Fe^{2+}$	-914
$4H_2 + SO_4^{2-} + H^+$	$HS^- + 4H_2O$	−152
$4H_2 + HCO_3^- + H+$	$CH_4 + 3H_2O$	−135
$4H_2 + HCO_3^- + H+$	$CH_3COO^- + 4H_2O$	−105
$CH_3COO^- + 8FeH + 4H_2O$	$2\ HCO_3^- + 8Fe^{2+} + 9H^+$	-809
$CH_3COO^- + SO_4^{2-}$	$2\ HCO_3^- + HS^-$	-47
$CH_3COO^- + H_2O$	$CH_4 + HCO_3^-$	-31

A 12.5% more favorable $\Delta G^{o'}$ was observed for sulfate reduction compared to methanogenesis when molecular hydrogen was used as the electron donor. So, methanogenesis is completely inhibited in habitats which have high sulfate concentrations.

2.4.2 Competition for Hydrogen

After completion of the first stage where hydrogen production took place, a considerable amount of hydrogen remains in the overhead space and also in the dissolved form. Molecular hydrogen is consumed under methanogenic condition (the second stage). For estimating the hydrogen utilization rate by hydrogenotrophic methanogens, the hydrogen consumption rate should be slow enough so that the reaction is not limited by hydrogen transfer from the gaseous phase to the liquid phase.

The competition for hydrogen under anaerobic conditions could be examined by observing the apparent K_m value for hydrogen utilization. The methanogens and methanogenic habitats have apparent K_m values of 4–8 μM hydrogen (550–1100 Pa). The values were even lower in the case of sulfate-reducing bacteria—about 2 μM (Table 2.2).

Table 2.2 Apparent K_m values for hydrogen uptake by pure cultures and methanogenic habitats

	Apparent K_m		
Organism/habitat	**μM**	**Pa**	**Reference**
Methanospirillum hungatei	5	670	(Robinson and Tiedje, 1984)
Methanosarcinia barkeri	13	1700	(Kristjansson et al., 1982)
Methanobacterium thermoautotrophicum	8	1700	(Kristjansson et al., 1982)
Desulfovibrio vulgaris	6	800	(Robinson and Tiedje, 1982)
Desulfovibrio formicicum	2	250	(Kristjansson et al., 1982)
Rumen fluids	4–9	860	(Robinson and Tiedje, 1982)
Sewage sludge	4–9	740	(Robinson and Tiedje, 1982)

The higher K_m values probably represent intrinsic limitations of the uptake hydrogenases for using hydrogen at lower partial pressures. In a bioreactor, under a higher loading rate, the effect of partial pressure of hydrogen could be observed.

Another approach that could be used to understand the competition among the anaerobes for hydrogen is by correlating the free energy available with the threshold partial pressure of hydrogen. Many hydrogenotrophic anaerobes were examined for hydrogen thresholds values (Cord-Ruwisch et al., 1988). An inverse correlation between the free energy available for the reaction and the threshold was observed (Table 2.3). The threshold value follows an order acetogen > methanogens > sulfate reducers. This implies that sulfate-reducing microorganisms can decrease the partial pressure of hydrogen to such a low level that even methanogens cannot use it. These threshold values are reaction specific and not organism specific. The thermodynamic effect of the hydrogen partial pressure on the hydrogen uptake rate can be explained by free-energy estimation using the Nemst equation. For a chemical reaction occurring at 25°C

$$aA + bB \rightarrow cC + dD. \tag{2.1}$$

The $\Delta G'$ values can be estimated (pH 7) in kilojoules, as stated in Eq. 2.2:

$$\Delta G' = \Delta G^{0'} + RT\ln\frac{(C)^c(D)^d}{(A)^a(B)^b} = \Delta G^{0'} + 5.7\log\frac{(C)^c(D)^d}{(A)^a(B)^b} \tag{2.2}$$

where A represents the molar concentration of reactant A, R is the ideal gas constant, and T is the absolute temperature in Kelvin.

For the methanogenesis process involving H_2-CO_2 reduction, free energy can be represented by the Nerst equation (Eq. 2.3). Assuming a HCO_3 concentration of 10 mM and a methane partial pressure of 0.5 atm, dependency on free energy ($\Delta G'$) can be calculated as

$$\Delta G' = -131 + 5.7\log\frac{(CH_4)}{(HCO_3^{1-})} - 5.7\log(H_2)^4 = -123 - 22.8\log(H_2) \tag{2.3}$$

Plotting a log of hydrogen partial pressure versus free energy ($\Delta G'$) gives a straight line. An identical slope was observed for acetogenesis and sulfate reduction. In both cases, 4 mol of hydrogen is used per reaction. However, they intersect $\Delta G'$ = 0 at different hydrogen partial pressures, as shown in Fig. 2.3.

Table 2.3 Thresholds for different hydrogenotrophic microorganisms (Cord-Ruwisch et al., 1988)

Organisms	Electron-accepting reaction	$\Delta G^{o'}$ (kJ/mol H_2)	H_2 threshold (Pa)
Acetobacterium woodii	CO_2 acetate	-26.1	52
Methanospirullum hungatei	CO_2 CH_4	-33.9	3
Methanobrevibacter smithii	CO_2 CH_4	-33.9	10
Desulfovibrio desulfuricans	CO_2 H_2S	-38.9	0.9

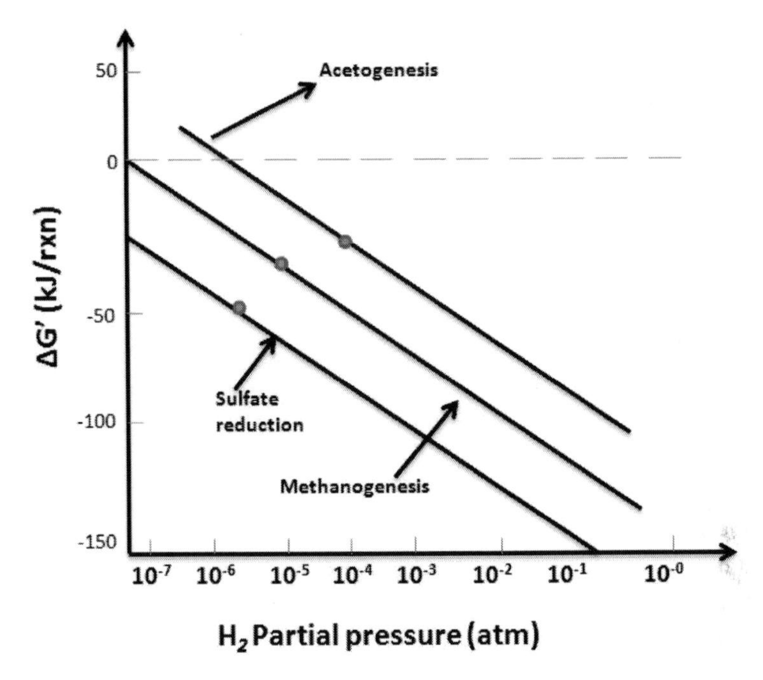

Figure 2.3 Effect of hydrogen partial pressure on the free energy of methanogenesis, sulfate reduction, or acetogenesis using hydrogen. The dots show typical hydrogen thresholds for the various microbial groups.

2.4.3 Competition for Acetate

Very little information is available related to the physiological properties of hydrogenotrophic methanogenes. The physicochemical properties that would be favorable for growth of one hydrogenotrophic species over another are still not clear. Many reports are available on the role of acetate in methane production. It was observed that higher acetate concentrations promote the growth of *Methanosarcina* sp. On the contrary, lower acetate concentration is preffered by *Methanothrix* sp. The microbial profile showed dominance of *Methanothrix* when the acetate concentration decreased beyond 1 mM in a thermophilic anaerobic digester (Wiegant, 1986; Zinder et al., 1984). To describe the competition for acetate, both Michaelis–Menton and threshold models have been used (as explained for hydrogenotrophic microbes). For acetoclastic methanogens such as *Methanosarcina*, the minimum thresholds for acetate utilization are typically in the range of 0.5 mM and higher. The minimum thresholds for acetate utilization of *Methanothrix* sp. are in the micromolar range. Methogenesis is favored at slightly alkaline pH ranges (7.2–8.0). Acetate dissociates poorly at pH ranges of 7.2–8.0, and thus this undissociated acetic acid is responsible for the threshold values.

2.4.4 Obligate Interspecies Hydrogen/Formate Transfer

Methanogens were originally believed to be able to grow on propionate, butyrate, and alcohols longer than methanol. One of the contrasting features of anaerobic methanogenesis is the symbiotic association among a vivid variety of microorganisms. In one such symbiotic association, one group of microorganisms would oxidize ethanol to acetate and *Methanobacterium* will use electrons from acetate and hydrogen to reduce CO_2 to CH_4. A phenomenon of syntrophisim widely observed between a hydrogen-producing microorganism and a hydrogen-oxidizing organism that coexist by breaking a single substrate. Moreover, it was also observed that the physical juxtaposition between hydrogen consumers and producers could facilitate hydrogen transfer (Thiele and Zeikus, 1988; Conrad et al., 1985). *Syntrophomonas wolfei*, when co-cultured with *Methanospirillum hungatei*, grows faster in the presence of formate, but when it was co-cultured with *Methanobacterium bryantii* (cannot utilize formate) it grew (McInerney et al., 1981). This emphasizes the fact that physiological differences between the two genera could also play a vital role in the two-stage biomethanation system. An interesting observation was made in the case of thermophilic methanogenesis. Under thermophilic conditions, acetate and propionate-oxidizing microorganisms could couple with *Methanobacterium thermoautotrophicum* (which could use formate). The partial pressure of hydrogen at a high temperature would be higher and high temperature facilitates diffusion. Thus, formate concentration may not be as important under thermophilic conditions.

2.4.5 Interspecies Acetate Transfer

Under anaerobic conditions, accumulation of acetate is countered by acetoclastic methanogens, thereby playing a vital role in maintaining pH homeostasis inside the reactor. In syntrophic reactions, acetate is a major product, whose accumulation could influence the overall thermodynamics of methane production. Syntrophic degradation of butyrate yields 2 mol of acetate and hydrogen, respectively. Considering the thermodynamic, a tenfold change in acetate concentration will have the same effect as hydrogen on the overall rate of the reaction. A co-culture of *Syntrophomonas wolfei–Methanospirillum hungatei* when bioagumented with *Methanosarcina barkeri* can facilitates butyrate degradation (Beaty and McInerney, 1989). In general, acetate concentration is usually higher compared to dissolved hydrogen concentration. Therfore, acetate turnover is probably not infered as hydrogen turnover. A potential example of obligate interspecies acetate transfer is the acetone-degrading methanogenic-enriched culture (Platen and Schink, 1987). When acetone was provided as the sole carbon and energy source, the mixed culture showed dominance of filamentous *Methanothrix* sp., and it was catabolized to aceto-acetate. The aceto-acetate then yielded 2 mol of acetate, which was consequently converted to methane. External feeding of acetate to the system leads to inhibition of acetone degradation. Addition of BES inhibits not only methanogenesis but also acetone degradation. The efficiency of acetone-degrading microbes showed dependency on acetate degradation, even though the ΔG for conversion of acetone to acetate was –34.2 kJ/reaction.

2.5 Conclusions

Two-stage integration of dark-fermentative hydrogen production with photofermentative hydrogen production has shown lots of drawbacks such as pigment accumulation and a low rate of hydrogen production in the second stage. Integration of biohydrogen production with biomethantion under the biohythane concept has posed new challenges and opportunities in terms of large-scale operation. This chapter deals with microbial insights of both processes. Understanding microbial diversity and its characteristics related to the biohythane process has led to many interesting findings. Acidogenic hydrogen grows faster than methanogens and eventually produces volatile fatty acids in the spent media. Major genuses related to acidogenic biohydrogen production are *Enterobater* sp., *Clostridium* sp., *Citrobacter* sp., *Thermoanaeobacterium* sp., *Caldicellulosiruptor* sp., etc. These volatile fatty acids are ideal feedstock for methanogenic microorganisms. Widely known methanogens are *Methanobacter* sp., *Methanospirillum* sp., Methanococcus sp., and *Methanosarcina* sp. The interaction of acidogens and methanogens at the substrate level and the oxidation-reduction potential level has potentially shown to influence the whole process. An intra- and interspecies acetate transfer has been one of the unique features among the interacting methanogens.

References

Abo-Hashesh M, Wang R, Hallenbeck PC (2011). Metabolic engineering in dark fermentative hydrogen production; theory and practice, *Bioresour Technol*, **102**, 8414–8422.

Angenent LT, Karim K, Al-Dahhan MH, Wrenn BA, Domíguez-Espinosa R (2004). Production of bioenergy and biochemicals from industrial and agricultural wastewater, *Trends Biotechnol*, **22**, 477–485.

Arooj M, Han S, Kim S, Kim D, Shin H (2008). Effect of HRT on ASBR converting starch into biological hydrogen, *Int J Hydrogen Energy*, **33**, 6509–6514.

Asakawa S (2003). Methanoculleus bourgensis, Methanoculleus olentangyi and Methanoculleus oldenburgensis are subjective synonyms, *Int J Syst Evol Microbiol*, **53**, 1551–1552.

Balch WE, Fox GE, Magrum LJ, Woese CR, Wolfe RS (1979). Methanogens: reevaluation of a unique biological group, *Microbiol Rev*, **43**, 260–296.

Beaty PS, McInerney MJ (1989). Effects of organic acid anions on the growth and metabolism of syntrophomonas wolfei in pure culture and in defined consortia, *Appl Environ Microbiol*, **55**, 977–983.

Bélaich JP, Bruschi M, Garcia JL (eds) (1990). *Microbiology and Biochemistry of Strict Anaerobes Involved in Interspecies Hydrogen Transfer*. Springer US, Boston, MA.

Benemann JR, Berenson JA, Kaplan NO, Kamen MD (1973). Hydrogen evolution by a chloroplast-ferredoxin-hydrogenase system, *Proc Natl Acad Sci USA*, **70**, 2317–2320.

Biavati B, Vasta M, Ferry JG (1988). Isolation and characterization of "Methanosphaera cuniculi" sp. nov., *Appl Environ Microbiol*, **54**, 768–771.

Blackwood AC, Ledingham GA, Neish AC (1956). Dissimilation of glucose at controlled pH values by pigmented and non-pigmented strains of *Escherichia coli, J Bacteriol*, **72**, 497–499.

Bleicher K, Zellner G, Winter J (1989). Growth of methanogens on cyclopentanol/CO_2 and specificity of alcohol dehydrogenase, *FEMS Microbiol Lett*, **59**, 307–312.

Boone DR, Johnson RL, Liu, Y (1989). Diffusion of the interspecies electron carriers H_2 and formate in methanogenic ecosystems and its implications in the measurement of K_m for H_2 or formate uptake, *Appl Environ Microbiol*, **55**, 1735–1741.

Bryant MP, Boone DR (1987). Emended description of strain MST(DSM 800T), the type strain of Methanosarcina barkeri, *Int J Syst Bacteriol*, **37**, 169–170.

Brynjarsdottir H, Scully SM, Orlygsson J (2013). Production of biohydrogen from sugars and lignocellulosic biomass using *Thermoanaerobacter* GHL15, *Int J Hydrogen Energy*, **38**, 14467–14475

Burggraf S, Stetter KO, Rouviere P, Woese CR (1991). Methanopyrus kandleri: an archaeal methanogen unrelated to all other known methanogens, *Syst Appl Microbiol*, **14**, 346–351.

Conrad R, Phelps TJ, Zeikus JG (1985). Gas metabolism evidence in support of the juxtaposition of hydrogen-producing and methanogenic bacteria in sewage sludge and lake sediments, *Appl Environ Microbiol*, **50**, 595–601.

Cord-Ruwisch R, Seitz HJ, Conrad R (1988). The capacity of hydrogenotrophic anaerobic bacteria to compete for traces of hydrogen depends on the redox potential of the terminal electron acceptor. *Arch Microbiol*, **149**, 350–357.

Das D, Veziroglu TN (2001). Hydrogen production by biological processes: a survey of literature, *Int J Hydrogen Energy*,**26**, 13–28.

Fascetti E (1998). Photosynthetic hydrogen evolution with volatile organic acids derived from the fermentation of source selected municipal solid wastes, *Int J Hydrogen Energy*, **23**, 753–760.

Finkelstein M, McMillan JD, Davison BH (eds) (2002). *Biotechnology for Fuels and Chemicals*. Humana Press, Totowa, NJ.

Gilbert JJ, Ray S, Das D (2011). Hydrogen production using Rhodobacter sphaeroides (O.U. 001) in a flat panel rocking photobioreactor, *Int J Hydrogen Energy,* **36**, 3434–3441.

Goud RK, Mohan SV (2012). Acidic and alkaline shock pretreatment to enrich acidogenic biohydrogen producing mixed culture: long term synergetic evaluation of microbial inventory, dehydrogenase activity and bio-electro kinetics, *RSC Adv,* **2**, 6336.

Hallenbeck PC, Abo-Hashesh M, Ghosh D (2012). Strategies for improving biological hydrogen production, *Bioresour Technol,* **110**, 1–9.

Hippe H, Caspari D, Fiebig K, Gottschalk G (1979). Utilization of trimethylamine and other N-methyl compounds for growth and methane formation by Methanosarcina barkeri, *Proc Natl Acad Sci USA,* **76**, 494–498.

Huber H, Thomm M, Knig H, Thies G, Stetter KO (1982). Methanococcus thermolithotrophicus, a novel thermophilic lithotrophic methanogen, *Arch Microbiol,* **132**, 47–50.

Huber, R, Langworthy, TA, Knig, H, Thomm, M, Woese, CR, Sleytr, UB, Stetter, KO (1986). Thermotoga maritima sp. nov. represents a new genus of unique extremely thermophilic eubacteria growing up to 90°C, *Arch Microbiol,* **144**, 324–333.

Hungate RE (1969). A roll tube method for cultivation of strict anaerobes, *Methods Microbiol,* **3**, 117–132.

Jo JH, Lee DS, Park D, Park JM (2008). Biological hydrogen production by immobilized cells of Clostridium tyrobutyricum JM1 isolated from a food waste treatment process, *Bioresour Technol,* **99**, 6666–6672.

Jung G (2002). Hydrogen production by a new chemoheterotrophic bacterium *Citrobacter* sp. Y19, *Int J Hydrogen Energy,* **27**, 601–610.

Kádár Z, de Vrije T, van Noorden GE, Budde MAW, Szengyel Z, Réczey K, Claassen PAM (2004). Yields from glucose, xylose, and paper sludge hydrolysate during hydrogen production by the extreme thermophile *Caldicellulosiruptor saccharolyticus, Appl Biochem Biotechnol,* **114**, 497–508.

Kalia VC, Jain SR, Kumar A, Joshi AP (1994). Frementation of biowaste to H_2 by *Bacillus licheniformis, World J Microbiol Biotechnol,* **10**, 224–227.

Kalia VC, Purohit HJ (2008). Microbial diversity and genomics in aid of bioenergy, *J Ind Microbiol Biotechnol,* **35**, 403–419.

Kamalaskar LB, Dhakephalkar PK, Meher KK, Ranade DR (2010). High biohydrogen yielding *Clostridium sp.* DMHC-10 isolated from sludge of distillery waste treatment plant, *Int J Hydrogen Energy,* **35**, 10639–10644.

Kotay SM, Das D (2007). Microbial hydrogen production with *Bacillus coagulans* IIT-BT S1 isolated from anaerobic sewage sludge, *Bioresour Technol,* **98**, 1183–1190.

Kristjansson JK, Schnheit P, Thauer RK (1982). Different Ks values for hydrogen of methanogenic bacteria and sulfate reducing bacteria: an explanation for the apparent inhibition of methanogenesis by sulfate, *Arch Microbiol,* **131**, 278–282.

Kumar N, Das D (2000). Enhancement of hydrogen production by *Enterobacter cloacae* IIT-BT 08, *Proc Biochem,* **35**, 589–593.

Kumar N, Ghosh A, Das D (2001). Redirection of biochemical pathways for the enhancement of H_2 production by *Enterobacter cloacae, Biotechnol Lett,* **23**, 537–541.

Kurr M, Huber R, Knig H, Jannasch HW, Fricke H, Trincone A, Kristjansson JK, Stetter KO (1991). *Methanopyrus kandleri* gen. and sp. nov. represents a novel group of hyperthermophilic methanogens, growing at 110°C, *Arch Microbiol,* **156**, 239–247.

Lauerer, G, Kristjansson, JK, Langworthy, TA, König, H, Stetter, KO (1986). Methanothermus sociabilis sp. nov., a second species within the *Methanothermaceae* growing at 97°C, *Syst Appl Microbiol,* **8**, 100–105.

Lee YE, Jain MK, Lee C, Zeikus JG (1993). Taxonomic distinction of saccharolytic thermophilic anaerobes: description of *Thermoanaerobacterium xylanolyticum* gen. nov., sp. nov., and *Thermoanaerobacterium saccharolyticum* gen. nov., sp. nov.; reclassification of *Thermoanaerobium brockii, Clostridium, Int J Syst Bacteriol,* **43**, 41–51.

Levin D, Islam R, Cicek N, Sparling R (2006). Hydrogen production by *Clostridium thermocellum* 27405 from cellulosic biomass substrates, *Int J Hydrogen Energy,* **31**, 1496–1503.

Mah RA, Kuhn DA (1984). Transfer of the type species of the genus *Methanococcus* to the genus Methanosarcina, naming it Methanosarcina mazei (Barker 1936) comb. nov. et emend. and conservation of the genus Methanococcus (Approved Lists 1980) with *Methanococcus vannielii. Int J Syst Bacteriol,* **34**, 263–265.

Mathrani IM, Boone DR (1985). Isolation and characterization of a moderately halophilic methanogen from a solar saltern, *Appl Environ Microbiol,* **50**, 140–143.

McInerney MJ, Bryant MP, Hespell RB, Costerton JW (1981). *Syntrophomonas wolfei* gen. nov. sp. nov., an anaerobic, syntrophic, fatty acid-oxidizing bacterium, *Appl Environ Microbiol*, **41**, 1029–1039.

Nandi R, Sengupta S (2008). Microbial production of hydrogen: an overview, *Crit Rev Microbio*l, **24**(1), 61–84.

Palazzi E, Fabiano B, Perego P (2000). Process development of continuous hydrogen production by *Enterobacter aerogenes* in a packed column reactor, *Bioproc Eng*, **22**, 205–213.

Platen H, Schink B (1987). Methanogenic degradation of acetone by an enrichment culture, *Arch Microbiol*, **149**, 136–141.

Rachman MA, Nakashimada Y, Kakizono T, Nishio N (1998). Hydrogen production with high yield and high evolution rate by self-flocculated cells of Enterobacter aerogenes in a packed-bed reactor, *Appl Microbiol Biotechnol*, **49**, 450–454.

Rainey FA, Donnison AM, Janssen PH, Saul D, Rodrigo A, Bergquist PL, Daniel RM, Stackebrandt E, Morgan HW (1994). Description of *Caldicellulosiruptor saccharolyticus* gen. nov., sp. nov: an obligately anaerobic, extremely thermophilic, cellulolytic bacterium, *FEMS Microbiol Lett*, **120**, 263–266.

Raskin L, Stromley JM, Rittmann BE, Stahl DA (1994). Group-specific 16S rRNA hybridization probes to describe natural communities of methanogens, *Appl Environ Microbiol*, **60**, 1232–1240.

Robinson JA, Tiedje JM (1982). Kinetics of hydrogen consumption by rumen fluid, anaerobic digestor sludge, and sediment, *Appl Environ Microbiol*, **44**, 1374–1384.

Robinson JA, Tiedje JM (1984). Competition between sulfate-reducing and methanogenic bacteria for H_2 under resting and growing conditions, *Arch Microbiol*, **137**, 26–32.

Rouviere PE, Wolfe RS (1988). Novel biochemistry of methanogenesis, *J Biol Chem*, **263**, 7913–7916.

Schut GJ, Adams MWW (2009). The iron-hydrogenase of *Thermotoga maritima* utilizes ferredoxin and NADH synergistically: a new perspective on anaerobic hydrogen production, *J Bacteriol*, **191**, 4451–4457.

Smith G, Ewart G, Tucker W (1992). Hydrogen production by cyanobacteria, *Int J Hydrogen Energy*, **17**, 695–698.

Soboh B, Linder D, Hedderich R (2004). A multisubunit membrane-bound [NiFe] hydrogenase and an NADH-dependent Fe-only hydrogenase in the fermenting bacterium *Thermoanaerobacter tengcongensis*, *Microbiology*, **150**, 2451–2463.

Stickland LH (1929). The bacterial decomposition of formic acid, *Biochem J*, **23**, 1187–1198.

Taguchi F, Mizukami N, Hasegawa K, Saito-Taki T (2011). Microbial conversion of arabinose and xylose to hydrogen by a newly isolated *Clostridium* sp. no. 2., **40**(3), 228–233.

Taguchi, F, Yamada, K, Hasegawa, K, Taki-Saito, T, Hara, K (1996). Continuous hydrogen production by *Clostridium* sp. strain no. 2 from cellulose hydrolysate in an aqueous two-phase system, *J Ferment Bioeng*, **82**, 80–83.

Tamburic B, Zemichael FW, Maitland GC, Hellgardt K (2011). Parameters affecting the growth and hydrogen production of the green alga *Chlamydomonas reinhardtii*, *Int J Hydrogen Energy*, **36**, 7872–7876.

Tanisho S (1998). Effect of CO_2 removal on hydrogen production by fermentation, *Int J Hydrogen Energy*, **23**, 559–563.

Tanisho S, Ishiwata Y (1994). Continuous hydrogen production from molasses by the bacterium *Enterobacter aerogenes*, *Int J Hydrogen Energy*,**19**, 807–812.

Tanisho S, Suzuki Y, Wakao N (1987). Fermentative hydrogen evolution by *Enterobacter aerogenes* strain E.82005, *Int J Hydrogen Energy*, **12**, 623–627.

Thiele JH, Zeikus JG (1988). Control of interspecies electron flow during anaerobic digestion: significance of formate transfer versus hydrogen transfer during syntrophic methanogenesis in flocs, *Appl Environ Microbiol*, **54**, 20–29.

Tsygankov AA (2001). *BioHydrogen II*. Elsevier.

Ueno Y, Kawai T, Sato S, Otsuka S, Morimoto M (1995). Biological production of hydrogen from cellulose by natural anaerobic microflora, *J Ferment Bioeng*, **79**, 395–397.

Ueno Y, Otsuka S, Morimoto M (1996). Hydrogen production from industrial wastewater by anaerobic microflora in chemostat culture, *J Ferment Bioeng*, **82**, 194–197.

Valdez-Vazquez I, Poggi-Varaldo HM (2009). Hydrogen production by fermentative consortia. *Renew Sust Energy Rev*, **13**, 1000–1013.

Vanniel E (2002). Distinctive properties of high hydrogen producing extreme thermophiles, *Caldicellulosiruptor saccharolyticus* and *Thermotoga elfii*, *Int J Hydrogen Energy*, **27**, 1391–1398.

Vatsala T (1992). Hydrogen production from (cane-molasses) stillage by citrobacter freundii and its use in improving methanogenesis, *Int J Hydrogen Energy*, **17**, 923–927.

Venkata Mohan S (2009). Harnessing of biohydrogen from wastewater treatment using mixed fermentative consortia: Process evaluation towards optimization, *Int J Hydrogen Energy*, **34**, 7460–7474.

Venkata Mohan S, Lalit Babu V, Sarma PN (2008). Effect of various pretreatment methods on anaerobic mixed microflora to enhance biohydrogen production utilizing dairy wastewater as substrate, *Bioresour Technol*, **99**, 59–67.

Venkata Mohan S, Lalit Babu V, Srikanth S, Sarma PN (2008). Bio-electrochemical evaluation of fermentative hydrogen production process with the function of feeding pH, *Int J Hydrogen Energy*, **33**, 4533–4546.

Weizmann C, Rosenfeld B (1937). The activation of the butanol-acetone fermentation of carbohydrates by *Clostridium acetobutylicum* (Weizmann), *Biochem J*, **31**, 619–639.

Wiegant W (1986). Separation of the propionate degradation to improve the efficiency of thermophilic anaerobic treatment of acidified wastewaters. *Water Res*, **20**, 517–524.

Wiegel J, Ljungdahl LG (1981). Thermoanaerobacter ethanolicus gen. nov., spec. nov., a new, extreme thermophilic, anaerobic bacterium, *Arch Microbiol*, **128**, 343–348.

Wildgruber G, Thomm M, Knig H, Ober K, Richiuto T, Stetter KO (1982). Methanoplanus limicola, a plate-shaped methanogen representing a novel family, the methanoplanaceae, *Arch Microbiol*, **132**, 31–36.

Woese CR, Kandler O, Wheelis ML (1990). Towards a natural system of organisms: proposal for the domains Archaea, Bacteria, and Eucarya, *Proc Natl Acad Sci USA*, **87**, 4576–4579.

Yokoi H, Tokushige T, Hirose J, Hayashi S, Takasaki Y (1998). H_2 production from starch by a mixed culture of *Clostridium butyricum* and *Enterobacter aerogenes*, *Biotechnol Lett*, **20**, 143–147.

Zeidan AA, Van Niel EWJ (2009). Developing a thermophilic hydrogen-producing co-culture for efficient utilization of mixed sugars, *Int J Hydrogen Energy*, **34**, 4524–4528.

Zhang T, Liu H, Fang HHP (2003). Biohydrogen production from starch in wastewater under thermophilic condition, *J Environ Manage*, **69**, 149–156.

Zinder SH, Cardwell SC, Anguish T (1984). Methanogenesis in a thermophilic (58°C) anaerobic digestor: *Methanothrix* sp. as an important aceticlastic methanogen, *Appl Environ Microbiol*, **47**(4), 796–807.

Chapter 3

Biochemistry of the Biohythane Production Process

3.1 Introduction

The biochemistry of biohydrogen and biomethane production processes has been the point of interest for many researchers. A detailed description of microorganisms involved in dark fermentation and biomethane production has been given in Chapter 2. The metabolic pathways involved in both processes are governed by various genes, and wide ranges of metabolites are formed. The metabolic uniqueness of mesophilic, thermophilic, obligate, and facultative anaerobes has been discussed in detail in this chapter. The principal enzymes involved in biohydrogen production, such as Ni-Fe H_2ase and Fe-Fe H_2ase, play a pivotal role. Similarly, the biomethanation process is governed by various vital enzymes and molecules such as coenzyme M (CoM), methyl reductases, acetyl kinase, and phosphor-transacetyl kinase.

3.2 Biochemistry behind Dark-Fermentative Hydrogen Production

Dark fermentation comprises a series of complex biochemical reactions manifested by a diverse group of bacteria, resulting in

Biohythane: Fuel for the Future
Debabrata Das and Shantonu Roy

ISBN 978-981-4745-29-1 (Hardcover), 978-981-4745-30-7 (eBook)
www.panstanford.com

conversion of organic substrates to biohydrogen. These distinct microorganisms include fermentative/hydrolytic bacteria that can hydrolyze complex organic polymers to monomers, which on further reaction with hydrogen-producing acidogenic bacteria are converted to a mixture of short-chain organic acids and alcohols. Under anaerobic conditions, the tricarboxylic acid (TCA) cycle of the cell is blocked, and instead fermentative metabolic processes occur to regenerate ATP. During this process, acids and alcohols are produced as reduced metabolites, which lead to disposition of extracellular reductants. Now to maintain the redox potential of the cell, among the many reduced end metabolites produced, hydrogen is also produced.

During fermentation, complex organic compounds are hydrolyzed to simple carbohydrates like glucose. Over other alternatives, glucose is a much preferred carbon source for the microbes and after being processed through the glycolytic pathway yields pyruvate for regeneration of ATP. The pyruvate leads to the formation of acetic acid and butyric acid with concomitant production of hydrogen gas by virtue of two different biochemical reactions.

In both thermophilic bacteria (McCord et al., 1971) and obligate anaerobes (Clostridia) (Zeikus, 1977), one of the mechanisms which is very common is the conversion of pyruvate to acetyl coenzyme A (acetyl-CoA) by oxidation. This reaction is carried out by pyruvate ferredoxin oxidoreductase (PFOR) (Uyeda and Rabinowitz, 1971). The acetyl-CoA is then further converted to acetyl phosphate, thereby generating ATP and acetate (doesn't fit in with pyruvate oxidation). This pyruvate oxidation step occurs simultaneously with ferredoxin (Fd) reduction. The reduced ferredoxin is then oxidized by [Fe-Fe] hydrogenase, catalyzing the formation of hydrogen (Fig. 3.1). The overall reaction is shown in Eq. 3.1 and Eq. 3.2.

$$\text{Pyruvate} + \text{CoA} + 2\text{Fd (ox)} \rightarrow \text{Acetyl-CoA} + 2\text{Fd (red)} + CO_2 \quad (3.1)$$

$$2H^+ + \text{Fd (red)} \rightarrow H_2 + \text{Fd (ox)} \quad (3.2)$$

The oxidation of pyruvate to acetate results in the production of 4 mol of hydrogen per mol of glucose (Benemann, 1996), while oxidation to butyrate as the sole end product produces 2 mol of hydrogen per mol of glucose. Thus, it is essential for organisms following a mixed-acid pathway that a higher A/B (acetate to butyrate) ratio be maintained in order to obtain higher yields of hydrogen production (Khanna et al., 2011a). The reactions governing acetic acid and butyric acid production are shown in Eq. 3.3 and Eq. 3.4, respectively:

$$C_6H_{12}O_6 + 2H_2O \rightarrow 2CH_3COOH + 2CO_2 + 4H_2 \quad (3.3)$$

$$C_6H_{12}O_6 \rightarrow CH_3CH_2COOH + 2CO_2 + 2H_2 \quad (3.4)$$

The catalysis of pyruvate oxidation to acetyl-CoA and formate performed by pyruvate formate lyase (PFL) is the second type of mechanism which occurs in few facultative anaerobic bacteria, such as *Escherichia coli* (Knappe and Sawers, 1990) (Eq. 3.5, Fig. 3.1):

$$\text{Pyruvate} + \text{CoA} \rightarrow \text{Acetyl-CoA} + \text{Formate} \quad (3.5)$$

The formate hydrogen lyase (FHL) then cleaves this formate, resulting in the production of hydrogen and CO_2 (Eq. 3.6, Fig. 3.1). This pathway was first reported in 1932 by Stephenson and Stickland.

$$HCOOH \rightarrow CO_2 + H_2 \quad (3.6)$$

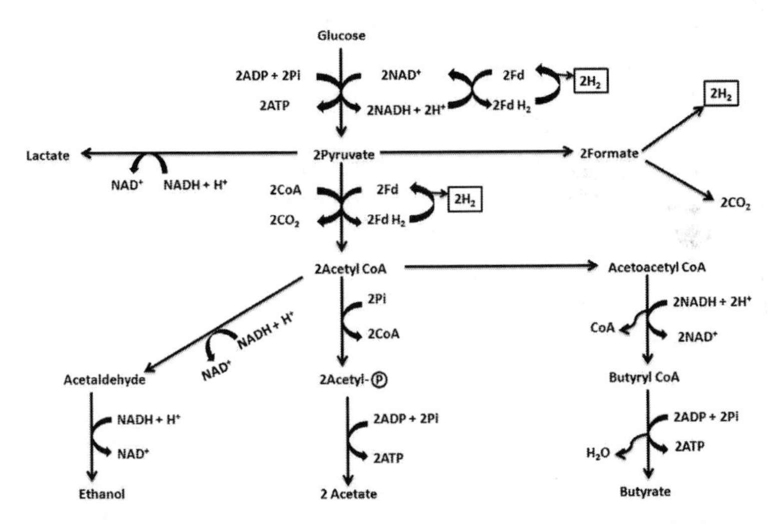

Figure 3.1 Metabolic pathways involved in dark-fermentative hydrogen production.

In case propionic acid, lactic acid, or ethanol is the end product of pyruvate oxidation, no hydrogen is produced. Besides, in the presence of alternate electron acceptors like nitrate, fumarate, etc., hydrogen production is hindered because it causes a biochemical shift of reactions from fermentation to anaerobic respiration, especially in facultative anaerobes like enteric bacteria. Thus, studies focusing on hydrogen production from facultative anaerobes should ensure that the production medium is devoid of any such electron acceptors. The much-gained attention toward dark fermentation for hydrogen production is due to its potential to utilize a wide variety of substrates. Furthermore, utilization of different kinds of wastewater as a substrate has broadened the scope of the process, hence providing dual benefits of waste bioremediation along with hydrogen production. For example, the use of industrial wastewater as a substrate suffices the basic yardsticks of substrate selection, viz., biodegradability, cost, and availability (Cai et al., 2009).

For biohydrogen production from wastewater via dark fermentation, researchers have explored the potential for the following wastewater types: palm oil mill wastewater (Pandu and Joseph, 2012), wheat straw wastes, molasses-based distillery effluents (Das et al., 2008), rice spent wash from wineries (Nath and Das, 2004), wastewater from processed food industries (Cheong and Hansen, 2006), cellulose- and pentose-rich paper mill effluents (Cheng et al., 2011), starchy wastewater from households (Chen et al., 2008b), wastewater generated from cattle-based industries (Chen et al., 2008a), and chemical wastewater (Cakir et al., 2010). These reports have majorly worked upon mixed cultures in acidophilic conditions after selective enrichment. Due to the nonsterile and heterogeneous environment of wastewater, utilization of mixed microflora is crucial and is more relevant for the dark fermentation process. Most of these wastes, however, require pretreatment before being used as a substrate source. For instance, plant biomass is rich in cellulose and must undergo physicochemical pretreatment before use in fermentors. The pretreatment method can range from treating it with an acid or alkali to the oxidative steam explosion method carried out at high temperatures. This pretreatment strategy of lignocellulosic material yields a mixture of pentose and hexose sugars, thus allowing microbial fermentation of these hydrolysis products. Hence while using plant biomass for hydrogen production, pretreatment plays a major role. However, fermentation of this mixture of hexose and pentose sugars is usually avoided as in the presence of glucose, catabolic repression causes lower conversion of pentose sugar and hence decreases the overall yield of hydrogen (Strobel, 1993; Aberu et al., 2010). Besides, the different fermentation pathways in organisms affect the efficiency of hydrogen production from sugars (Abreu et al., 2012).

Dark fermentation yields organic acids that can be further used as substrates by photofermentative bacteria and produce CO_2 and hydrogen. Thus, by combination of photofermentation and dark fermentation into a two-stage (hybrid) process, higher hydrogen yields can be obtained. This hybrid process can theoretically generate around 12 mol of hydrogen per mol of glucose (Nath and Das, 2008).

$$2CH_3COOH + 4H_2O \rightarrow 8H_2 + 4CO_2 \quad (3.7)$$

But photofermentation is marred by many operational constrains. The major bottleneck is low photosynthetic light conversion efficiency of photofermentative microorganisms. Moreover, a low rate of hydrogen production and the shading effect of pigments produced by the photofermentative microorganisms also undermine the potential of photofermentation.

Another hybrid system that propped up in recent times is two-stage biohydrogen followed by the biomethanation process. This could be achieved either in the same reactor or in a different bioreactor. During dark-fermentative hydrogen production, only 12%–18% of the total energy available from the feedstock could be extracted. To maximize gaseous energy extraction, another hybrid system has been developed in which hydrogen production would be followed by second-stage methanogenesis. Theoretically, 4 mol of hydrogen are obtained from 1 mol of glucose. The theoretical maximum was almost achieved; however, the energy trapped in 2 mol of acetic acid generated could not be recovered. The energy trapped in acetic acid could be recovered by acetoclastic methanogens as the seed to form methane.

$$C_6H_{12}O_6 + 2H_2O \rightarrow 4H_2 + 2CH_3COOH + 2CO_2$$

(Stage I: Biohydrogen production) (3.8)

$$2CH_3COOH \rightarrow 2CH_4 + + 2CO_2$$

(Stage II: Biomethane production) (3.9)

The biomethanation process is a well-established process. So, integration of the biohydrogen and biomethane production processes, known as the biohythane process, could prove to be a better option for gaseous energy recovery compared to the photofermentative process.

3.2.1 Enhancement of Hydrogen Production by Metabolic Engineering

Current advances in metabolic engineering with development of techniques like expression analysis, gene technology, and genome sequencing have improved the possibilities of engineering microorganisms for introduction, deletion, or modification of desired metabolic pathways. The results of quantitative and systematic analysis of metabolic pathways are integrated with genomic approaches and molecular biology via metabolic engineering. The alteration of the metabolic pathway could increase the production of a native or nonnative product (Stephanopoulos, 1988; Wiechert, 2002), such as biohydrogen and biomethane. The effect of limiting factors in biohydrogen production pathways can be diminished by metabolic engineering to divert and concentrate the flow of electrons toward hydrogen-evolving pathways. The approach can also be used for increasing substrate consumption and making oxygen-tolerant hydrogenases, which will ultimately improve the efficiency of the process. In dark-fermentative hydrogen production, the metabolic engineering approach can be employed at different stages for overall process improvement. Metabolic engineering can be employed either for introducing new pathways for hydrogen production or for altering existing pathways so that hydrogen production can be improved in terms of hydrogen yields and/or rates.

3.2.1.1 Metabolic engineering approach for improvement of hydrogen-producing obligate anaerobes

In strict anaerobes, hydrogen is produced by the enzyme [Fe-Fe] hydrogenase that mediates transfer of electrons from ferredoxin (Fd) to a proton (H^+ ion). The metabolic pathway and possible target enzymes for metabolic engineering to enhance hydrogen production in obligate anaerobes are shown in Fig. 3.2. In such organisms, pyruvate produced in glycolysis is converted to acetyl-CoA and CO_2 through PFOR. This oxidation of pyruvate requires ferredoxin, which later transfers its electron to a proton facilitated by [Fe-Fe] hydrogenase to evolve hydrogen. The whole process results in a maximum yield of 2 mol of hydrogen per mol of glucose metabolized. However, strict anaerobes are known to produce higher yields of hydrogen, that is, up to 4 mol of hydrogen per mol of glucose. Another 2 mol of hydrogen can be produced by the oxidation of NADH formed during glycolysis. The NADH undergoes oxidation and transfers its electron to ferredoxin (Fd) through NADH ferredoxin oxidoreductase (NFOR). Thus, the overall maximum yield of 4 mol of hydrogen per mol of glucose consumed can be obtained in strict anaerobes via metabolic engineering.

However, the reason for lower hydrogen yields is also due to its dependence on the other competing pathways and metabolites, for example, acetate, butyrate, lactate, or ethanol produced during fermentation. The maximum yield of hydrogen is achieved when acetate is the end product. Thermophilic fermentation is an interesting approach to direct the reaction toward hydrogen production because of more favorable thermodynamic conditions (Kadar et al., 2004; van Niel et al., 2002). High temperatures shift the equilibrium point of the hydrogen pathways in the direction of hydrogen production by a factor of up to 4.5, which results in higher hydrogen yields (Kongjan and Angelidaki, 2011; Veit et al., 2008). It has been reported that under thermophilic conditions, a higher hydrogen production rate (1050 ± 63 mmol h^{-1} at 60°C) is exhibited (Ahn et al., 2005) and a lesser variety of fermentation end products are obtained (Schonheit and Schafer, 1995). Moreover, the H^+ ion produced inside the cell during fermentation is expelled outside the cell. This generates a proton motive force that further influences hydrogen production and ATP generation (Hakobyan et al., 2012).

Different approaches applied to increase hydrogen production in obligate anaerobes have been summarized in Table 3.1. To increase hydrogen production in strict anaerobes overexpression of [Fe-Fe] hydrogenase has been reported in some studies. The [Fe-Fe] hydrogenases are known to evolve hydrogen and are found in strict anaerobes such as *Clostridia*, *Thermotoga*, and *Desulfovibrio* sp. The hydrogenase gene from different microorganisms has been characterized, and genetic modifications of this gene have been performed to increase hydrogen production. In strict anaerobes, hydA encoding the [Fe-Fe] hydrogenase can be overexpressed homologously as well as heterologously. Morimoto et al. reported that homologous overexpression of the *hydA* gene in *Clostridium paraputrificum* M-21 resulted in 1.7-fold increased hydrogen production compared to the wild strain with no lactic acid production and increased acetic acid production (Morimoto et al., 2005). Homologous overexpression of the [Fe-Fe] hydrogenase gene, *hydA* in *C. tyrobutyricum* JM1, resulted in a 1.7-fold and 1.5-fold enhancements in the hydrogenase activity and hydrogen yield, respectively, against the wild-type strain (Jo et al., 2010). Earlier attempts of heterologous overexpression of [Fe-Fe] hydrogenase were in vain, and later it was shown that for successful functional expression of hydrogenase, co-expression of three maturation genes, *hydE*, *hydF*, and *hydG*, is necessary. These maturation genes are required for maturation and insertion of the H-cluster in organisms where these genes are absent (Posewitz et al., 2008). Therefore, the reported work of hydrogenase overexpression where these accessory genes were not co-expressed (Karube et al., 1983; Subudhi et al., 2011). For co-expression of maturation genes along with *hydA*, cloning vectors having two cloning sites are available (Duet vectors). Co-expressing *hydE*, *hydF*, and *hydG* genes from *C. acetobutylicum* in *E. coli* has shown a stable hydrogen production. Co-expression of *C. acetobutylicum* maturation proteins along with various algal and bacterial [Fe-Fe] hydrogenase results in enzymes with similar specific activity as the ones purified from native sources. Hence, it can be inferred that for the catalytically active [Fe-Fe] hydrogenase to be biosynthesized, it is essential that the Hyd maturation proteins and catalytic domain of the hydrogenase be present (King et al., 2006). However, if somehow the host genome is encoded for hydrogen-producing proteins, accessory genes will not be required and heterologous expression will be possible.

During dark-fermentative hydrogen production, one of the major bottlenecks encountered is the accumulation of end metabolites that inhibit hydrogen production. Accessory metabolic pathways operating during fermentative hydrogen production also compete for NADH, diluting the pool of NADH required for the hydrogen production via the NFOR pathway, resulting in decreased hydrogen production. An alternative approach to increase hydrogen production is to block the existing competing pathways that utilize the pool of NADH. Integration mutagenesis was used to inactivate acetate kinase-encoding *ack* gene so that the acetate-forming pathway can be blocked in *C. tyrobutyricum* (Liu et al., 2006). The objective of this study was to increase butyrate and hydrogen production. The wild strain of *C. tyrobutyricum* produces acetate, butyrate, and hydrogen, while it was reported that the *ack*-deleted mutant strain produced 50% more hydrogen from glucose and increased butyrate production. Another study showed that in the *C. saccharoperbutylacetonicum* strain N1-4 down-regulation of the *hupCBA* gene cluster using antisense RNA strategy improved the hydrogen production by 3.1-fold compared to the wild strain. HupCBA proteins are responsible for uptake of hydrogen in *Clostridium* sp. (Nakayama et al., 2008).

Directed mutagenesis is an important approach in molecular biology, and most of the mutations in *Clostridium* sp. are still constructed through a homologous recombination. Recently, a ClosTron system has been developed, which allows directed construction of a stable mutant in *Clostridium* sp. using a bacterial group II intron (Heap et al., 2007). The *ltrB* gene present in *Lactococcus lactis* encodes for a mobile group II intron, which lays the foundation for development of the ClosTron system. This mutagenic system was developed to function in *Clostridium* sp. hosts. With the help of the RNA-mediated "retrohoming" mechanism, the group II mobile intron elements from the *ltrb* gene of *L. lactis* inserts into their specific target gene. Base pairing between the excised intron lariat RNA and target site DNA helps in target site recognition (Mohr et al., 2000). Intron target specificity can be reprogrammed by alteration of the DNA sequence encoding for the target part of the intron. Recently, several improvements have been made in the ClosTron system so that multiple mutation-carrying strains can be constructed. Automated ClosTron design bioinformatics tools are available online (http://clostron.com). The group II intron (TargeTron gene knockout system)-mediated mutagenesis was used to delete the lactate dehydrogenase–encoding *ldh* gene in the *C. pefringens* strain W11, and it was reported that in the *ldh*-deleted mutant strain W13, the hydrogen yield increased by 51% with almost zero lactate production (Wang et al., 2011). This work was further extended by inactivating the acetyl-CoA acetyltransferase gene (*atoB*) so that butyrate formation could be blocked by use of the TargeTron gene knockout system. The resultant mutant strain W15 showed decreased hydrogen production, revealing the importance of butyrate formation pathway for hydrogen production in *C. pefringens* (Yu et al., 2013). Group II intron-based mutation has successfully worked in *C. difficle, C. acetobutylicum, C. sporogenes, C. botulinum, C. beijerinckii*, and *C. pefrigenes*. This newly developed system has opened up a new avenue for engineering of pathways for improved production of biofuels, including hydrogen production.

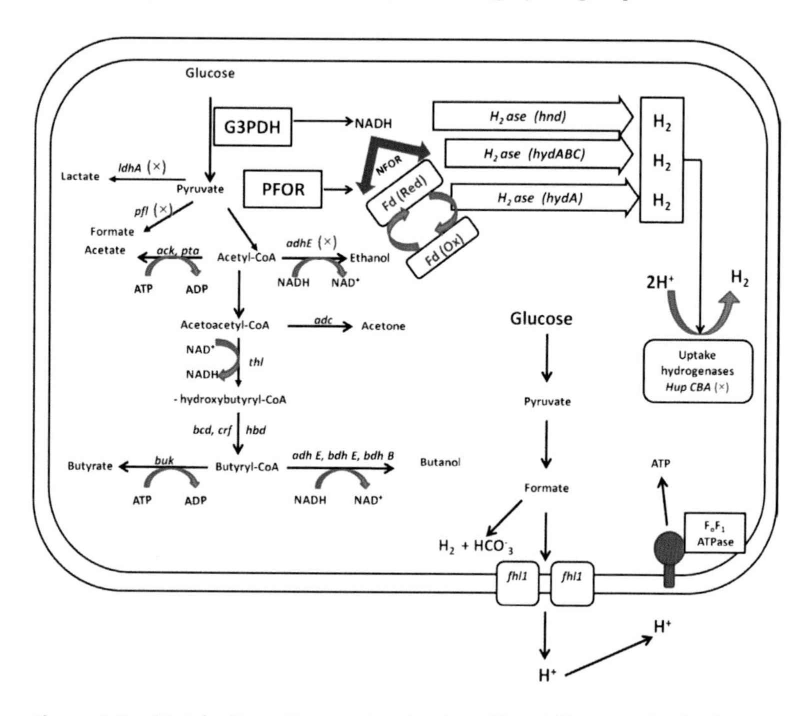

Figure 3.2 Metabolic pathways involved in *Clostridium* sp. for hydrogen production and possible genetic manipulation to increase hydrogen production. (X) represents the knockout of a particular gene. ATP, adenosine triphosphate; ADP, adenosine diphosphate; G3PDH, glyceraldehyde-3-phosphate dehydrogenase; NADH, nicotinamide adenine dinucleotide; Fd, ferredoxin; PFOR, pyruvate ferredoxin oxidoreductase; NFOR, NADH Fd oxidoreductase; CoA, coenzyme A; *ack*, acetate kinase; *ldh*, lactate dehydrogenase; *pfl*, pyruvate formate lyase; *adhE*, aldehyde-alcohol dehydrogenase; *thl*, thiolase; *bdh*, 2,3-butanediol dehydrogenase.

The other approach that can be followed for increasing biohydrogen production is to knock out the genes whose products direct the pool of pyruvate and NADH in pathways not in line with hydrogen production, such as *pfl*-encoding PFL that diverts the pyruvate pool toward formate formation, *adhE*-encoding aldehyde-alcohol dehydrogenase that by utilizing NADH produces ethanol from acetyl-CoA, *adhE*-encoding aldehyde-alcohol dehydrogenase, and *bdh* gene-encoding 2,3-butanediol dehydrogenase that facilitates the formation of butanol from butyrl-CoA by converting NADH to NAD^+. Some *Clostridium* species are well known to degrade cellulose to produce hydrogen. *C. cellulolyticum* and *C. populeti* have been found to produce more hydrogen with cellulose, while *C. acetobutylicum* gives higher hydrogen yields with cellobiose (Ren et al., 2007). It was speculated that the cellulose-degrading pathway could be expressed in other noncellulolytic clostridial strains such as *C. acetobutylicum*. This approach could be highly advantageous for production of hydrogen from lignocellulosic wastes.

3.2.1.2 Challenges in genetic manipulation in hydrogen-producing thermophilic obligate anaerobes

Very few studies are available for metabolic engineering in thermophiles to increase hydrogen production due to lesser availability of compatible vector systems for thermophiles. Contemporary molecular biology techniques available for mesophiles cannot be used for thermophiles. The commonly used vectors (pGEX, PET, pBSK) are not suitable for thermophilic studies because of their instability at higher temperatures. Moreover, most of the antibiotic selection markers are heat labile and very few heat-resistant antibiotics are available (Kanamycin). In recent studies, in the thermophilic strain *Thermoanaerobacterium aotearoense*, the lactate dehydrogenase gene encoded by *ldh* was subjected to knockout for redirecting the pool of NADH toward hydrogen production. The resulted *ldh*-deleted mutant strain showed a twofold increase in hydrogen production with respect to the wild strain (Li et al., 2010).

Although group II introns are well known to knock out the target genes in mesophilic clostridia, since the system is known to have temperature-based limitations, the use of such approaches is incompatible with thermophilic clostridia and is impractical.

Recently, a genetic system for making targeted gene knockouts in thermophilic *C. thermocellum* was developed by using a dual selection marker around the pyrF strain (Tripathi et al., 2010). The Δ pyrF strain is an uracil auxotroph, which can be restored to prototrophy by ectopic expression of pyrF from a plasmid, hence providing a positive genetic control. For selection against plasmid-expressed pyrF, a toxic uracil analog, 5-fluoroorotic acid (5-FOA), can be used. This creates a negative selection for plasmid loss. The genetic tools were validated for application purposes with respect to metabolic engineering by the creation of a *pta* mutant by deletion of the phosphotransacetylase-encoding gene *pta* in *C. thermocellum*. The resulting mutant strain was found to be devoid of any acetate production. A similar study was also reported for *Caldicellulosiruptor bescii*, where the lactate dehydrogenase gene (*ldh*) was deleted to redirect the electron flow from lactate production to acetate and hydrogen production. The *ldh* mutant strain showed increased hydrogen and acetate production (Cha et al., 2013).

Emergence of genetic selections based on the pyrF strain coupled with electroporation-based techniques has enabled us to practice easier transformation and genetic modifications in thermophilic microbes. Advancements in these areas, facilitated by the available genetic tools, has opened up new horizons for further development of mutant microorganisms to knock out the undesired metabolic pathways as well production of sustainable biofuels by enhanced deconstruction of lignocellulosic biomass.

3.2.1.3 Application of metabolic engineering in facultative anaerobes for improvement in hydrogen production

In facultative anaerobes, fermentative hydrogen production is mediated through a multienzyme complex known as the FHL system. FHL facilitates hydrogen production via breakdown of formate. FHL-mediated hydrogen production has been reported in various hydrogen-producing facultative anaerobes such as *Salmonella typhimurium* (Chippauro et al., 1977), *Klebsiella pneumonia, Rhodospirillum rubrum* (Voelskow and Schon, 1980), *Methanobacterium formiccium* (Baron and Ferry et al., 1989), *Enterobacter aerogenes, E. cloacae* IIT-BT 08 (Khanna et al., 2013), and *E. coli* (Peck and Gest, 1957). Facultative anaerobes such as *E. coli* and four types of nickel-iron [NiFe] hydrogenases—hydrogenase-1

(Hyd-1), hydrogenase-2 (Hyd-2), hydrogenase-3 (Hyd-3), and hydrogenase-4 (Hyd-4)—are known for hydrogen metabolism. Hyd-1 and Hyd-2 are uptake hydrogenases and are encoded by the *hya* and *hyb* operons, respectively (Menon et al., 1991; Menon et al., 1994). Hyd-3 is a bidirectional hydrogenase and is responsible for both production and consumption of hydrogen; however, the synthesis reaction predominates (Maeda et al., 2007a). Hyd-3 is encoded by the *hyc* operon and is responsible for hydrogen production via breakdown of formate mediated by the FHL complex at a slightly acidic pH. Hyd-4 is encoded by the *hyf* operon, and in normal cellular conditions it remains nonfunctional. Formate metabolism in *E. coli* is facilitated by the action of various genes and their products, as shown in Fig. 3.3.

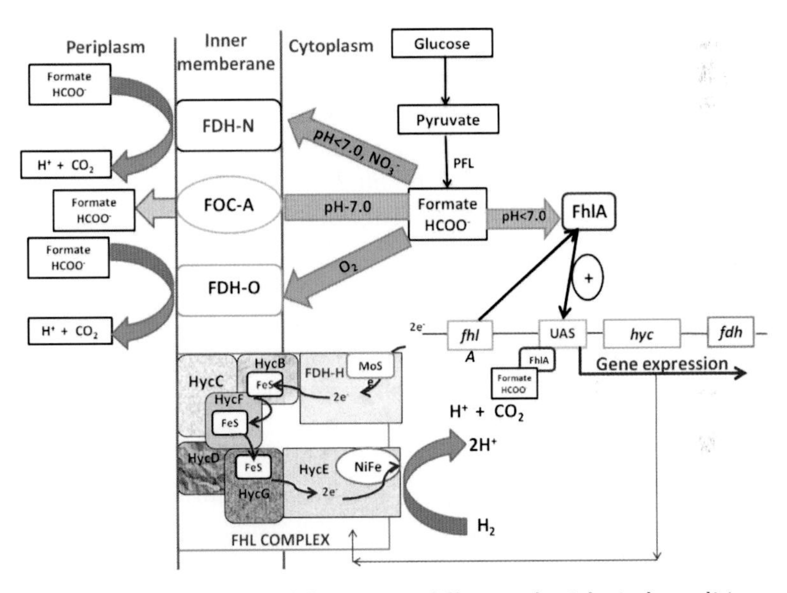

Figure 3.3 Metabolism of formate at different physiological conditions and maturation of the FHL complex involved in hydrogen production in facultative anaerobes. Flow of electrons is shown by red arrows. FDH, formate dehydrogenase; FOC-A, formate transporter; PFL, pyruvate formate lyase; FHL, formate hydrogen lyase.

Formate dehydrogenase-H (FDH-H) is encoded by the monocistronic *fdhF* gene. The FHL system, which is encoded by the *hyc* BCDEFGH genes, accepts electrons from FDH-H and further transfers this electron to a proton to evolve hydrogen. FDHs have small subunits, which act as docking targets of FDH-H. HycB exhibits high similarity of sequences with such small subunits. Among other proteins, HycC and HycD are membrane-integral proteins with a putative anchoring function. HycE and HycG are large and small subunits of hydrogenase-3, respectively. The function of HycF is unknown, but based on the sequence signature, it appears to be an electron carrier. FHL is generated by Hyc-H. HycA and HycI are involved in regulation and maturation of the complex, respectively.

The maturation of hydrogenases is governed by five different proteins encoded by the *hyp* operon. The ultimate gene of this system is *fhlA*, which codes for the central regulator of the formate regulon. The expression of the FHL system is activated by FHLA, a transcriptional activator that in conjunction with a ligand; formate binds to UAS of the *fdhF* gene and activates transcriptional promoter as a consequence, synthesizing FDH-H and other components of the FHL system (Leonhartsberger et al., 2002). The FHL repressor protein (HycA) encoded by *hycA* binds to FHLA so that the FHLA-formate complex is unable to bind to the UAS, stopping the transcription of FHL (Boehm et al., 1990).

Hydrogenase-1 and hydrogenase-2 are uptake hydrogenases, which for activation need to be compulsorily transported by the twin arginine translocation (tat) protein system. These proteins are usually located in the periplasmic space of the cell. However, proteins oriented on the cytoplasmic site, like hydrogenase-3 and FDH-H, are not to be transported.

In facultative anaerobes, pyruvate formed during glycolysis is catalyzed by PFL and converted to acetyl-CoA and formate. Hydrogen is then produced by formate by the FHL complex. In facultative anaerobes, single glucose molecules metabolize into two pyruvate molecules. Later this pyruvate molecule is converted to formate and hence the maximum yield theoretically is 2 mol of hydrogen per mol of glucose. The yield of hydrogen is affected by several factors such as the competitive metabolic pathway that lowers the pool of pyruvate for production of some other metabolite (formation of succinate, lactate) rather than formate. These competing pathways utilizing pyruvate as a reducing agent lower the production of hydrogen drastically (Sinha et al., 2011). The metabolic pathways that can be manipulated in facultative anaerobes are shown in Fig 3.4.

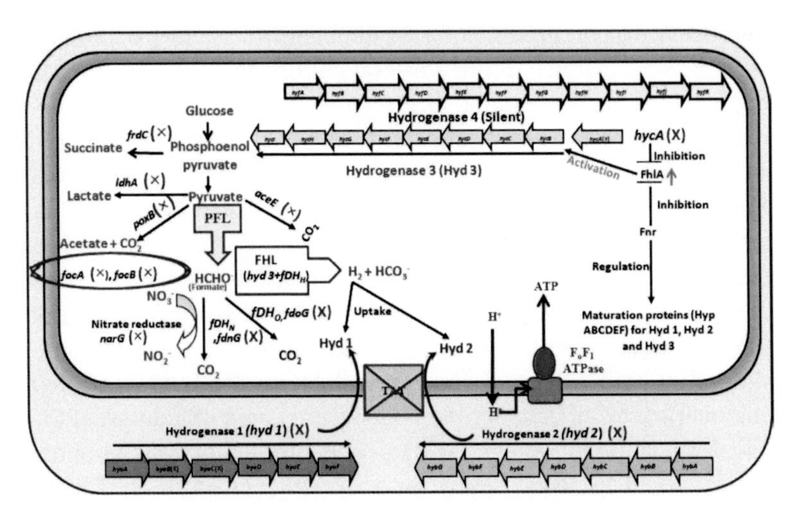

Figure 3.4 Metabolic pathway for hydrogen production in facultative anaerobes. ace, pyruvate dehydrogenase; ldh, lactate dehydrogenase; pox B, pyruvate dehydrogenase; foc, formate transporter; narG, nitrate reductase; fDH, formate dehydrogenase; PFL, pyruvate hydrogen lyase; FHL, formate hydrogen lyase; Fhl A, formate hydrogen lyase activator; hyd, hydrogenase; ATP, adenosine triphosphate; ADP, adenosine diphosphate; (X) represents the knockout of a particular gene; (↑) shows overexpression of a particular gene.

One approach that can be used to enhance hydrogen production is the production of stable hydrogenases via expression in a heterologous host or by knocking out the genes responsible for uptaking hydrogenase. Enhanced hydrogen yields were reported which went up to 41-fold in *E. coli* by cloning the bidirectional hydrogenase (encoded by hox EFUYH) of *Cyanobacterium synechocystis* sp. PCC 6803 (Maeda et al., 2007a). The hydrogenase-1 and hydrogenase-2 uptake was suppressed by the expression of bidirectional hydrogenase in *E. coli* by unknown mechanisms. Another way to increase the hydrogen production is to remove hydrogenase-1 and hydrogenase-2 in modified strains (Mathews and Wang, 2009) or by disturbing the maturation of uptake hydrogenases. The effect of deleting the genes encoding the tat system on hydrogen production in the *E. coli* strain MC4100 was enhanced hydrogen production in two mutant strains defective in tat transport (*tatC* deleted and *tatA-E* deleted). It was reported that the hydrogen production rate doubled from 0.88 ± 0.28 ml H_2 mg^{-1} dcw L^{-1} culture in the parental culture to 1.70 ± 0.15 ml H_2 mg^{-1} dcw L^{-1} culture and 1.75 ± 0.18 ml H_2 mg^{-1} dcw L^{-1} culture, respectively, in the *tatC*-deleted and *tatA-E*-deleted strains. In one such study, the correct assembly of the uptake hydrogenases and FDHs (FDH-O and FDH-N) was hampered on deletion of the tat system, thereby suggesting that a subsequent loss of basal levels of respiratory-linked hydrogen and formate oxidation was the reason behind the increase in formate-dependent hydrogen evolution (Penfold et al., 2006). The literature suggests that when the large subunit of hydrogenase-3, HycE, was subjected to random protein engineering by an error-prone polymerase chain reaction (epPCR) and DNA shuffling, enhanced hydrogen production was obtained (Maeda et al., 2008a). Seven variants with enhanced HycE were obtained. On analysis of results, the most desirable resulting epPCR variant comprised eight mutations (L523Q, M444I, V433L, T366S, A291V, I171T, Y50F, and S2T) and when compared to the wild-type HycE exhibited around 17 times' higher hydrogen production and an eightfold higher yield of hydrogen using formate. Also, a 23-fold higher hydrogen-producing strain was developed from three different highly active HycE variants by deoxyribonucleic acid shuffling. On truncation of the 74-amino-acid carboxy terminal of the large subunit of hydrogenase-3, the same variant also resulted in a ninefold higher yield on formate. Saturation mutagenesis at T366 of HycE also led to increased hydrogen production via truncation at this position. Hence, a 30-fold increase in hydrogen production was observed on deletion of 204 amino acids at the carboxy terminus. Hydrogen production was also improved by knocking out the gene encoding for a small subunit of the uptake hydrogenase (*hybO*) in *E. aerogenes*. A mutant strain IAM 1183-O (Δ hybO) was created by the pYM-Red recombination system, which showed comparatively greater hydrogen production around 18% and a hydrogen yield of about 20% (Zhao et al., 2009). Although *E. coli* is one of the most useful microorganisms as a target for metabolic engineering, from the viewpoint of hydrogen production, a lack of NADH-dependent hydrogenase in the organism creates a barrier for alteration of metabolism of hydrogen in the organism. Several types of NAD(P)H-dependent hydrogenases have been characterized, ranging from soluble [Ni-Fe] hydrogenases (Ralstonia) to [Fe-Fe] hydrogenase, such as the Hnd (hndA, hnd B, hnd C, and hnd D) NADP-reducing hydrogenase of *Desulfovibrio fructosovarans* (Malki et al., 1998). However, till now, the heterologous expression of this hydrogenase in *E. coli* being available hasn't been supported by any credible report. Of late, successful expression of a cytoplasmic, NADP-dependent hydrogenase of a hyperthermophile *Pyrococcus furiosus* in *E. coli* showed promising results (Sun et al., 2010). Remarkably, the native *E. coli* maturation system could also generate a functional hydrogenase when provided subunits and a single hydrogenase-specific protease from *P. furiosus* required for C-terminal cleavage during insertion of the Ni-Fe-active site.

Among the many reasons *E. coli* is used so frequently for genetic modification is one that it has been one of the most well-studied organisms and has a highly characterized genome. Another fact that makes *E. coli* such a desirable host is because it can survive on a variety of carbon sources, making it very adaptable. In addition, a metabolic pathway database, EcoCyc, is also available so that any unknown metabolic pathway can be predicted (Keseler et al., 2005). To enhance hydrogen production, recombinant strains of *E. coli* harboring mutation in several genes were developed, for example, mutations were made in the large subunit of uptake hydrogenase-1 and hydrogenase-2 (*hyaB* and *hybC*, respectively) (Maeda et al., 2008b; Maeda et al., 2007b), in lactate dehydrogenase (*ldh*), in the FHL repressor (*hycA*) (Yoshida et al., 2005) in the FHL activator (*fhlA*), in fumarate reductase (*frdBC*) (R-63, 64), in the tat system (*tatA-E*) (Mathews and Wang, 2009), in FDH-N and FDH-O by altering the alpha subunit (*fdnG* and *fdoG*, respectively), in the alpha subunit of nitrate reductase A (narG), in proteins that transport formate (*focA* and *focB*), in pyruvate oxidase (*poxB*), and in pyruvate dehydrogenase (aceE). In *E. coli*, genetic recombination was employed to elevate hydrogen production efficiency of the resultant SR13 strain. The construction of strain S13 that overexpressed FHL was done by combining inactivation of the FHL repressor (*hycA*) and overexpression of FHL activator (fhlA). It was shown that in the mutant S13 strain, the net hydrogen productivity increased by 2.8 folds compared to the wild type strain; also, transcriptomics results showed transcription of a large subunit of hydrogenase. hycE was increased by 7 folds and that for large subunit of FDH, *fdhF* increased by 6.5 folds (Yoshida et al., 2005).

In another study enhancement in hydrogen, yield was demonstrated by blocking the competing lactate (via deleting ldh) and succinate (via deleting frdBC) pathway. The hydrogen production yield was found to be 1.87 mol hydrogen per mol of glucose in the ldhA and frdBC deleted *E. coli* strain SR15 while it was 1.08 mol hydrogen per mol of glucose in the wild type strain W3110 (Yoshida et al., 2006). In another study, the *hycA* inactivation in the *E. coli* strain HD701 showed higher hydrogen production rates in comparison to the parental strain MC4100. The highest rate of hydrogen evolution by the strain HD701 at a glucose concentration of 100 mM was 31 mL h^{-1} (OD unit)$^{-1}$, while it was much lower, around 16 mL h^{-1} (OD unit)$^{-1}$, in the original strain MC4100 when grown under the same conditions (Penfold et al., 2003). The testing of the strain HD701 was also extended to the use of industrial wastes which had a high sugar content. The results showed improved performance over the parental strain in every case.

The P1 transduction method for development of stable mutation in the *E. coli* K-12 library was used so that metabolic flux could be directed toward production of hydrogen. Alteration in the regulation of FHL was brought about by inactivation of the repressor *hycA* and by overexpression of the activator *fhlA*, and deletion of *hyaB* and *hybC* stopped the hydrogen uptake by the cell; glucose metabolism was redirected toward formate production by incorporating deletions of the *poxB, focB, focA, narG, fdoG* and *fdng*, and *aceE* genes. Succinate and lactate production was ceased by deleting *frdC* and *ldhA* genes. BW25113 *aceE ldhA fdog hycA hybC hyaB* was one such strain among the different metabolically engineered ones that resulted in 4.6-fold increased hydrogen production from glucose with twofold increased yield (from 0.65 to 1.3 mol hydrogen per mol glucose). In a similar study, the theoretical yield of production of hydrogen from formate was obtained by constructing a strain carrying various mutations, which gave a 141-fold increase. It was observed in this mutant strain that the hydrogen yield from glucose increased by 50% (Madea et al., 2008)

In another study, the hydrogen promoting gene (HPP) from *E. cloacae* IIT-BT08 was cloned and overexpressed in *E. cloacae* CICC10017 so that the overall hydrogen yield could be improved. In the recombinant strain, the hydrogenase activity was found to be 534.78 ± 18.51 mL (g-DW h)$^{-1}$, while the hydrogen yield was observed to be 2.55 ± 0.1 mol per mol of glucose. Both the hydrogenase activity and hydrogen yield in the recombinant strain were twofold higher than the wild-type strain. The recombinant strains also exhibited increased production of acetate and butyrate and lesser production of ethanol (Song et al., 2011). The regulators of transcription can also be modified to regulate operon expressions in genes necessary for the biosynthesis of FHL. This can be another possible approach for improving hydrogen production. In one of the studies, a systemic modification strategy was employed for multiple, discrete, metabolic segments that include global regulatory, transport, and auxiliary components necessary for FHL biosynthesis, processing, and assembly. It was observed that modification of transcriptional regulators and metabolic enzymes involved in dissimilation of pyruvate and formate improved the specific hydrogen yield from glucose. The engineered *E. coli* strain ZF1 (disrupted in the formate transporter gene *focA*) and ZF3 (disrupted in a global transcriptional regulator gene *narL*) produced 14.9 and 14.4 μmole of hydrogen (mg of dry cell weight)$^{-1}$, respectively, compared to 9.8 μmoles of hydrogen (mg of dry cell weight)$^{-1}$ produced by the wild-type *E. coli* strain W3110 (Fan et al., 2009).

3.3 Biochemistry behind Biomethane Production

The spent media generated after dark-fermentative hydrogen production are rich in short-chain fatty acids (SCFAs) such as acetate, butyrate, and propioniate. These SCFAs could be potential substrates for methanogenic microorganisms. Moreover, the dissolved hydrogen and CO_2 in the spent media could also serve as substrates for a distinct group of methanogens (hydrogenotrophic methanogens). Interactions among different groups of acetoclastic and hydrogenotrophic methanogens are greatly influenced by the action of many metabolic enzymes.

3.3.1 Methanogenesis from CO_2 and Hydrogen: Bioenergetics of Hydrogenotrophic Methanogens

During acidogenic dark fermentation, CO_2 and hydrogen are produced as gaseous products. A considerable amount of dissolved hydrogen and CO_2 remains in the spent media. These dissolved gases could prove to be energy sources in the metabolic pathway of a special group of methanogens (Fig. 3.5). Hydrogenotrophic methanogenic bacteria depend on CO_2 and hydrogen for energy (Balch et al., 1979), whereas acetoclastic methanogens like *Methanothrix* sp. can perform only acetate metabolism to produce methane (Huser et al., 1982); *Methanolobus tindarius* can only utilize methylamine and methanol (Konig and Stetter, 1982); and *Methanosphaera stadtmaniae* is responsible for reduction of methanol with hydrogen (Miller and Wolin, 1985).

$$CO_2 + 4H_2 \rightarrow CH_4 + 2H_2O \; (\Delta G^o = -131 \text{ kJ mol}^{-1}) \quad (3.10)$$

Under methanogenic conditions, the partial pressure of hydrogen ranges between 1 Pa and 10 Pa. The free-energy change ($\Delta G'$) values for CO_2 and hydrogen were in the range of –20 kJ mol^{-1} to –40 kJ mol^{-1}. However, at least 50 kJ mol^{-1} of free energy is required for in vivo synthesis of ATP from ADP and inorganic phosphate (Thauer and Morris, 1984). Thus, under methanogenic conditions, for 1 mol of CH_4 produced, less than 1 mol of ATP is generated. An exhaustive literature review suggested the existence of chemiosmotic mechanism-mediated coupling of the exergonic formation of CH_4 and the endergonic phosphorylation of ADP (Blaut et al., 1990; Gottschalk and Blaut, 1990).

3.3.1.1 Role of C_1 unit carriers and electron carriers in reduction of CO_2

Methanofuran (MFR), tetrahydromethanopterin (H_4MPT), and HS-CoM are the three major C_1 unit carriers (Fig. 3.6). Thy collectively comprise coenzyme-bound C_1 intermediates that are found in all methanogens. These play a major role in the CO_2 to CH_4 reduction (Wolfe, 1991).

MFR is a C_4-substituted furfurylamine, characteristically present in all methanogenic bacteria (Leigh et al., 1985) and also in *Archaeoglobus fulgidus* (Moller-Zinkhan et al., 1989). Usually, there are five different MFR derivatives that exist in nature, which differ in their respective R-groups. Other than its presence in methanogenic bacteria, H_4MPT can also be found in nonmethanogenic archaebacteria (Moller-Zinkhan et al., 1989), which show significant similarities with tetrahydrofolate. It differs with respect to the two

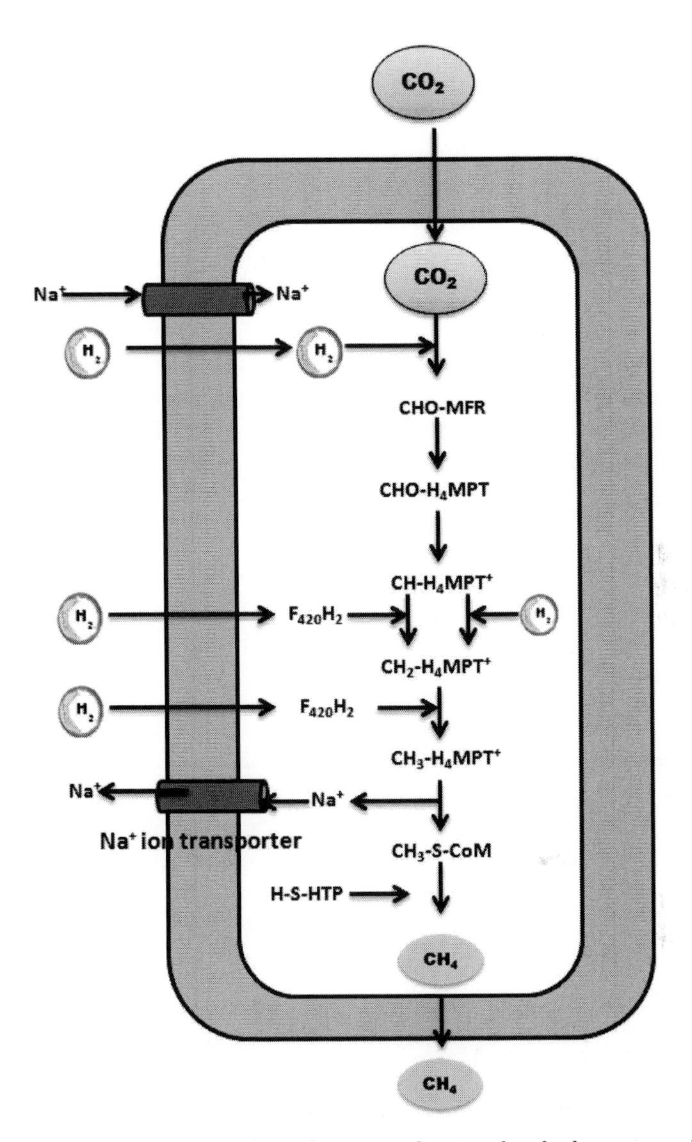

Figure 3.5 Mechanism of methane production by hydrogenotrophic methanogens. CHO-MFR, *N*-formyl methanofuran; CHO-H_4MPT, N^5-formyl tetrahydromethanopterin; CH-H_4MPT^+, N^5-N^{10}-methenyl tetrahydromethanopterin; CH_2-H_4MPT, N^{10}-methylene tetrahydromethanopterin; CH_3-H_4MPT^+, N^5-methyl tetrahydromethanopterin; CH_3-S-CoM, methyl-coenzyme M; H-S-HTP, *N*-7-mercaptoheptanoyl-O-phospho-L-threonine; $F_{420}H_2$-reduced coenzyme F_{420}.

distinct methyl groups present at positions 7 and 9 and by the nature of its substituent R-group. Among the different chemical derivatives of H_4MPT, tetrahydrosarcinapterin (H_4SPT), which is found in *Methanosarcina* spp., differs from H_4MPT only by the presence of an additional glutamyl moiety (van Beelen et al., 1984). Till date, the simplest enzyme known is CoM found exclusively in methanogenic bacteria (Balch and Wolfe, 1979). The functional groups of these three coenzymes, that is, the amino group on methanofuran, the mercapto group of CoM, and the nitrogen present on positions 5 and 10 of tetrahydromethanopterin, serve as the binding sites for the C_1 intermediates. The conversion of CO_2 to methane in methanogenic bacteria involves a large number of electron carriers such as *N*-7-mercaptoheptanoyl-O-phospho-L-threonine (H-S-HTP) and coenzyme F_{420} (Eirich et al., 1978) (Fig. 3.7).

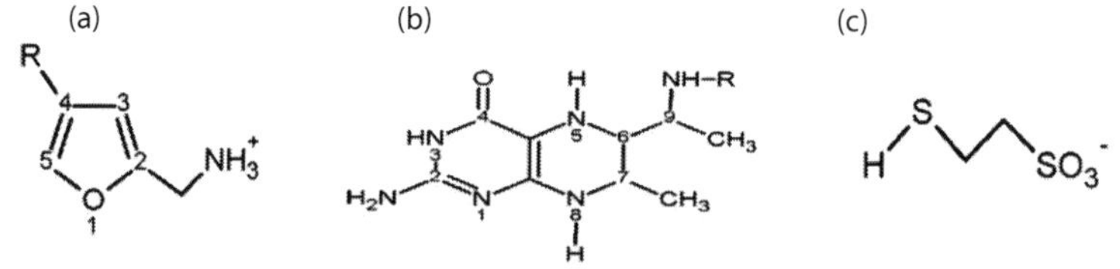

Figure 3.6 (a) MFR, (b) H_4MPT, and (c) H-S-CoM.

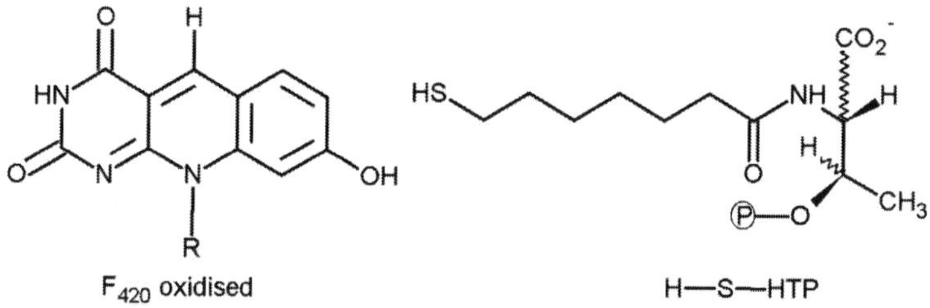

Figure 3.7 Structure of coenzyme F_{420} and *N*-7-mercaptoheptanoyl-O-phospho-L-threonine.

Coenzyme F_{420}'s substituent R can contain multiple glutamyl moieties ranging from two glutamyl to five glutamyl moieties—hence named (F_{420}-2), (F_{420}-3), (F_{420}-4), and (F_{420}-5). Methanogens also contain polyferredoxin with 12 [4Fe-4S] clusters, ferredoxins, and other iron-sulfur proteins, the functions of which haven't been properly explored yet. Methanogens also have cytochromes that are capable of oxidizing methyl groups (of acetate, methanol, or methylamine) to CO_2. However, their activity during the growth of methanogens on hydrogen and CO_2 is yet to be completely elucidated.

3.3.1.2 C_1 intermediates and reactions involved in CO_2 reduction

The CO_2 reduction to methane is characterized by the following coenzyme-bound C_1 intermediates: NS-methyl-H_4MPT (CH_3-H_4MPT), *N*-formyl-MFR (CHG-MFR), NIO-methylene-H_4MPT (CH_2-H_4MPT), NS-formyl-H_4MPT (CHG-H4MPT), NS, NIO-methenyl-H_4MPT (CH-H_4MPT^+), and methyl-CoM (CH_3-S-CoM) (Keltjens et al., 1990).

The corresponding partial reactions can be written as

$$CO_2 + MFR + H_2 \rightarrow CHO\text{-}MFR + H_2O + H^+ (\Delta G^o = +16 \text{ kJ mol}^{-1}) \quad (3.11)$$

$$CHO\text{-}MFR + H_4MPT \rightarrow CHO\text{-}H_4MPT + MFR(\Delta G^o = -4.4 \text{ kJ mol}^{-1}) \quad (3.12)$$

$$CHO\text{-}H_4MPT + H^+ \rightarrow CH \equiv H_4MPT^+ + H_2O(\Delta G^o = -4.6 \text{ kJ mol}^{-1}) \quad (3.13)$$

$$CH \equiv H_4MPT^+ + H_2 \rightarrow CH_2 = H_4\,MPT + H^+(\Delta G^o = -5.5 \text{ kJ mol}^{-1}) \quad (3.14)$$

$$CH_2 = H_4MPT + H_2 \rightarrow CH_3H_4MPT(\Delta G^o = -17.2 \text{ kJ mol}^{-1}) \quad (3.15)$$

$$CH_3H_4MPT + HSCoM \rightarrow CH_3SCoM + H_4MPT(\Delta G^o = -29.7 \text{ kJ mol}^{-1}) \quad (3.16)$$

$$CH_3\,SCoM + H_2 \rightarrow CH_4 + HSCoM(\Delta G^o = -85.7 \text{ kJ mol}^{-1}) \quad (3.17)$$

Equilibrium constants determined from experiments were used to calculate the free-energy changes for reactions represented by Eqs. 3.11 to 3.17. In the absence of experimental values of the constants, analogous reactions were used to calculate the free-energy change for particular reactions. Collectively, the total free-energy change contributed by all of these reactions is –130.4 kJ mol^{-1}. The difference in this free-energy change from that calculated by considering standard free energies for the formation of CH_4 is only

0.6 kJ mol^{-1}. Among the above-mentioned reactions, a few reactions follow energy conservation, for example, Eqs. 3.16 and 3.17 are exergonic reactions that follow the principle of energy conservation caused by the chemiosmotic mechanism, while on the other hand, reversed electron transport drives the endergonic reaction (Eq. 3.11) toward the forward direction.

Three different organisms belonging to the class of methanogenic bacteria have been investigated for enzymes that are responsible for the biocatalysis of reactions (Balch et al., 1979). These include *Methanobacterium thermoautotrophicum*, *Methanosarcina barkeri*, strain Fusaro, and strain MS (Bryant and Boone, 1987), and *Methanopyrus kandleri* with optimal growth temperatures of 65°C (order *Methanobacteriales*), 37°C (order *Methanomicrobiales*), and 98°C, respectively. *Methanopyrus kandleri* is also the sole member of its group so far; hence, the investigation has majorly covered most of the different methanogenic bacteria. It has been discovered that *A. julgidus*, a sulfate-reducing Achaean also contains some of these enzymes. It has also been found out that this strain is phylogenetically related to the group *Methanomicrobiales*.

3.3.1.3 Role of molecular hydrogen for methanogenic activation

Coenzyme F_{420}-reducing hydrogenase and coenzyme F_{420}-nonreducing hydrogenase (Jacobson et al., 1982) are the two [Ni-Fe] hydrogenases generally present in methanogenic bacteria growing on hydrogen and CO_2. The latter enzyme is also known as methylviologen-reducing hydrogenase. However, it hasn't been completely studied and its electron acceptor is still unknown. Both the hydrogenases mentioned can catalyze viologen dye reduction and hence the name methylviologen-reducing hydrogenase is somewhat misleading. Other than the above-mentioned hydrogenases, a third hydrogenase is actively found in most methanogens. This particular hydrogenase is called hydrogen-forming methylene tetrahydromethanopterin dehydrogenase, which contains nickel and/or iron-sulfur clusters (Zimgibl et al., 1990). This particular hydrogenase is unique and very different from all other known hydrogenases.

3.3.1.3.1 *Role of coenzyme F_{420}-reducing [Ni-Fe] hydrogenase*

The reversible reduction of coenzyme F_{420} with hydrogen is carried out by the [Ni-Fe] enzyme. Coenzyme F_{420} is a two-electron hydride acceptor with an E^{o} value of –360 mV (Gloss and Hausinger, 1987) and is stereospecific in nature (Yamazaki et al., 1985).

The above-mentioned reaction is accompanied by a free-energy change ($\sim G^{o}$) of –11 kJ/mol. It has been reported that hydrogen-/CO_2-grown methanogens contain a very active coenzyme F_{420}–dependent methylene-H4MPT dehydrogenase and a coenzyme F_{420}–dependent methylene-H_4MPT reductase (Schworer and Thauer, 1991). The reaction shown in Fig. 3.8 reiterates the role of reduced coenzyme F_{420} for reduction of CH-H_4MPT^+ to CH_2-H_4MPT (Eq. 3.14) and for reduction of CH_2-H_4MPT to CH_3-H_4MPT (Eq 3.15). Several immunogold-labeling techniques suggest that the F_{420}-reducing hydrogenase, a peripheral membrane protein, is localized at the inner surface of the cytoplasmic membrane (Lunsdorf et al., 1991). Most of the activity, however, can be recovered in the soluble cell fraction. The enzyme has been extracted and characterized from *Methanobacterium thermoautotrophicum, Methanobacteriumformicicum, Methanosarcina barkeri, Methanococcus voltae, Methanococcus vannielii,* and *Methanospirillum hungatei* (Fox et al., 1987; Baron and Ferry, 1989; Yamazaki, 1982). It was found to be reversibly inactivated by oxygen (Choquet and Sprott, 1991).

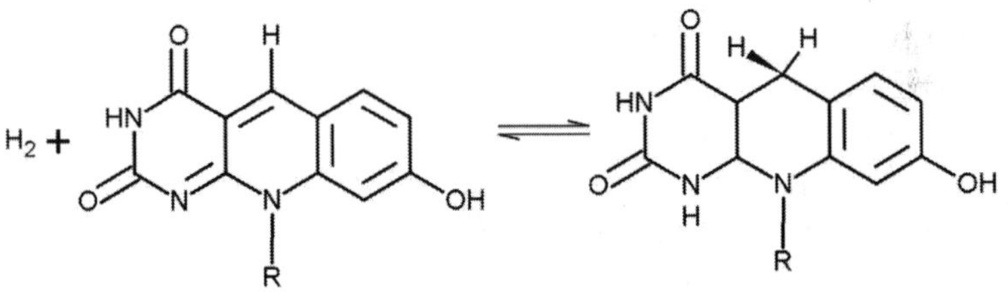

Figure 3.8 Reaction mechanism of reduction of F_{420} by molecular hydrogen.

The enzyme purified from *Methanobacterium thermoautotrophicum* comprises three subunits designated α (47 kDa), β (31 kDa), and γ (26 kDa). The trimeric protein exists as $\alpha_1\beta_1\gamma_1$, and it contains approximately 1 mol nickel, 1 mol FAD, and 13–14 mol nonheme iron and acid-labile sulfur (Fox et al., 1987). The iron-sulfur clusters are organized in [4Fe-4S] clusters. These clusters include the 47 kDa subunit, which contains nickel. The presence of characteristic nickel-binding motifs in the amino terminus (Arg-X_1-Cys-X_2-Cys-X_3-His) and the carboxy terminus (Asp-Pro-CysX_2-Cys-X_2-His) was also reported in this cluster (Alex et al., 1990). It was found out that the amino acid sequence of this particular 47 kDa subunit exhibited 30% homology with the large subunit of the F_{420}-nonreducing hydrogenase obtained from *Methanobacterium thermoautotrophicum* and also with other nickel-containing hydrogenases (Alex et al., 1990). It is also established that oxidation of hydrogen comprises one-electron redox steps, justified by the fact that the F_{420}-reducing hydrogenase also catalyzes reduction of viologen dyes with hydrogen. This also corroborates with the presence of iron-sulfur clusters and describes the presence of flavin in the enzyme, which is required as a one-electron/two-electron switch for the two-electron reduction of coenzyme F_{420} (Walsh, 1979).

3.3.1.3.2 *Role of coenzyme F_{420}–nonreducing [Ni-Fe] hydrogenase*

Information regarding the physiological electron acceptor of the [Ni-Fe] hydrogenase is inadequate. This electron acceptor can catalyzes the reduction of viologen dyes. It also lacks flavins and contains Fe-S clusters responsible for the one-/two-electron switch. Thus it could be speculated that it probably accepts one electron at a time.

$$H_2 + 2\,X \rightarrow 2\,X^- + 2\,H^+ \qquad (3.18)$$

The $E^{o'}$ of the H^+/H_2 couple is –414 mV. When methanogens are grown in the normal conditions, the hydrogen partial pressure which prevails in their growth environment generates $E^{o'}$ of the H^+/H_2 couple between –270 mV and –300 mV. Therefore, the electron acceptor X for the [Ni-Fe] hydrogenase should have $E^{o'}$ near –300 mV. It is also suggested that the catalysis of reactions (Eq. 3.11) and (Eq. 3.17) is done by the F_{420}-nonreducing hydrogenase, which is based on the findings that in vivo the reduction of CO_2 to formyl-MFR and of methyl-CoM to CH_4 with hydrogen are independent of coenzyme F_{420}. Immunogold labeling hasn't been used yet to localize this enzyme in intact cells. However, upon cell breakage most of the enzyme activity is recovered in the soluble cell fraction. The ambiguity in the nature of polyferredoxin points toward a possibility that this iron-sulfur protein might be the physiological electron acceptor of the [Ni-Fe] hydrogenase (Hedderich et al., 1992).

3.3.1.3.3 *Selenocysteine-containing hydrogenases*

The carboxy terminal nickel-binding site of a subunit of both F_{420}-reducing hydrogenase and F_{420}-nonreducing hydrogenase from *Methanococcus voltae* contains selenocysteine instead of a cysteine (Asp-Pro-selenocysteine-X_2-Cys-X_2-His) (Muth et al., 1987). Interestingly, the organism additionally contains the genes for the two [Ni-Fe] hydrogenases without selenocysteine (Halboth and Klein, 1992). However, further details about the conditions of their expression are still unknown. The [Ni-Fe] hydrogenase expression from *Methanococcus vannielii* has also exhibited traces of selenium (Yamazaki, 1982).

3.3.1.3.4 *H_1-forming methylene tetrahydromethanopterin dehydrogenase*

This particular hydrogenase is responsible for catalyzing the reversible reduction of CH-H_4MPT$^+$ to CH_2-H_4MPT in the presence of hydrogen (Fig. 3.9).

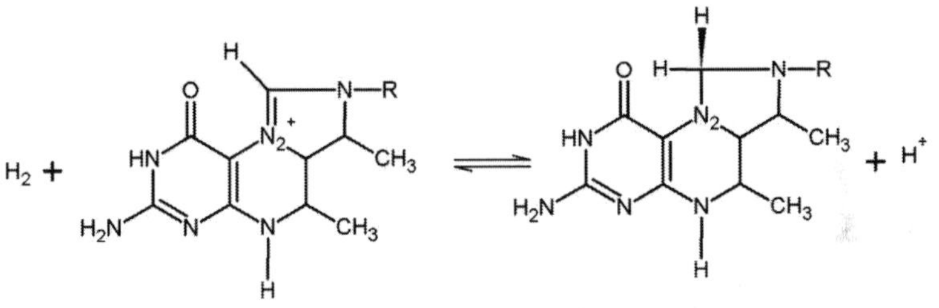

H_2 + [structure] ⇌ [structure] + H^+

Figure 3.9 Reaction mechanism of H_1-forming methylene tetrahydromethanopterin dehydrogenase catalyzing the reversible reduction of CH-H_4MPT$^+$ to CH_2-H_4MPT in the presence of hydrogen.

In comparison to the two [Ni-Fe] hydrogenases, which catalyze the reduction of viologen dyes and perform H_2/H^+ exchange, this enzyme performs neither such operation. Among all methanogens, hydrogen-forming methylene-H_4MPT dehydrogenase is found in the orders *Methanobacteriales* and *Methanococcales* and also in *Methanopyrus kandleri*. However, it is absent in the order *Methanomicrobiales*.

A very active, soluble hydrogenase that has a high tendency to be inactivated in the presence of air has been purified from *Methanobacterium thermoautotrophicum*, *Methanobacterium wolfei*, and *Methanopyrus kandleri* (Zirngibl et al., 1990; Zirngibl et al., 1992; Ma et al., 1991a). This hydrogenase is a homodimer with an apparent molecular mass of 43 kDa. It does not contain nickel or iron-sulfur clusters. However, zinc has been found to be present in significant amounts in the enzyme. On analysis of the enzyme, it was found that the primary structure lacked nickel-binding site sequence motifs that are commonly associated with [Ni-Fe] hydrogenases and of iron-sulfur clusters of bacterial ferredoxin.

3.3.1.4 Utilization of CO_2 for methane production via formation of *N*-formyl methanofuran

CH_4 formation from CO_2 is associated with no formation of free formate. Whatever formate is formed as an intermediate is subsequently oxidized to CO_2 by the methanogens. CO_2 formed is then reduced to *N*-formyl methanofuran, which is an *N*-substituted formamide (Leigh et al., 1985).

The free-energy change associated with the reaction shown in Fig. 3.10 is estimated to be +16 kJ/mol (Keltjens and van der Drift, 1986), which is based on a ΔG^o value of +3.45 kJ/mol for the reduction of CO_2 with hydrogen (both in the gaseous state) to formate. Moreover, the formation of formyl methanofuran from formate and methanofuran has a ΔG^o value of +12.5 kJ/mol. Hence, it can be seen that the reduction of CO_2 to *N*-formyl methanofuran is an endothermic reaction with respect to the low hydrogen partial pressures prevailing in the natural habitats of methanogens. When subjected to these conditions, the free-energy change observed is between +34 and +40 kJ/mol. It could be suggested that the reduction proceeds via carbamate formation from CO_2 and methanofuran in a spontaneous reaction (Ewing et al., 1980).

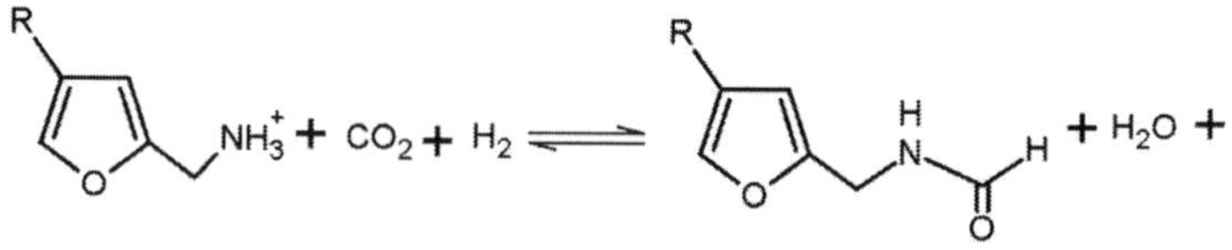

$$R\text{-furan-CH}_2\text{-}NH_3^+ + CO_2 + H_2 \rightleftharpoons R\text{-furan-CH}_2\text{-NH-CHO} + H_2O + H^+$$

Figure 3.10 Mechanism of reduction of CO_2 to *N*-formyl methanofuran.

The free-energy change associated with Fig. 3.11 is estimated to be +6.4 kJ/mol, which is calculated from ΔG^o = +1.6 kJ/mol for HCO_3^- + NH_3-NH_2COO^- + H_2O and ΔG^o = +4.8 kJ/mol for CO_2 (gaseous) + H_2O-HCO_3 + H^+ (Thauer et a., 1977).

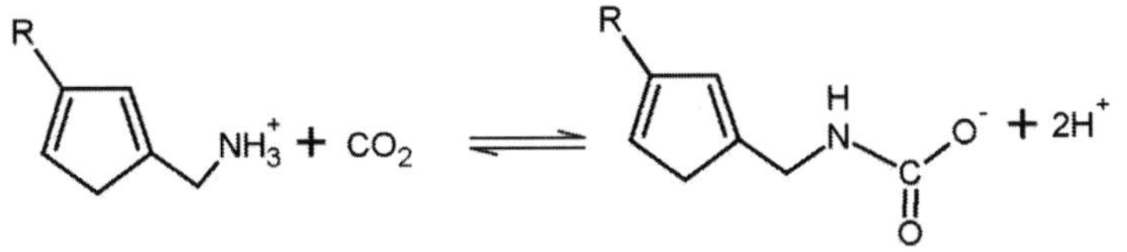

Figure 3.11 Reduction of CO_2 to *N*-carbamoyl methanofuran.

3.3.1.4.1 *CO_2 reduction to N-formyl methanofuran with hydrogen*

The catalytic formation of *N*-formyl methanofuran from methanofuran, CO_2, and hydrogen is majorly governed by three enzyme complexes:

- The F_{420}-nonreducing [Ni-Fe] hydrogenase catalyzing the reduction of an unknown electron acceptor X
- The reversible reduction of (CO_2 + MFR) to CHO-MFR catalyzed by formyl methanofuran dehydrogenase via an unknown electron donor Y
- An oxidoreductase complex which catalyzes the reversed electron transport from reduced X to oxidized Y

The mentioned reactions are described below:

$$H_2 + 2\,X \rightarrow 2\,X^- + 2\,H^+ \tag{3.19}$$

$$CO_2 + MFR + 2\,Y^- + H^+ \rightarrow CHO\text{-}MFR + H_2O + 2\,Y \tag{3.20}$$

$$2\,X^- + 2\,Y \rightarrow 2\,X + 2\,Y^- \tag{3.21}$$

The estimation of the redox potential E' of X and Y is –300 mV and –500 mV, respectively. The redox potential of the CHO-MFR couple is –497 mV. Hence it can be said that the reduction of Y by X^- must be an endergonic reaction that must be energy driven. Several literature studies have suggested that this driving force is probably the sodium motive force (Kaesler and Schonheit, 1989).

Since the mechanism of reversed electron transport involves ion gradients as the driving force, it was proposed that this complex forms an integral part of the cytoplasmic membrane. The purification and characterization of formyl methanofuran dehydrogenase revealed that it undergoes reversible reduction of artificial one-electron acceptors, very similar to that of viologen dyes with *N*-formyl methanofuran (Borner et al., 1989). The enzyme contains either molybdenum or tungsten and is rapidly inactivated in the presence of air. It has been purified from *Methanobacterium wolfei, Methanosarcina barkeri,* and *Methanobacterium thermoautotrophicum* and partially purified from *A. fulgidus* (Borner et al., 1991).

3.3.1.4.2 *Role of molybdenum-containing formyl methanofuran dehydrogenases*

The formyl methanofuran dehydrogenases found in *Methanobacterium thermoautotrophicum* comprise two different subunits of an apparent molecular mass of 60 kDa and 45 kD. Approximately 1 mol of molybdenum, 1 mol of molybdopterin dinucleotide, and 4 mol of nonheme iron and acid-labile sulfur are present per mol of the $\alpha_1\beta_1$ dimer. It was also observed that molybdopterin guanine dinucleotide (Fig. 3.12), molybdopterin adenine dinucleotide, and molybdopterin hypoxanthine dinucleotide were mixed together to form pterin dinucleotide (Borner et al., 1991).

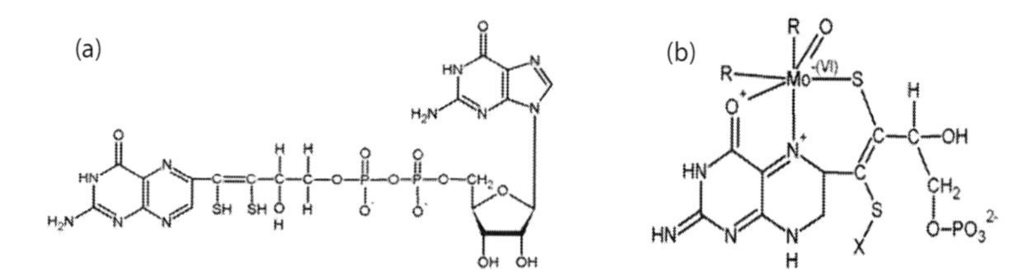

Figure 3.12 (a) Structure of molybdopterin guanine dinucleotide in the oxidized form and (b) structure of the complex with molybdenum in the active site of formyl methanofuran dehydrogenase.

3.3.1.4.3 *A tungsten-containing formyl methanojuran dehydrogenase in* **Methanobacterium wolfei**

Two formyl methanofuran dehydrogenases have been reported in *Methanobacterium wolfei.* The first enzyme, Enzyme I, represents a molybdenum protein and is preferentially formed when the growth medium contains molybdenum. Three subunits together make up this enzyme and are of molecular masses 63 kDa, 51 kDa, and 31 kDa, respectively. Every mol of the $\alpha_1\beta_1\gamma_1$ trimer on an estimate contains 4–6 mol of nonheme iron, 0.3 mol of molybdopterin guanine dinucleotide, 0.3 mol of molybdenum, and acid-labile sulfur. Enzyme II is a tungsten protein and like the enzyme mentioned above contains three subunits. This protein is only formed when the growth medium contains tungstate. This enzyme contains acid-labile sulfur, 4–6 mol of nonheme iron, 0.4 mol of molybdopterin guanine dinucleotide, and 0.4 mol of tungsten per mol of $\alpha_1\beta_1\gamma_1$. The three subunits of this enzyme have apparent molecular weights of 64 kDa, 51 kDa, and 35 kDa, respectively. The enzymes exhibit significantly different catalytic properties and activities, for example, the tungsten enzyme cannot use *N*-furfurylformamide and formate as substrates besides *N*-formyl methanofuran, whereas the molybdenum enzyme can do so (Schmitz et al., 1992).

3.3.1.5 Formyl group transfer from formyl methanofuran to H_4MPT

The formation of N^5-formyl-H_4MPT from formyl methanofuran and H_4MPT is catalyzed by formyl methanofuran:H_4MPT formyl transferase (Fig. 3.13) (Donnelly and Wolfe, 1986). This enzyme can also use *N*-furfurylformamide as a substrate but with much lower catalytic efficiency.

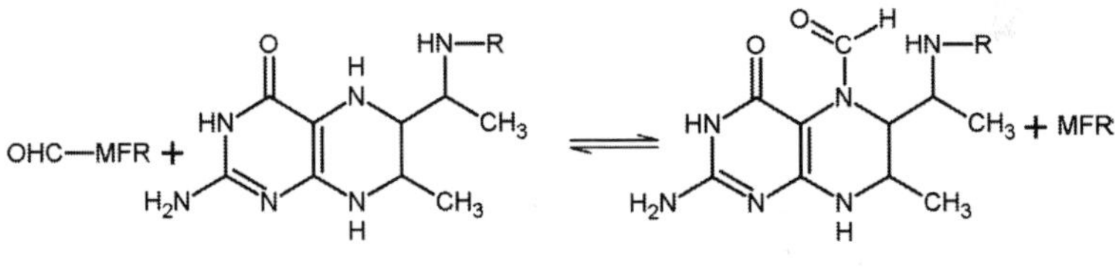

OHC—MFR + [H4MPT structure] ⇌ [N5-formyl-H4MPT structure] + MFR

Figure 3.13 Role of formyl methanofuran:H_4MPT formyl transferase in conversion of formyl methanofuran and H_4MPT to N^5-formyl-H_4MPT.

It was found out in certain reports that in some methanogens N^{10}-formyl-H_4MPT is formed rather than N^5-formyl-H_4MPT. The formation of the N^5-isomer is of special interest in eubacteria and in eukaryotes. The product N^{10}-formyl tetrahydrofolate is the intermediate which is formed in an ATP-dependent reaction instead of a formyl transferase reaction between free formate and tetrahydrofolate. The formyl transferase is a soluble enzyme stable in the presence of oxygen. It's widely present in *Methanobacterium thermoautotrophicum, Methanosarcina barkeri,* and *Methanopyrus kandleri.*

3.3.1.6 Conversion of N^5-formyl-H_4MPT to N^{10}-methenyl-H_4MPT

The reversible hydrolysis of N^5-N^{10}-methenyl-H_4MPT to N^5-formyl-H_4MPT is catalyzed by methenyl-H_4MPT cyclohydrolase (Donnelly et al., 1985).

In alkaline conditions, N^5-N^{10}-methenyl-H_4MPT is spontaneously hydrolyzed to N^{10}-formyl-H_4MPT (Fig. 3.14). The presence of anions increases the rate of hydrolysis. Interestingly, in eubacteria and eukaryotes, N^{10}-methenyl tetrahydrofolate cyclohydrolase mediates the formation of the N^{10}-formyl isomer from N^5-N^{10}-methenyl tetrahydrofolate.

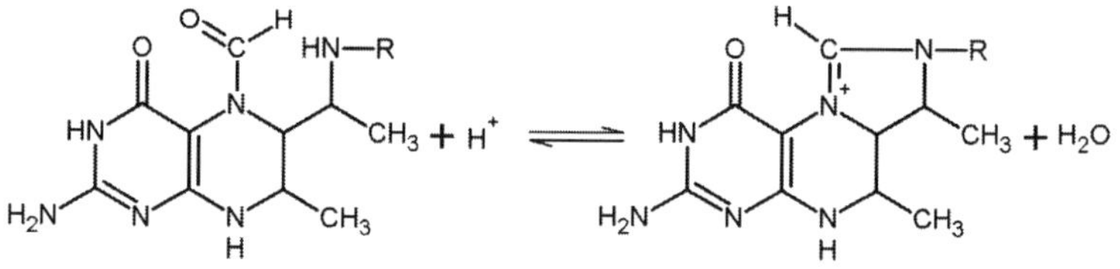

Figure 3.14 Hydrolysis of N^5-N^{10}-methenyl-H_4MPT to N^5-formyl-H_4MPT.

3.3.1.7 Conversion of N^5-N^{10}-methen-H_4MPT to N^5-N^{10}-methene-H_4MPT

All hydrogen-/CO_2-grown methanogenic microbes have coenzyme F_{420}–dependent methylene-H_4MPT dehydrogenase, which catalyzes the reversible reduction of N^5-N^{10}-methenyl-H_4MPT with reduced coenzyme F_{420} to N^5-N^{10}-methylene-H_4MPT. The reduction is probably stereospecific and proceeds via hydride transfer (Fig. 3.15).

The two enzymes mediate the reduction of N^5-N^{10}-methenyl-H_4MPT with hydrogen by the coenzyme F_{420}–reducing [Ni-Fe] hydrogenase:

$$H_2 + F_{420} \rightarrow F_{420}H_2 (\Delta G^{0'} = 11 \text{kJ/mol}) \quad (3.22)$$

$$CH \equiv H_4MPT + F_{420}H_2 \rightarrow CH_2 = H_4MPT + H^+ + F_{420} \; (\Delta G^{0'} = 5.5 \text{kJ/mol}) \quad (3.23)$$

The above-mentioned reactions can be catalyzed by the hydrogen-forming methylene-H_4MPT dehydrogenase in a single step. This enzyme is present in almost all methanogenic bacteria, with an exception of those belonging to the order *Methanomicrobiales.* This enzyme is a polypeptide with an apparent molecular mass of 32 kDa and forms a homopolymer. It also lacks a chromophoric prosthetic group, just like the N^5-N^{10}-methylene tetrahydrofolate dehydrogenase from eubacteria and eukaryotes.

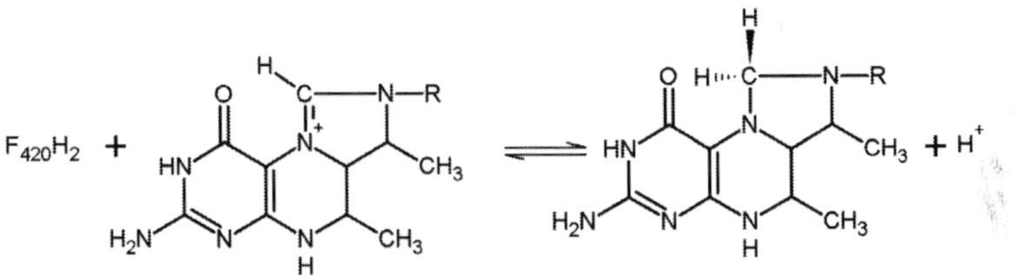

Figure 3.15 Mechanism of N^5-N^{10}-methen-H_4MPT to N^5-N^{10}-methene-H_4MPT.

3.3.1.8 N^5-N^{10}-methene-H_4MPT reduction to N^5-methyl-H_4MPT

F_{420}-dependent methylene-H_4MPT reductase coenzyme is responsible for catalyzing the conversion of N^5-N^{10}-methylene-H_4MPT to N^5-methyl-H_4MPT with reduced coenzyme F_{420} as the electron donor. The reaction mentioned proceeds stereospecifically via hydride transfer (Fig. 3.16).

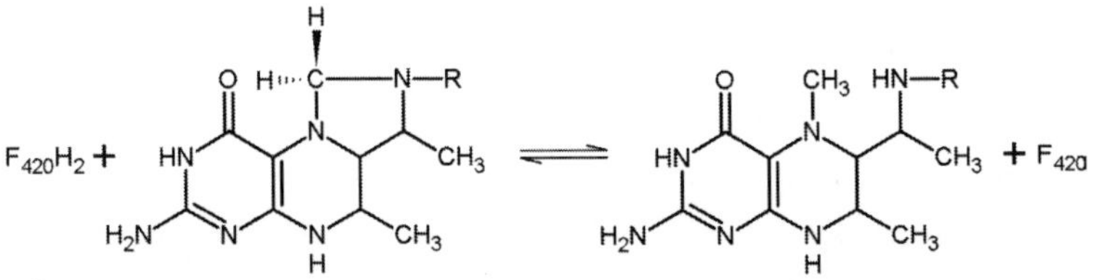

Figure 3.16 Formation of N^5-methyl-H_4MPT.

F_{420}-dependent reductase enzyme has been majorly purified from the *Methanobacterium thermoautotrophicum* strain Marburg, *Methanosarcina barkeri, Methanopyrus kandleri,* and *A. fulgidus* (Schmitz et al., 1991; Ma and Thauer, 1990). This enzyme is a soluble enzyme and is relatively stable in air. The structural studies have revealed that it is a homotetramer or homohexamer of a polypeptide with an apparent molecular mass of 35 kDa.

3.3.1.9 Methyltransfer from N^5-methyl-H_4MPT to coenzyme M

The formation of methyl-CoM from N^5-methyl-H_4MPT and CoM is an ATP-dependent step (Sauer, 1986). This reaction was found to accumulate a methylated corrinoid under strong reducing conditions and the absence of CoM. This product was demethylated upon addition of coenzyme H (CoH). The corrinoid was identified as 5-hydroxybenzimidazolyl cobamide (Pol et al., 1982) (Fig. 3.17), and it was proposed that methyltransfer from methyl-H_4MPT to CoM proceeds in two steps. The first step is the transfer of the methyl group of CH_3-H_4MPT to a corrinoid protein in the reduced form [Co(I)], which is followed by further transfer of this methyl group to CoM:

$$CH_3 - H_4MPT + [Co(I)] \rightarrow CH_3 - [Co(III)] + H_4MPT \quad (3.24)$$

$$CH_3 - [Co(III)] + H - S - CoM \rightarrow CH_3 - S - CoM + [Co(I)] \quad (3.25)$$

The explanation of requirement for reducing conditions and ATP for activity was given assuming that in the presence of traces of O_2, the corrinoid protein is oxidized to the Co(II) form. This oxidized form can be reduced back to the Co(I) form in an ATP-dependent reaction (Fischer et al., 1992). However, the exact role of ATP is still unknown.

$$[Co(I)] + O_2 \rightarrow [Co(II)] + O^-_2 \quad (3.26)$$

$$[Co(II)] + e^- \xrightarrow[ATP]{} [Co(I)] \quad (3.27)$$

Fractionation studies have revealed that for methyl transfer from methyl-H_4MPT to CoM, only the membrane fraction is required and that the polypeptides involved are integral membrane proteins (Fischer et al., 1992), thereby evidencing that methyl-H_4MPT:molecule M methyltransferase is an integral membrane macromolecule.

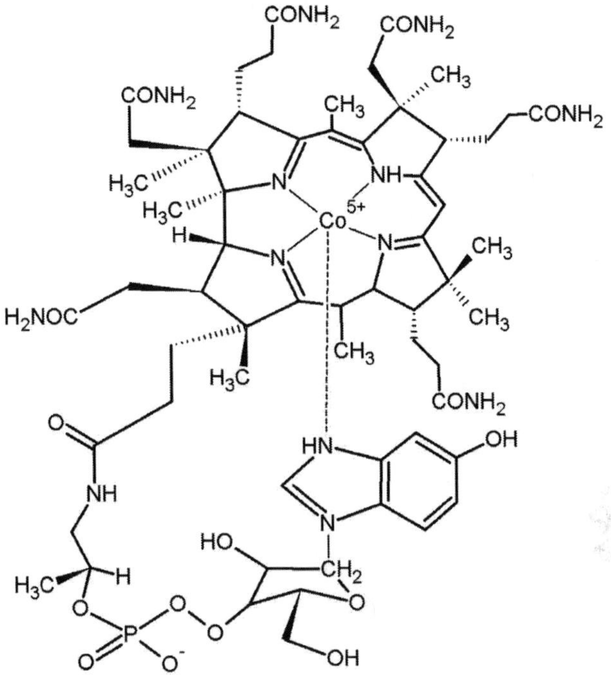

Figure 3.17 Structure of 5-hydroxybenzimidazolyl cobamide (Pol et al., 1982).

3.3.1.10 Reduction of methyl–coenzyme M to methane

The reduction of CO_2 with hydrogen to methyl-CoM returns through steps that are analogous to reactions identified in a nonmethanogenic microorganism. If a reaction was to be distinctive to methanogenic microbes, the reduction of methyl-CoM to the methane series could be considered to be fairly distinctive. Three novel coenzymes (H-S-CoM, H-S-HTP, and molecule F_{430}) and four catalyst complexes are involved in the reaction. The methyl-CoM enzyme and the heterodisulfide enzyme are mentioned in greater detail elsewhere.

3.3.1.10.1 *Four enzyme complexes involved in methyl–coenzyme M reduction with hydrogen to CH_4*

The reduction of methyl-CoM to methane with hydrogen, which is the last step in methanogenesis, involves four enzymes or enzyme complexes that catalyze the following reactions:

$$\mathrm{CH_3-S-CoM + H-S-HTP \rightarrow CH_4 + CoM-S-S-HTP} \quad (\Delta G^{0'} = -45\,\mathrm{kJ/mol}) \tag{3.28}$$

$$\left.\begin{array}{l} \mathrm{H_2 + 2\,X \rightarrow 2\,X^- + 2\,H^+} \\ \mathrm{2\,X^- + 2Z \rightarrow 2\,X + 2\,Z^-} \\ \mathrm{2\,Z^- + 2\,H^+ + CoM\text{-}S\text{-}S\text{-}HTP \rightarrow} \\ \qquad \mathrm{HS\text{-}CoM + H\text{-}S\text{-}HTP + 2Z} \end{array}\right\} (\Delta G^{0'} = -40\,\mathrm{kJ/mol}) \tag{3.29}$$

The heterodisulfide component of CoM (H-S-CoM) is CoM-S-S-HTP. The other part of CoM is H-S-HTP. The role of X is still unknown. It might act as an electron acceptor of the molecule F_{420}-nonreducing [Ni-Fe] hydrogenase, and its chemical reaction potential has been projected to be close to –300 mV. Another still unknown electron donor of the heterodisulfide enzyme is Z. The redox potential E^o of Z was found to be close to –200 mV. The reduction of Z (–200 mV) by X^{-1} (–300 mV) is an exergonic reaction. The reduction of CoM-S-S-HTP with hydrogen cannot be explained via a chemiosmotic mechanism. The probable site of energy conservation is proposed to be the electron transport chain between X and Z. The methyl-CoM enzyme is a heterodisulfide enzyme.

3.3.1.10.2 *Reaction catalyzed by methyl–coenzyme M reductase*

Earlier, it was believed that the methyl-CoM enzyme catalyzes the reduction of methyl-CoM to methane and a free molecule M. However, the mode of action of the catalyst was unclear. The methyl-CoM enzyme might catalyze the reduction of methyl-CoM with H-S-HTP to methane and CoM-S-S-HTP (Fig. 3.18) (Ellermann et al., 1998).

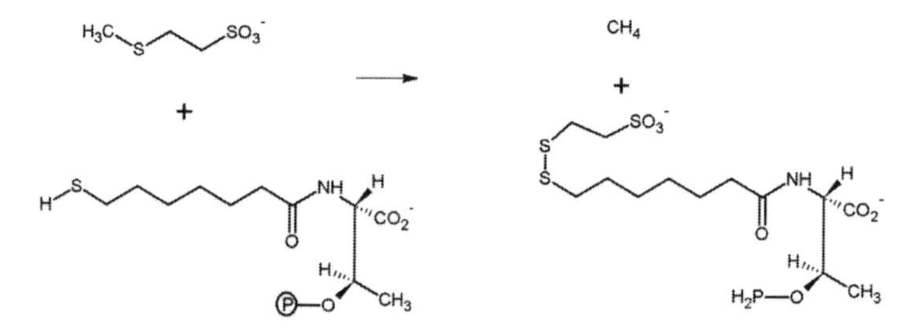

Figure 3.18 Mechanism of reduction of methyl–coenzyme M with H-S-HTP to methane and CoM-S-S-HTP.

It has been observed that methyl-CoM reductase exhibits extreme specificity for both methyl-CoM and H-S-HTP. Among the different methyl-CoM analogues tested, only ethyl-CoM, selenomethyl-CoM, monoftuoromethyl-CoM, and diftuoromethyl-coenzyme are slowly reduced the analogues *N*-6-mercaptohexanoyl-O-phospho-L-threonine. The *N*-8-mercapto-octanoyl-O-phospho-L-threonine acts as an inhibitor for the reaction and cannot substitute for H-S-HTP as an electron donor (Ellermann et al., 1988).

Despite the fact that H-S-HTP may be extremely specific and an effective electron donor for methyl-CoM reduction, evidence suggests that the physiological electron donor of methyl-CoM enzyme may presumably be a bigger molecule that comprises a sugar moiety covalently bonded through a mixed anhydride linkage to H-S-HTP (Fig. 3.19) (Sauer et al., 1990).

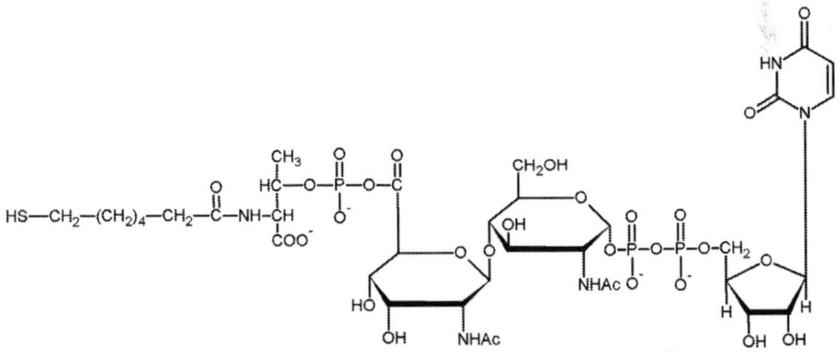

Figure 3.19 Proposed structure of electron donor of methyl-coenzyme M (Sauer et al., 1990).

3.3.2 Acetoclastic Methanogenesis

Earlier, research in the field of acetate conversion to methane had several hypotheses for description of the general mechanism of the process. In 1936, the evidence of reduction of CO_2 to CH_4 along with the generation of four hydrogen molecules with complete oxidation of acetate to two molecules of CO_2 was found (Barker, 1936). Later, with the use of ^{14}C-labeled acetate, it was established that the methyl group and the carboxyl carbon were major contributors to methane production. This observation ruled out the CO_2 reduction theory (Buswell and Sollo, 1948). It was also found out in further studies that the hydrogen (deuterium) atoms of the methyl group are directly transferred to CH_4. Additional studies on the conversion of the methyl group of other substrates led to the conclusion that the ultimate step in methanogenesis from all substrates is reductive demethylation of a common precursor X-CH_3 (Pine and Vishniac, 1957). Reduction of acetate to CH_4 by the supply of exogenous hydrogen was another theory that developed over time. However, this mechanism was ruled out by the finding that acetate is a sole source for growth and methanogenesis by pure cultures.

In most of the acetate-utilizing bacterial anaerobes, cleavage of acetyl-CoA is followed by reduction of an exogenous electron acceptor and oxidation of the methyl and carbonyl groups to CO_2. The acetotrophic methane-producing Archaea group also cleaves acetate. However, in this case, methane is formed by reduction of the methyl group with electrons derived from oxidation of the carbonyl group to CO_2. Thus, it can be concluded that the conversion of acetate to CH_4 and CO_2 is a fermentative process. The phylogenetic difference between the two domains creates doubts for a valid comparison of biochemical mechanisms involved in the cleavage and activation of acetate. Figure 3.20 depicts the pathway as it is currently understood in methane production from acetate. It's generally observed in *Methanosacina* sp. and *Methanothrix* sp.

As described, the first step in the pathway is the activation of acetate to acetyl-CoA, which is then followed by cleavage of the carbon–carbon and carbon–sulfur bonds (decarbonylation). It is catalyzed by the nickel/iron-sulfur component of the CO dehydrogenase (CODH) enzyme complex. The oxidation of the carbonyl group to CO_2 and the reduction of a ferredoxin are primarily carried out by the nickel/iron-sulfur component of the enzyme. The transfer of the methyl group to the corrinoid/iron-sulfur component within the complex and finally to HS-CoM is catalyzed by at least two methyl transferases. The CH_3-S-CoM is demethylated to methane with electrons derived from the sulfur atoms of CH_3-S-CoM and HS-HTP, which results in the formation of the heterodisulfide CoM-S-S-HTP. The electrons derived from reduced ferredoxin carry out the reduction of the heterodisulfide to the corresponding sulfhydryl forms of the cofactors. However, the process requires further investigation as the electron transport chain mechanism from ferredoxin to the heterodisulfide is still unknown.

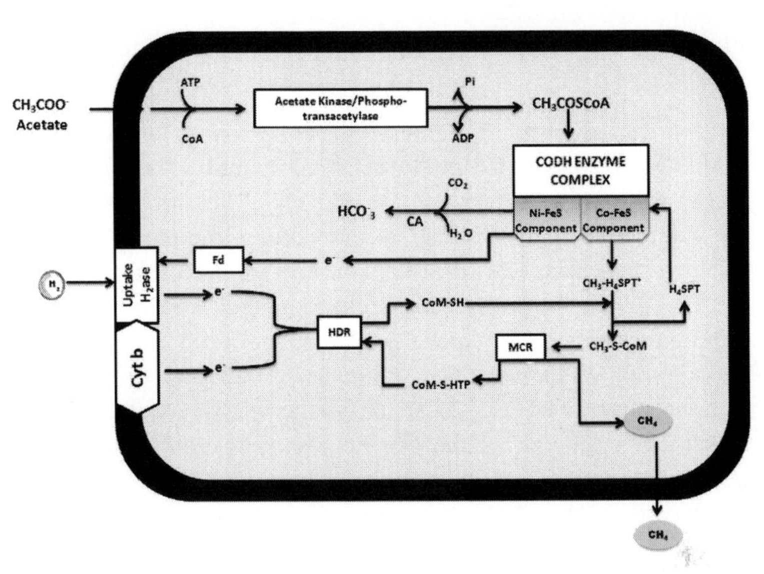

Figure 3.20 Proposed pathway for the conversion of acetate to CO_2 and CH_4 in *Methanosarcina* sp. H_4SPT, tetrahydrosarcinapterin; CA, carbonic anhydrase; MCR, active methylreductase; HDR, heterodisulfide (CoM-S-S-HTP) reductase; Fd, ferredoxin; CODH, carbon monoxide dehydrogenase; cyt b, cytochrome b; H_2ase, hydrogenase.

3.3.2.1 Activation of acetate to acetyl-CoA

Acetyl phosphate and acetyl-CoA were found out to be the potentially activated forms of acetate for methane formation. Few microbes such as *Methanosarcina thermophila* synthesize acetate kinase and phosphotransacetylase only when grown on acetate (Lundie and Ferry, 1989), a result which is consistent with acetate activation by Eqs. 3.29 and 3.30:

$$CH_3COO^- + ATP \rightarrow CH_3CO_2PO_3^{2-} + ADP\ (\Delta G^{0'} = 13\,kJ/mol) \tag{3.29}$$

$$CH_3CO_2PO_3^{2-} + CoA \rightarrow CH_3COSCoA + Pi\ (\Delta G^{0'} = -9\,kJ/mol) \tag{3.30}$$

$$CH_3COSCoA + H_4SPT \rightarrow CoA + CO + CH_3 - H_4SPT\ (\Delta G^{0'} = 62\,kJ/mol) \tag{3.31}$$

$$CH_3 - H_4SPT + HS\text{-}CoM \rightarrow CH_3\text{-}S\text{-}CoM + H_4SPT\ (\Delta G^{0'} = -29\,kJ/mol) \tag{3.32}$$

$$CH_3\text{-S-CoM} + \text{HS-HTP} \rightarrow CH_4 + \text{CoM-S-S-HTP}(\Delta G^{0'} = -43 \text{kJ/mol}) \quad (3.33)$$

Interestingly, the K_m values of acetate kinase are lower for *Methanothrix* sp. (K_m = 0.86 mM acetate) compared to *Methanosarcina* sp. (K_m = 22 mM acetate) (Jetten et al., 1990). This is consistent with the ability of *Methanothrix* sp. to outcompete the *Methanosarcina* for low concentrations of acetate in the environment (Jetten et al., 1990). For both *Methanosarcina* sp. and *Methanothrix* sp., expenditure of one ATP equivalent is made for each mole of acetate metabolized. *Methanosarcina thermophila* falls into the category of strict anaerobes but has an air-stable enzyme acetate kinase, which has a subunit of 53,000 Da. Also, phosphotransacetylase, another air-stable monomeric protein present in *Methanosarcina thermophila* (Mr = 42,000) is purified from the soluble fraction (Lundie and Ferry, 1989). The presence of potassium or ammonium ions in concentrations of around 50 mM increases the activity of this enzyme by sevenfold. Moreover, ions such as phosphate, arsenate, and sulfate cause inhibition of this enzyme. Similar to acetate kinase obtained from *Methanosarcina thermophila, Methanothrix soehngenii* also has an air-stable enzyme acetyl-CoA synthetase. The special characteristic of these enzymes is the presence of the ATP-binding site.

3.3.2.2 Cleavage of the carbon–sulfur and carbon–carbon bonds in acetyl-CoA

CODH plays a central role in the pathway of methanogenesis from acetate and acts as a catalyst during acetyl-CoA. CODHs are commonly found in anaerobes. Complete oxidation of acetate molecules to CO_2 is very common in anaerobes, and this particular process helps in the reduction of different electron acceptors found in the cell. The cleavage of acetyl-CoA by CODH gives methyl and carbonyl groups, which are consequently oxidized to CO_2. In the energy-yielding Ljungdahl–Wood pathway in homoacetogenic bacteria, CODH (acetyl-CoA synthase) is employed to catalyze acetyl-CoA synthesis by reducing CO_2. In methanogenic Archaea cells, carbon synthesis from CO_2 is also carried out by the CODH enzyme. In the pathway, 1 mole of CO_2 is reduced to the methyl level in the form of CH_3-THF. The other 1 mol of CO_2 is reduced to the redox level of CO, which ultimately binds to the CODH enzyme. The methyl group of CH_3-THF is transferred to the corrinoid/iron-sulfur (corrinoid/Fe-S) protein. Then the methyl group is transferred to the CODH enzyme. CODH then catalyzes the synthesis of acetyl-CoA from the CODH-bound methyl, CO, and CoA moieties. The observation of increased CO-oxidizing activity in acetate-grown cells of *Methanosarcina barkeri* was the first indicator of the involvement of CODH in methanogenesis from acetate (Krzycki et al., 1982).

3.3.2.3 Electron transport and bioenergetics during acetate activation

Conversion of acetate to CH_4 and CO_2 yields a small amount of energy, given that acetate activation consumes an equivalent of one ATP.

$$CH_3COO^- + H^+ \rightarrow CH_4 + CO_2(\Delta G^{0'} = -36\,\text{kJ/mol}) \qquad (3.34)$$

But over time these microorganisms have developed mechanisms that are very efficient at conservation of energy. Though substrate-level phosphorylation is not so evident in the reactions in microorganisms, there exists evidence in support of the chemiosmotic mechanism for ATP synthesis. A proton-motive force of −120 mV is generated in whole cells of *Methanosarcina barkeri*, which can degrade acetate. The dependency of electron transport on the membrane-bound proteins in the process of electron transfer from carbonyl of acetyl-CoA to CoM-S-S-HTP is majorly responsible for generating the above-mentioned proton-motive force. Three b-type cytochromes with low midpoint potentials ranging from −330 to −182 mV were reported in acetate-grown cells of *Methanosarcina* and *Methanothrix soehngenii* (Kuhn and Gottschalk, 1983; Kuhn et al., 1983). Although the CO_2-reducing pathway would not be able to proceed if coenzyme F_{420} would not act as an electron carrier (a 5-deazaflavin), F_{420} does not act as an electron acceptor for CODH. However, F_{420} may perform reductive biosynthesis by providing electrons for oxidation of the methyl group of acetate to CO_2. Acetate-grown *Methanosarcina barkeri* can also convert F_{420} to F_{390} in an ATP-dependent reaction, similar to the CO_2-reducing methanogenic species (van de Wijngaard et al., 1991).

In *Methanosarcina thermophila*, a reconstitution of the CODH complex, cytochrome b-linked hydrogenase, and ferredoxin was done for a CO-oxidizing:hydrogen-evolving system (Terlesky and

Ferry, 1988). Although acetate-grown *Methanosarcina thermophila* is unable to utilize H_2-CO_2 for methanogenesis, they do exhibit hydrogen-dependent heterodisulfide reductase activity (Clements et al., 1993). Thus, it can be postulated that the CoM-S-S-HTP is reduced by the electrons provided by oxidation of hydrogen. Moreover, the membrane-bound electron carriers had the ability to couple acetate oxidation (carbonyl group) and hydrogen evolution. According to this hypothesis, a possibility of synthesizing ATP by a chemiosmotic mechanism arises. In *Methanosarcina barkeri*, which are grown on acetate, there exists a coupling between CO oxidation to CO_2, hydrogen, and proton translocation. The findings of methane production by feeding on acetate in *Methanosarcina barkeri* have led to the proposal that a membrane-bound electron transport system is operative in whole cells and requires hydrogen to direct the electron flow in case membranes are not present. Sodium plays a critical role in methane formation from acetate, and the sodium ion gradient generated happens simultaneously along with this process in *Methanosarcina barkeri*. The transmembrane protein CH-H_4MSPT:HS-CoM methyltransferase, which was recently found in acetate-grown *Methanosarcina barkeri* (Fischer et al., 1992), catalyzes an exergonic reaction. This exergonic reaction is coupled to a primary electrogenic sodium extrusion in the CO_2-reducing and methanol utilization pathways. However, the possibility of the reaction being coupled to sodium extrusion during methanogenesis from acetate is yet to be explored and not much information is available about it.

3.4 Challenges and Opportunities toward Commercialization of Biohythane Technology

For the transition of a fossil fuel–based economy to a hydrogen energy–based economy, several challenges need to be addressed, such as:

- Inherent bottleneck of low hydrogen yield: The hydrogen yields obtained so far from the above-described processes are too low for commercial application. To improve the yields, identification of diverse pathways, chief enzymes, and electron acceptors involved in hydrogen production is crucial, information about which is still lacking till date. Moreover, the reaction is limited by thermodynamic constraints, making it energetically unfavorable. Till date, among the different enzymes which take part during dark fermentation hydrogenase I and [Fe-Fe] hydrogenase from *C. pasteurianum* (CpI) and *Desulfovibrio desulfuricans* (D*d*HydAB) are the only two [Fe-Fe] enzymes whose crystal structures have been studied. Due to this, there is no proper understanding of the catalytic processes and development of engineered hydrogenases. Also, compared to bioethanol or methane production, biohydrogen production is limited by certain thermodynamic constraints that make the process technologically and economically more challenging. Moreover, the efficiency of the process is severely disrupted by hydrogen partial pressure that contributed to the existing problem of lower yield. However, integration of processes might provide a solution to overcome this problem.

- Expensive costs pertaining to processing of biomass feedstock: It is essential that for improving the feasibility of the process, inexpensive methods for growing, harvesting, transporting, and pretreating energy crops and/or biomass waste products need to be developed. Dark fermentation has an added advantage of utilizing biomass waste for generation of clean energy; however, this waste needs to be pretreated before it can be used effectively. For example, for biochemical conversion of lignocellulosic biomass into biohydrogen, alteration of the cellulosic biomass structure is essential to make cellulose accessible to enzymes that can convert complex carbohydrate polymers into simple fermentable sugars (Mosier et al., 2005). Among all the processing steps, pretreatment is recognized as a crucial and expensive step and several recent review articles provide a general overview of the process (Alvira et al., 2009; Carvalheiro et al., 2008). Several physical, chemical, and biological pretreatment processes have been reported in the literature (Harmsen et al., 2010); however, these methods result in sugar loss and increased operational costs. Hence, to address this major bottleneck, an expeditious investigation into development of efficient and cost-effective processes is required.

- Inefficient substrate degradation efficiency: Incomplete substrate conversion during dark fermentation limits the total cumulative biohydrogen and biomethane production. So for improvement in process yields and improved conversion of substrates, new metabolic engineering approaches are being adopted to alter existing and introduce novel pathways. In general, chemical oxygen demand (COD) reduction during dark-fermentative hydrogen production is in the range 40%–50%. A considerable amount of unutilized substrate still remains in the system. Implementation of the biomethanation process further improves COD reduction to 30%–40%. So improved cumulative COD reduction could be achieved under the biohythane process.
- Need of the hour: potent and robust strains: At present, we still are facing a lack of microorganisms that are robust enough and can be used for industrial production of hydrogen, producing more than 4 mol of hydrogen per mol of glucose. Thus use of metabolic approaches for creation of an able organism is essential for achieving significantly high yields for biohydrogen production via dark fermentation. The ideal outcome of this newly created microorganism would be to produce high hydrogen yields from an inexpensive feedstock. To achieve this, we need to alter the electron flux and overcome the metabolic limitations which affect the hydrogen production, for which a number of high-priority technical breakthroughs and R&D activities are needed. These include genetic tools that can help (e.g., use of genomic database, genome shuffling, use of a single host organism for expression of genes involved in production of hydrogen). Few obligate extreme thermophilic microbes such as *Thermotoga* sp. are well known to produce hydrogen near the theoretical limit. The operational constraints associated with fermentation at extremely high temperatures mar the prospect of using such microbes in a large-scale setup. In the case of the biomethanation process, most of the microbes associated with the process are slow-growing obligate anaerobes. Stringent operational requirements for obligate anaerobes such as

creation of an oxygen-free environment and dominance of sulfur-reducing bacteria create constraints in the process. Moreover, very few potent strains have been identified related to biomethanation.

- Engineering challenges: Difficulty in sustaining a long-term, steady hydrogen production rate, avoiding interspecies hydrogen transfer in nonsterile conditions, and scaling up and separation/purification of hydrogen are the key engineering bottlenecks that need to be addressed. The reactor configurations used at present for industrial hydrogen production do not perform optimally, that is, they lack robust, reliable performance and have low sustained hydrogen yields. Therefore, extensive research is called for to improve reactor designs and techniques for process parameters, such as mixing, pH, temperature control, and cell harvesting. Development of a hybrid system encompassing the need of both hydrogen producers and methanogens harbors great challenges but also provides an opportunity to improve gaseous energy extraction.
- Oxygen sensitivity of key metabolic enzymes in the biohythane process: Sensitivity to oxygen remains a critical issue during hydrogen process development. For the past few decades, several researchers have attempted to develop oxygen-insensitive hydrogenase. However, the results so far have been discouraging. Oxygen irreversibly destroys [Fe-Fe] hydrogenases, while it reversibly inactivates the catalytic function of [Ni-Fe] hydrogenases. Spectroscopic techniques have revealed that iron and nickel bind to oxygen species, forming a complex (Brecht et al., 2003). Under turnover conditions, this compound keeps the formal hydride binding position occupied and is removed reductively, thereby exposing the catalytic site for activity of hydrogen. Recently, Bingham et al. (2012) elucidated upon the finding of a *C. pasteurianum* mutant strain (CpI) in which [Fe-Fe] hydrogenase I had higher sensitivity toward hydrogen. A randomly mutated CpI library along with cell-free protein synthesis screening was used as a basis for identification of a mutant which exhibited significantly higher tolerance toward oxygen. For sensitivity toward oxygen, three mutations were

found to be majorly responsible. These exhibited higher tolerance on being subjected to polymerase chain reaction (PCR) mutagenesis. They observed that under the resting-state condition of the enzyme, compared to the wild-type strain, the mutant hydrogenase retained significant enzyme activity even after oxygen exposure. However, surprisingly, after oxygen exposure under the catalyzing state of the enzyme during hydrogen production, the mutant hydrogenase showed no improvement in oxygen tolerance. Similar studies like this depict the complicated catalytic nature and oxygen sensitivity of hydrogenase enzymes.

Many models and explanations have been used to describe the possibility of overcoming oxygen instability in hydrogenases. According to the literature, diffusion of oxygen to the active site pocket through the protein and oxygen sensitivity of [2Fe] H subcluster were identified as the two rate-limiting steps of oxygen inhibition (Roseboom et al., 2006). Also the oxygen inactivation mechanism is actively controlled by accessory [FeS] clusters (Lemon and Peters, 1999). It was explained that the distal Fe of the [2Fe] H subcluster binds with the oxygen molecule, which is farthest from the Fe^{2+} of the [4Fe4S]H cluster, resulting in the formation of reactive oxygen species (ROS), thereby inactivating the [4Fe4S]H subcluster. This distal Fe is supposed to act as a hydrogen-binding site as well as a site for reversible binding of carbon monoxide (Peters et al., 1998).

Certain other research groups had tried using protein film electrochemistry for studying the inactivation of oxygen. In this process, hydrogenase was adsorbed onto an electrode during exposure to oxygen and its activity was directly measured by electron transfer through the electrode under oxidizing or reducing potentials. These experiments had indicated that oxygen inactivation was averted by reversible carbon monoxide inhibition and also provided further evidence of the distal Fe atom binding to the oxygen (Pandey et al., 2008). The literature also suggests that oxygen diffusion occurs at the hydrogenase active site that causes enzyme inactivation. Considering this, as described by Stripp et al. (2009), a method was proposed to overcome oxygen sensitivity by limiting diffusion of oxygen molecules to the active site. Using this research as the basis, Boyer et al. (2008) developed a screen that in comparison to the wild-type protein showed higher retention of activity. The mutated hydrogenases were activated in the screen and expressed in reactions following the principle of cell-free protein synthesis. To further study the proteins, oxygen exposure was provided and corresponding hydrogenase activities were recorded. Bingham et al. reported in 2012 that the protein structure could be modified in such a way that more protection is offered to the active site. This would eventually ensure proper exclusion of oxygen. This process is independent of the fact whether the enzyme is actively involved in catalyzing the hydrogen conversion reaction at that time.

Many vital enzymes involved in methanogenesis are oxygen sensitive. Enzymes such as acetate kinase, CoM, and methyl reductase are well-known enzymes that play a vital role in biomethanation. Oxygen sensitivity of these enzymes could prove to be disadvantageous if the spent media used for biomethanation get oxygen contamination.

- Suppression of methanogens in mixed consortia during enrichment of acetogenic hydrogen producers: Compared to pure strains, mixed consortia showed promising hydrogen production. However, in certain cases, hydrogen production was not observed as it was rapidly consumed by the methanogens. Thus, for successful biological hydrogen production, inhibition of hydrogen-consuming microorganisms such as methanogens is a prerequisite. Pretreatment of parent culture is one of the strategies that is used for selecting the required microflora. To develop hydrogen-producing seeds, the fundamental basis relies on physiological differences between hydrogen-producing bacteria and hydrogen-consuming bacteria (methanogenic bacteria). Methanogens have an inability to survive under extreme environments such as high temperature, extreme acidity, and alkalinity, while spore-forming hydrogen-producing bacteria such as clostridia can withstand and survive in such environments. Traditional dilution methods can also be used to solve this problem. Another strategy to suppress methanogens from a mixed culture is by maintaining a short hydraulic retention time (HRT) (2–10 h) as hydrogen-producing bacteria grow faster than methanogens (Fang and Liu, 2002; Fan et al., 2006; Lin and Jo, 2003).

- End metabolite formation competes with hydrogen production: End metabolites formed during anaerobic fermentation, like acetate and butyrate, are associated with hydrogen production and yield 4 and 2 mol of hydrogen per mol of glucose, respectively. However, the production of hydrogen is independent of certain metabolites such as ethanol, lactate, and propionate. These metabolites drive the reducing equivalents or protons away from hydrogen production, thereby decreasing the overall yield. Consequently, the present research focus has shifted to redirect the metabolic pathways by deleting or blocking other end metabolites.
- Pertaining issue of thermodynamic limitations: For making biohydrogen production commercially feasible, the thermodynamic limitations of the processes need to be overcome. The theoretical yield from glucose is restricted to 4 mol of hydrogen per mol of glucose if the acetate pathway is followed; however, around 10 mol of hydrogen per mol of glucose are required to make this technology commercially compatible to methane or ethanol fermentation (Benemann, 2008). In 1977, Thauer and his coworkers proposed the Thauer limit, according to which there is a restricted limit of 4 mol of hydrogen per mol of glucose. Even after nearly 50 years of research, the Thauer limit is still in existence, which prevents large-scale commercialization of this technology. About 75%–80% of the total cellular energy is consumed in releasing reduced end products, for example, propionic acid, alcohol, and butyric acid, which reduce the overall hydrogen yield and hence only a small amount is recovered as gaseous energy. The already commercialized processes of biomethane and bioethanol production show about 90% energy conversion from glucose. The Thauer limit sets thermodynamic limitations on the process, where the maximum free energy (–215 kJ under STP) is obtained during the conversion of glucose to 2 mol of acetate and 4 mol of molecular hydrogen. ATP generation by the cell is one major

process where this energy is utilized; however, ATP generation is not limited to acetate-based pathways and, for instance, the butyrate pathway is also quite capable to produce ATP. The decision of which pathway is to be taken rests solely upon the free-energy availability. During the process, a few reducing equivalents are produced in the form of NADH or ferredoxin, which during hydrogenase activity to produce hydrogen are employed as electron carriers. The midpoint redox potentials of NAD^+/NADH and oxidized ferredoxin/reduced ferredoxin are −320 and −398 mV, respectively. These reducing equivalents must be recycled to maintain the continuity of the process, which can be accomplished by several reactions such as reduction of pyruvate by NADH for ethanol production, though the governing factor for the reactions to occur is the standard free-energy change ($\Delta G^{o\prime\prime}$). It is observed that bacteria undergo mixed-acid fermentation when glucose is not a limiting substrate, owing to mixed-acid fermentation-induced improved growth. This gives a much faster growth when compared to the growth where only acetate is produced. However, the latter growth is much more efficient when it comes to ATP production. The redox potential of the couple H^+/H_2 is formed by the reducing action of NADH on protons, resulting in a potential difference of −414 mV, which is higher than that of the NAD+/NADH couple. Thus, molecular hydrogen formation by NADH oxidation is thermodynamically constrained, whereas ferredoxin oxidation to produce molecular hydrogen is more favorable. At high temperatures, when production of hydrogen is targeted using thermophiles, the Thauer limit of 4 mol of hydrogen per mol of glucose is nearly met. In such a case, a large entropic factor favors high hydrogen production accounting for its higher redox potential, which allows use of NADPH as a reductant, as described above. Stoichiometrically, 12 mol of hydrogen can be obtained per mol of glucose, but this entire pathway is associated with a very low Gibbs free-energy change of 26 kJ per mole of glucose (~2 kJ/mol hydrogen, at 25°C, 1 bar of hydrogen and physiological pH, and bicarbonate concentration). Thus, because of insufficiency of energy for the reaction to take place, it doesn't take place. In a scientific study, pentose phosphate pathway enzymes were put to good use for conversion of glucose-6-phosphate to 11.6 mol of hydrogen. This experiment was conducted in 2000 by Woodward et al. This experiment was performed on hyperthermophile *Pyrococcus furiosus,* where NADPH was generated by the action of these enzymes (of the pentose phosphate pathway) and this further reduced the hydrogenase, leading to hydrogen production (Woodward et al., 2000). Though the operations on glucose-6-phosphate is carried out at low partial pressures, improved feasibility at high partial pressures, and a quicker reaction rate can only be obtained if a substrate with energy content higher than glucose-6-phosphate is used. More research hence needs to go into usage of improved substrate molecules.

3.5 Conclusions

The metabolic pathways involved in the biohythane process are complex in nature. The complexity of metabolism comes from the diversity of microbes present in the two-stage process. The biohydrogen production potential of 4 mol per mol of glucose has been the major bottleneck for the single-stage gaseous energy recovery process. Low activity of the methanogenic enzymatic pathway makes the overall process slow. Thus the metabolic interplay between the two processes has been a point of interest in recent times. Facultative, dark-fermentative hydrogen producers follow different pathways compared to obligate anaerobes. Facultative microbes use a unique enzyme complex called the FHL complex to produce hydrogen from formate. Obligate anaerobes show an entirely different pathway for biohydrogen production. They have [Fe-Fe] hydrogenase, which can donate an electron to the H^+ ion and convert it to molecular hydrogen. For this NADH is the reducing equivalent that is required by the [Fe-Fe] hydrogenase. In the case of methane production, two distinct domains of microbes were identified on the basis of their source of electron donors. In the two-stage process, the roles of hydrogenotrophic methanogens and acetogenic methanogens are crucial. The dissolved hydrogen remaining after first-stage hydrogen production could be a potential source of electron donors to convert CO_2 to methane. Similarly, the acetate and butyrate-rich spent media can provide electrons to reduce CO_2 to methane. This chapter encompasses all the metabolic features of biohydrogen and biomethane-producing bacteria.

References

Abreu AA, Alves JI, Pereira MA, Karakashev D, Alves MM, Angelidaki I (2010). Engineered heat treated methanogenic granules: a promising biotechnological approach for extreme thermophilic biohydrogen production, *Bioresour Technol,* **101**, 9577–9586.

Ahn Y, Park EJ, Oh YK, Park S, Webster G, Weightman AJ (2005). Biofilm microbial community of a thermophilic trickling biofilter used for continuous biohydrogen production, *FEMS Microbiol Lett,* **249**, 31–38.

Alex LA, Reeve JN, Orme Johnson WH, Walsh CT (1990). Cloning, sequence determination, and expression of the genes encoding the subunits of the nickel containing 8-hydroxy-5-deazaflavin reducing hydrogenase from *Methanobacterium thermoautotrophicum* ΔH, *Biochemistry,* **29**, 7237–7244.

Alvira P, Tomas Pejo E, et al. (2009). Pretreatment technologies for an efficient bioethanol production process based on enzymatic hydrolysis: a review, *Bioresour Technol,* **101**, 4851–4861.

Balch WE, Wolfe RS (1979). Specificity and biological distribution of coenzyme M (2 mercaptoethanesulfonic acid), *J Bacteriol,* **137**, 256–263.

Balch WE, Fox GE, Magru LJ, Woese CR, Wolfe RS (1979). Methanogens: reevaluation of a unique biological group, *Microbiol Rev,* **43**, 260–296.

Barker HA (1936). On the biochemistry of the methane fermentation, *Arch Microbiol,* **7**, 404–419.

Baron SF, Ferry JG (1989). Reconstitution and properties of a coenzyme F420 mediated formate hydrogen lyase system in *Methanobacterium formiccium, J Bacteriol,* **171**, 3854–3859.

Baron SF, Ferry JG (1989). Purification and properties of the membrane-associated coenzyme F 420-reducing hydrogenase from *Methanobacterium jormicicum, J Bacteriol,* **171**, 3846–3853.

Benemann JR (1996). Hydrogen biotechnology: progress and prospects, *Nat Biotechnol,* **14**, 1101–1103.

Bingham AS, Smith PR, Swartz JR (2012). Evolution of an [FeFe] hydrogenase with decreased oxygen sensitivity, *Int J Hydrogen Energy,* **37**, 2965–2976.

Blaut MV, Miiller Gottschalk G (1990). Energetics of methanogens. In: Sokatch JR, Ornston LN (eds), *The Bacteria*, Vol XII, 505–537. Academic Press, San Diego, USA.

Boehm R, Sauterm M, Boeck A (1990). Nucleotide sequence and expression of an operon in an *Escherichia coli* coding for formate hydrogen lyase components, *Mol Microbiol*, **4**, 231–244.

Borner G, Karrasch M, Thauer RK (1991). Molybdopterin adenine dinucleotide and molybdopterin hypoxanthine dinucleotide in fonnylmethanofuran dehydrogenase from *Methanobacterium thermoautotrophicum* (Marburg), *FEBS Lett*, **290**, 31–34.

Borner G, Karrasch M, Thauer RK (1989). Fonnylmethanofuran dehydrogenase activity in cell extracts of *Methanobacterium thermoautotrophicum* and of *Methanosarcina barkeri*, *FEBS Lett*, **244**, 21–25.

Boyer M, Stapleton JA, Kuchenreuther JM, Wang C, Swartz JR (2008). Cell-free synthesis and maturation of hydrogenases, *Biotechnol Bioengen*, **99**, 59–67.

Brecht M, van Gastel M, Buhrke T, Friedrich B, Lubitz W (2003). Direct detection of a hydrogen ligand in the [NiFe] center of the regulatory H_2-sensing hydrogenase from *Ralstonia eutropha* in its reduced state by HYSCORE and ENDOR spectroscopy, *J Am Chem Soc*, **125**, 13075–13083.

Breitung J, Thauer RK (1990). Fonnylmethanofuran: tetrahydromethanopterin formyltransferase from *Methanosarcina barkeri*, identification of NS-fonnyltetrahydro methanopterin as the product, *FEBS Lett*, **275**, 226–230.

Bryant MP, Boone DR (1987). Emended description of strain MS^T (DSM $8oo^T$), the type strain of *Methanosarcina barkeri*, *Inst J Syst Bacteriol*, **37**, 169–170.

Buswell AM, Sollo FW (1948). The mechanism of the methane fennentation, *J Am Chem Soc*, **70**, 1778–1780.

Cai JL, Wang GC, Li YC, Zhu DL, Pan GH (2009). Enrichment and hydrogen production by marine anaerobic hydrogen producing microflora, *Chinese Sci Bull*, **54**, 2656–2661.

Cakir A, Ozmihci S, Kargi F (2010). Comparison of biohydrogen production from hydrolyzed wheat starch by mesophilic and thermophilic dark fermentation, *Int J Hydrogen Energy*, **35**, 13214–13218.

Carvalheiro F, et al. (2008). Hemicellulose biorefineries: a review on biomass pretreatments. *J Sci Indus Res*, **67**, 849–864.

Cha M, Chung D, Elkins JG, Guss AM, Westpheling J (2013). Metabolic engineering of *Caldicellulosiruptor bescii* yields increased hydrogen production from lignocellulosic biomass, *Biotechnol Biofuels,* **6**, 85.

Chen CY, Saratale GD, Lee CM, Chen PC, Chang JS (2008). Phototrophic hydrogen production, *Int J Hydrogen Energy*, **33**, 6886–6895.

Chen SD, Lee KS, Lo YC, Chen WM, Wu JF, Lin JY, Chang JS (2008). Batch and continuous biohydrogen production from starch hydrolysate by *Clostridium* species, *Int J Hydrogen Energy,* **33**, 1803–1812.

Cheng J, Su H, Zhou J, Song W, Cen K (2011). Hydrogen production by mixed bacteria through dark and photo-fermentation, *Int J Hydrogen Energy,* **36**, 450– 457.

Cheong DY, Hansen CL (2006). Bacterial stress enrichment enhances anaerobic hydrogen production in cattle manure sludge, *Appl Microbiol Biotechnol,* **72**, 635–643.

Chippauro M, Pasca MC, Casse F (1977). Formate hydrogenlyase system in *Salmonella typhimurium* LT2, *Eur J Biochem*, **72**, 149–155.

Choquet CG, Sprott GD (1991). Metal chelate affinity chromatography for the purification of the F420-reducing (Ni,Fe) hydrogenase of *Methanospirillum hungatei*, *J Microbiol Methods*, **13**, 161–169.

Clements AP, White RH, Ferry JG (1993). Structural characterization and physiological function of component B from *Methanosarcina thermophila*, *Arch Microbiol*, **159**, 296–300.

Das D, Khanna N, Veziroglu TN (2008). Recent developments in biological hydrogen production processes, *Chem Ind Chem Eng Quart*, **14**, 57–67.

Donnelly MI, Wolfe RS (1986). The role of formylmethanofuran: tetrahydromethanopterin formyltransferase in methanogenesis from carbon dioxide, *J Biol Chem*, **261**, 16653–16659.

Donnelly MI, Escalante-Semerena JC, Rinehart KL Jr, Wolfe RS (1985). Methenyl-tetrahydromethanopterin cyclohydrolase in cell extracts of *Methanobacterium*, *Arch Biochem Biophys*, **242**, 430–439.

Eirich LD, Vogels GD, Wolfe RS (1979). Distribution of coenzyme F_{420} and properties of its hydrolytic fragments, *J Bacteriol*, **140**, 20–27.

Ellermann JR, Hedderich Bocher R, Thauer RK (1988). The final step in methane formation: investigations with highly purified methyl-CoM reductase (component C) from *Methanobacterium thermoautotrophicum* (strain Marburg), *Eur J Biochem*, **172**, 669–677.

Fan Z, Yuan L, Chatterjee R (2009). Increased hydrogen production by genetic engineering of *Escherichia coli*, *PLoS One*, **4**(2), 4432.

Fischer RP, Gartner A, Yeliseev Thauer RK (1992). NS-methyltetrahydromethanopterin:coenzyme M methyltransferase in methanogenic archaebacteria is a membrane protein, *Arch Microbiol*, **158**, 208–217.

Fox JA, Livingston DJ, Onne-Johnson WH, Walsh CT (1987). 8-hydroxy-5-deazaflavin-reducing hydrogenase from *Methanobacterium thermautotrophicum*, purification and characterization, *Biochemistry*, **26**, 4219–4227.

Gloss LM, Hausinger RP (1987). Reduction potential characterization of methanogen factor 390, *FEMS Microbiol Lett*, **48**, 143–145.

Gottschalk G, Blaut M (1990). Generation of proton and sodium motive forces in methanogenic bacteria, *Biochim Biophys Acta*, **1018**, 263–266.

Hakobyan L, Gabrielyan L, Trchounian A (2012). Relationship of proton motive force and the F_0F_1-ATPase with bio-hydrogen production activity of *Rhodobacter sphaeroides*: effects of diphenylene iodonium, hydrogenase inhibitor, and its solvent dimethylsulphoxide, *J Bioenerg Biomembr*, **44**(4), 495–502.

Harmsen P, Huijgen W, Bermudez L, Bakker R (2010). Literature review of physical and chemical pretreatment processes for lignocellulosic biomass, *Biosynenergy Rep*, **1184**, 1–54.

Heap JT, Pennington OJ, Cartman ST, Carter GP, Minton NP (2007). The ClosTron: a universal gene knock-out system for the genus *Clostridium*, *J Microbiol Methods*, **70**, 452–464.

Hedderich R, Albracht SPJ, Linder D, Koch J, Thauer RK (1992). Isolation and characterization of polyferredoxin from *Methanobacterium thermoautotrophicum*, *FEBS Lett*, **298**, 65–68.

Huser BA, Wuhrmann K, Zehnder AJB (1982). *Methanothrix soehngenii* gen. nov. sp. nov., a new acetotrophic non-hydrogen-oxidizing methane bacterium, *Arch Microbiol*, **132**, 1–9.

Jacobson FS, Daniels L, Fox JA, Walsh CT, Orme-Johnson WH (1982). Purification and properties of an 8-hydroxy-5-deazaflavin-reducing hydrogenase from *Methanobacterium thermoautotrophicum*, *J Bioi Chem*, **257**, 3385–3388.

Jetten MSM, Starns AJM, Zehnder AJB (1990). Acetate threshold values and acetate activating enzymes in methanogenic bacteria, *FEMS Microbiol Ecol*, **73**, 339–344.

Jo JH, Jeon CO, Lee SY, Lee DS, Park JM (2010). Molecular characterization and homologous overexpression of [FeFe]-hydrogenase in *Clostridium tyrobutyricum* JM1, *Int J Hydrogen Energy*, **35**(3), 1065–1073.

Kadar Z, De Vrijek T, van Noorden GE, Budde MAW, Szengyel Z, Reczey K (2004). Yields from glucose, xylose, and paper sludge hydrolysate during hydrogen production by the extreme thermophile *Caldicellulosiruptor saccharolyticus*, *Appl Biochem Biotech*, **113**, 497–508.

Kaesler B, Schonheit P (1989). The role of sodium ions in methanogenesis, Formaldehyde oxidation to CO_2 and $2H_2$ in methanogenic bacteria is coupled with primary electrogenic Na^+ translocation at a stoichiometry of 2-3 $Na+/CO_2$, *Eur J Biochem*, **184**, 223–232.

Karube I, Urano N, Yamada T, Hirochika H, Sakaguchi K (1983). Cloning and expression of the hydrogenase gene from *Clostridium butyricum* in *Escherichia coli*, *FEBS Lett*, **158**, 119–122.

Keltjens JT, van der Drift C (1986). Electron transfer reactions in methanogens, *FEMS Microbiol Rev*, **87**, 327–332.

Keltjens JT, te Brommelstroet BW, Kengen SWM, van der Drift C, Vogels GD (1990). 5,6,7,8 tetrahydromethanopterin-dependent enzymes involved in methanogenesis, *FEMS Microbiol Rev*, **87**, 327–332.

Keseler IM, Collado Vides J, Gama-Castro S, Ingraham J, Paley S, Paulsen IT, Peralta-Gil M, Karp PD (2005). EcoCyc: a comprehensive database resource for *Escherichia coli*, *Nucleic Acids Res*, **33**, 334–337.

Khanna N, Ghosh AK, Huntemann M, Deshpande S, Han J, Das D, et al. (2013). Complete genome sequence of *Enterobacter sp* IIT-BT 08: a potential microbial strain for high rate hydrogen production, *Stand Genomic Sci*, **9**, 359–369.

Khanna N, Kotay SM, Gilbert JJ, Das D (2011). Improvement of biohydrogen production by *Enterobacter cloacae* IIT-BT 08 under regulated pH, *J Biotechnol*, **152**, 15–30.

King PW, Posewitz MC, Ghirardi ML, Seibert M (2006). Functional studies of [FeFe] hydrogenase maturation in an *Escherichia coli* biosynthetic system, *J Bacteriol*, **188**, 2163–2172.

Knappe J, Sawers G (1990). A radical-chemical route to acetyl-CoA: the anaerobically induced pyruvate formate-lyase system of *Escherichia coli*, *FEMS Microbiol Lett*, **75**, 383–398.

Kongjan P, Angelidaki I (2010). Extreme thermophilic biohydrogen production from wheat straw hydrolysate using mixed culture fermentation: effect of reactor configuration, *Bioresour Technol*, **101**, 7789–7796.

Konig H, Stetter KO (1982). Isolation and characterization of *Methanolobus tindarius*; sp. Nov, a coccoid methanogen growing only on methanol and methylamines, *Zbl Bakt Hyg1 Abt Orig*, **C3**, 478–490.

Krzycki J, Wolkin ARH, Zeikus JG (1982). Comparison of unitrophic and mixotrophic substrate metabolism by an acetate-adapted strain of *Methanosarcina barkeri*, *J Bacteriol*, **149**, 247–254.

Kuhn W, Gottschalk G (1983). Characterization of the cytochromes occurring in *Methanosarcina* species, *Eur J Biochem*, **135**, 89–94.

Kuhn W, Fiebig K, Hippe H, Mah RA, Huser BA, Gottschalk G (1983). Distrubtion of cytochromes in methanogenic bacteria, *FEMS Microbiol Lett*, **20**, 407–410.

Leigh JA, Rineart Jr KL, Wolfe RS (1985). Methanofuran (carbon dioxide reduction factor), a formyl carrier in methane production from carbon dioxide in *Methanobacterium*, *Biochemistry*, **24**, 995–999.

Lemon BJ, Peters PW (1999). Binding of exogenously added carbon monoxide at the active site of the iron-only hydrogenase (CpI) from *Clostridium pasteurianum*, *Biochemistry*, **38**, 12969–12973.

Leonhartsberger S, Korsa I, Bock I (2002). The molecular biology of formate metabolism in enterobacteria, *J Mol Microbiol Biotechnol*, **4**(3), 269–276.

Li S, Lai C, Cai Y, Yang X, Yang S, Zhu M, (2010). High efficiency hydrogen production from glucose/xylose by the ldh-deleted *Thermoanaerobacterium* strain, *Bioresour Technol*, **101**, 8718–8724.

Liu X, Zhu Y, Yan ST (2006). Construction and characterization of ack deleted mutant of *Clostridium tyrobutyricum* for enhanced butyric acid and hydrogen production, *Biotechnol Prog*, **22**, 1265–1275.

Lundie LL, Ferry JG (1989). Activation of acetate by *Methanosarcina thermophila*. Purification and characterization of phosphotransacetylase, *J Bioi Chem*, **264**, 18392–18396.

Ma K, Thauer RK (1990). Purification and properties of N^5,N^{10}-methylenetetrahy dromethanopterin reductase from *Methanobacterium thermoautotrophicum* (strain Marburg), *Eur J Biochem*, **191**, 187–193.

Ma K, Linder D, Stetter KO, Thauer RK (1991). Purification and properties of N^5,N^{10}-methylenetetrahydromethanopterin reductase (coenzyme F420-dependent) from the extreme thermophile *Methanopyrus kandleri*, *Arch Microbiol*, **155**, 593–600.

Maeda T, Sanchez-Torres V, Wood T (2008). Protein engineering of hydrogenase 3 to enhance hydrogen production, *Appl Microbiol Biotechnol*, **79**, 77–86.

Maeda T, Sanchez Torres V, Wood TK (2007). *Escherichia coli* hydrogenase 3 is a reversible enzyme possessing hydrogen uptake and synthesis activities, *Appl Microbiol Biotechnol*, **76**, 1035–1042.

Maeda T, Sanchez Torres V, Wood TK (2007). Enhanced hydrogen production from glucose by a metabolically-engineered *Escherichia coli*, *Appl Microbiol Biotechnol*, **77**, 879–890.

Maeda T, Sanchez Torres V, Wood TK (2008). Metabolic engineering to enhance bacterial hydrogen production, *Microbiol Biotechnol*, **1**(1), 30–39.

Maeda T, Vardar G, Self WT, Wood TK (2007). Inhibition of hydrogen uptake in *Escherichia coli* by expressing the hydrogenase from the cyanobacterium *Synechocystis sp.* PCC 6803, *BMC Biotechnol*, **7**, 25.

Malki S, Saimmaime I, De Luca G, Rousse M, Dermoun Z, Belaich JP (1995). Characterization of an operon encoding an NADP-reducing hydrogenase in *Desulfovibrio fructosovorans*, *J Bacteriol*, **177**, 2628–2636.

Mathews J, Wang G (2009). Metabolic pathway engineering for enhanced biohydrogen production, *Int J Hydrogen Energy*, **34**, 7404–7416.

McCord J, Keele M, Fridovich I (1971). An enzyme-based theory of obligate anaerobiosis: the physiological function of superoxide dismutase, *Proc Natl Acad Sci USA*, **68**, 1024–1027.

Menon NK, Chatelus CY, Dervartanian M, Wendt JC, Shanmugam KT, Peck Jr HD, Przybyl AE (1994). Cloning, sequencing, and mutational analysis of the hyb operon encoding *Escherichia coli* hydrogenase 2, *J Bacteriol*, **176**, 4416–4423.

Menon NK, Robbins J, Wendt JC, Shanmugam KT, Przybyla AE (1991). Mutational analysis and characterization of the *Escherichia coli* hya operon, which encodes [NiFe] hydrogenase 1, *J Bacteriol*, **173**, 4851–4861.

Miller TL, Wolin MJ (1985). *Methanosphaera stadtmaniae* gen, nov, sp nov.: a species that forms methane by reducing methanol with hydrogen, *Arch Microbiol*, **141**, 116–122.

Mohr G, Smith D, Belfort M, Lambowitz AM (2000). Rules for DNA target-site recognition by a lactococcal group II intron enable retargeting of the intron to specific DNA sequences, *Genes Dev*, **14**, 559–573.

Moller-Zinkhan D, Bomer G, Thauer RK (1989). Function of methanofuran, tetrahydromethanopterin, and coenzyme F420 in *Archaeoglobus fulgidus*, *Arch Microbiol*, **152**, 362–368.

Morimoto K, Kimura T, Sakka K, Ohmiya K (2005). Overexpression of a hydrogenase gene in *Clostridium paraputrificum* to enhance hydrogen gas production, *FEMS Microbiol Lett*, **246**(2), 229–234.

Mosier N, Wyman C, Dale B, Elander R, Lee YY, Holtzapple M, Ladisch M (2005). N5, N10-methylenetetrahydromethanopterin reductase (coenzyme F420-dependent) from the extreme thermophile *Methanopyrus kandleri*, *Arch Microbiol*, **155**, 593–600.

Nakayama S, Kosaka T, Hirakawa H, Matsuura K, Yoshino S, Furukawa K (2008). Metabolic engineering for solvent productivity by downregulation of the hydrogenase gene cluster hupCBA in *Clostridium saccharoperbutylacetonicum* strain N1-4, *Appl Microbiol Biotechnol*, **78**, 483–493.

Nath K, Das D (2008). Effect of light intensity and initial pH during hydrogen production by an integrated dark and photofermentation process, *Int J Hydrogen Energy*, **34**, 7497–7501.

Noll KM, Rinehart Jr KL, Tanner RS, Wolfe RS (1986). Structure of component B (7-mercaptoheptanoylthreonine phosphate) of the methylcoenzyme M methylreductase system of *Methanobacterium thermoautotrophicum*, *Proc Natl Acad Sci USA*, **83**, 4238–4242.

Pandey A, Pandey A (2008). Reverse micelles as suitable microreactor for increased biohydrogen production, *Int J Hydrogen Energy*, **33**, 273–278.

Pandu K, Joseph S (2012). Comparisons and limitations of biohydrogen production processes : a review, *Int J Adv Eng Technol*, **2**, 342–356.

Peck HD, Gest H (1957). Formic dehydrogenase and hydrogen lyase enzyme complex in coli- aerogens bacteria, *J Bacteriol*, **73**, 706–721.

Penfold DW, Forster C, Macaskie E (2003). Increased hydrogen production by *Escherichia coli* strain HD701 in comparison with the wild-type parent strain MC4100, *Enzyme Microb Technol*, **33**, 185–189.

Penfold DW, Sargent F, Macaskie LE (2006). Inactivation of the *Escherichia coli* K-12 twin-arginine translocation system promotes increased hydrogen production, *FEMS Microbiol Lett*, **262**(2), 135–137.

Peters JW, Lanzilotta WN, Lemon BJ, Seefeldt LC (1998). X-ray crystal structure of the Fe-only hydrogenase (Cpl) from *Clostridium pasteurianum* to 1.8 angstrom resolution, *Science*, **282**, 1853–1858.

Pine MJ, Vishniac W (1957). The methane fermentations of acetate and methanol, *J Bacteriol*, **73**, 736–742.

Pol A, van der Drift C, Vogels GD (1982). Corrlnoids from *Methanosarcina barkeri*: structure of the a-ligand, *Biochem Biophys Res Commun*, **108**, 731–737.

Posewitz MC, King PW, Smolinski SL, Zhang LP, Seibert M, Ghirardi ML (2004). Discovery of two novel radical S-adenosylmethionine proteins required for the assembly of an active Fe-hydrogenase, *J Biol Chem*, **279**, 25711–25720.

Thauer RK (1992). H_2-forming methylenetetrahydromethanopterin dehydrogenase, a novel type of hydrogenase without iron-sulfur clusters in methanogenic archaea, *Eur J Biochem*, **208**, 511–520.

Ren Z, Ward TE, Logan BE, Regan JM (2007). Characterization of the cellulolytic and hydrogen-producing activities of six mesophilic *Clostridium* species, *J Appl Microbiol*, **103**, 2258–2266.

Roseboom W, Lacey AL, Fernandez VM, Hatchikian EC, Albracht SPJ (2006). The active site of the [FeFe]-hydrogenase from *Desulfovibrio desulfuricans* II, Redox properties, light sensitivity, and CO-ligand exchange as observed by infrared spectroscopy, *J Biol Inorganic Chem*, **11**, 102–118.

Sauer FD, Blackwell BA, Kramer LKG, Marsden BL (1990). Structure of a novel cofactor containing N-(7-mercaptoheptanoyl)-O-3-phosphothreonine, *Biochemistry*, **29**, 7593–7600.

Sauer FD (1986). Tetrahydromethanopterin methyltransferase, a component of the methane synthesizing complex of *Methanobacterium thermoautotrophicum*, *Biochem Biophys Res Commun*, **136**, 542–547.

Schmitz RA, Linder D, Stetter KO, Thauer RK (1991). N^5,N^{10} methylenetetrahydromethanopterin reductase (coenzyme F_{420}-dependent) and formylmethanofuran dehydrogenase from the hyperthermophile *Archaeoglobus fulgidus*, *Arch Microbiol*, **156**, 427–434.

Schmitz RA, Richter M, Linder D, Thauer RK (1992). A tungsten containing active formylmethanofuran dehydrogenase in the thermophilic archaeon *Methanobacterium wolfei*, *J Biochem*, **207**, 539–565.

Schonheit P, Schafer T (1995). Metabolism of hyperthermophiles, *World J Microbiol Biotechnol*, **11**, 26–57.

Schworer B, Thauer RK (1991). Activities of formylmethanofuran dehydrogenase, methylenetetrahydromethanopterin dehydrogenase, methylenetetrahydromethanopterin reductase, and heterodisulfide reductase in methanogenic bacteria, *Arch Microbiol*, **155**, 459–465.

Sinha P, Pandey A (2011). An evaluative report and challenges for fermentative biohydrogen production, *Int J Hydrogen Energy*, **36**, 4760–4778.

Song W, Cheng J, Zhao J, Carrieri D, Zhang C, Zhou J, Cen K (2011). Improvement of hydrogen production by over-expression of a hydrogen promoting protein gene in *Enterobacter cloacae*, *Int J Hydrogen Energy*, **36**, 6609–6615.

Stripp ST, Goldet G, Brandmayr C, Sanganas O, Vincent KA, Haumann M, Armstrong FA, (2009). How oxygen attacks [FeFe] hydrogenases from photosynthetic organisms, *Proc Natl Acad Sci USA*, **106**, 1–6.

Subudhi S, Lal B (2011). Fermentative hydrogen production in recombinant *Escherichia coli* harbouring a [FeFe]-hydrogenase gene isolated from *Clostridium butyricum, Int J Hydrogen Energy,* **36**, 14024–14030.

Sun J, Hopkins RC, Jenney Jr F, McTernan PM, Adams MW (2010). Heterologous expression and maturation of an NADP-dependent [NiFe] hydrogenase: a key enzyme in biofuel production, *PLoS One,* **5**(5), 105–126.

Tard C, et al. (2005). Synthesis of the H-cluster framework of iron-only hydrogenase, *Nature*, **433**, 610–613.

Terlesky KC, Ferry JG (1988). Ferredoxin requirement for electron transport from the carbon monoxide dehydrogenase complex to a membrane-bound hydrogenase in acetate-grown *Methanosarcina thermophil, J Biol Chem*, **263**, 4075–4079.

Thauer RK, Morris JG (1984). Metabolism of chemotrophic anaerobes: old views and new aspects. In: Kelly DP, Carr NG (eds), The Microbe 1984, Part II, Prokaryotes and Eukaryotes, *Soc Gen Microbiol Symb*, **36**, 123–168. Cambridge University Press, Cambridge, England.

Tripathi SA, Olson DG, Argyros DA, Miller BB, Barrett TF, Murphy DM, et al. (2010). Development of pyrF-based genetic system for targeted gene deletion in *Clostridium thermocellum* and creation of a pta mutant, *Appl Environ Microbiol*, **76**(19), 6591–6599.

Uyeda K, Rabinowitz JC (1971). Pyruvate-ferredoxin oxidoreductase, *J Biol Chem*, **246**, 3111–3119.

van de Wijingaard WM, Vermey HP, van der Drift C (1991). Formation of factor 390 by cell extracts of *Methanosarcina barkeri*, *J Bacteriol*, **173**, 2710–2711.

van Beelen P, Labro JFA, Keltjens JT, Geerts WJ, Vogels GD, Laarhoven WH, Guijt W, Haasnoot CAG (1984). Derivatives of methanopterin, a coenzyme involved in methanogenesis, *Eur J Biochem*, **139**, 359–365.

van Niel EWJ, Budde MAW, de Haas GG, van der Wal FJ, Claasen PAM, Stams AJM (2002). Distinctive properties of high hydrogen producing extreme thermophiles, *Caldicellulosiruptor saccharolyticus* and *Thermotoga elfii*, *Int J Hydrogen Energy,* **27**(11–12), 1391–1398.

Veit A, Akhtar MK, Mizutani T, Jones PR (2008). Constructing and testing the thermodynamic limits of synthetic NAD(P)H: H_2 pathways, *Microb Biotechnol*, **1**, 382–394.

Voelskow H, Schon G (1980). Hydrogen production of *Rhodospirillum rubrum* during adaptation to anaerobic dark conditions, *Arch Microbiol*, **125**, 245–249.

Walsh C (1979). *Enzymatic Reaction Mechanisms*, WH Freeman and Company, San Francisco.

Wang R, Zong W, Changli Q, Wei Y, Yu R, Zhou Z (2011). Isolation of *Clostridium pefrigens* strain W11 and optimization of its biohydrogen production by genetic modification, *Int J Hydrogen Energy*, **36**, 12159–12167.

Wiechert W (2002). Modeling and simulation: tools for metabolic engineering, *J Biotechnol*, **94**, 37–63.

Wolfe RS (1991). My kind of biology, *Annu Rev Microbiol*, **45**, 1–35.

Yamazaki S (1982). A selenium-containing hydrogenase from *Methanococcus vannielii*. Identification of the selenium moiety as a selenocysteine residue, *J Biol Chem*, **257**, 7926–7929.

Yoshida A, Nishimura T, Kawaguchi H, Inui M, Yukawa H (2005). Enhanced hydrogen production from formic acid by formate hydrogen lyase overexpressing *Escherichia coli* strains, *Appl Environ Microbiol*, **71**, 6762–6768.

Yoshida A, Nishimura T, Kawaguchi H, Inui M, Yukawa H (2006). Enhanced hydrogen production from glucose using ldh- and frd-inactivated *Escherichia coli* strains, *Appl Microbiol Biotechnol*, **73**, 67–72.

Yu R, Wang R, Bi T, Sun W, Zhou Z (2013). Blocking the butyrate formation pathway impairs hydrogen production in *Clostridium pefringens*, *Acta Biochim Biophys Sin*, **45**, 408–415.

Zeikus JG (1977). The biology of methanogenic bacteria, *Bacteriol Rev*, **41**, 514–541.

Zhao H, Ma K, Lu Y, Zang C, Wang L, Xing XH (2009). Cloning and knockout of formate hydrogen lyase and H_2 uptake hydrogenase genes in *Enterobacter aerogenens* for enhanced hydrogen production, *Int J Hydrogen Energy*, **34**, 186–194.

Zimgibl C, Hedderich R, Thauer RK (1990). N5,N10-methylenetetrahydromethanopterin dehydrogenase from *Methanobacterium thermoautotrophicum* has hydrogenase activity, *FEBS Lett*, **261**, 112–116.

Chapter 4

Mathematical Modeling and Simulation of Biohydrogen Production Processes

4.1 Introduction

Global economic development is primarily driven by energy provided by fossils fuels. Overdependence and nonjudicious use of these fossil fuels has contributed toward global climate change, environmental degradation, and health problems. Use of hydrogen as fuel does offer a promising alternative to fossil fuels. Certain characteristics such as being carbon neutral, renewable, and highest energy density pitched its prospect as a future fuel. Hydrogen is compatible with electrochemical and oxidative combustion processes. The by-products of these processes are energy and water. Thus in a true sense, hydrogen is a carbon-neutral fuel. A hydrogen-based economy could be a boon, not only for the humans, but also for the environment.

At present, hydrogen is produced by various conventional processes such as electrolysis of water, thermocatalytic reformation of organic compounds, and biological routes. Commercial-scale hydrogen is produced mainly by electrolysis of water or by steam reformation of methane. The technologies available for biohydrogen production are direct biophotolysis, indirect biophotolysis, photofermentation, and dark fermentation. As previously discussed,

Biohythane: Fuel for the Future
Debabrata Das and Shantonu Roy

ISBN 978-981-4745-29-1 (Hardcover), 978-981-4745-30-7 (eBook)
www.panstanford.com

dark and fermentative hydrogen production has shown the highest rate of production among other biological routes. In this chapter, the mathematical modeling of biohydrogen production technologies has been discussed in detail.

A model can be used to explain a system and also to study the effects of different parameters on the same. More importantly, a mathematical model finds its application in making predictions about the behavior of the system. It is imperative to have a predicted outcome if a process is going to be scaled up from the laboratory to an industry. The quality of a mathematical model depends on how well theoretical predictions agree with the observations obtained from repeated experiments. In no case, can the disagreement between the theoretical values and experimental values be greater than 5%. If the disagreement is greater than 5%, the model is unsuitable and needs to be modified.

4.2 Development of Mathematical Models to Correlate Substrate and Biomass Concentration with Time

4.2.1 Monod Model for Cell Growth Kinetics

The Monod equation is a mathematical model that deals with the growth kinetics of microorganisms. Jacques Monod (1949) proposed a relation between microbial growth rates in an aqueous environment and the concentration of a limiting substrate, given by the equation

$$\mu = \mu_{max} \frac{S}{K_S + S} \tag{4.1}$$

where μ is the specific growth rate of microorganisms (h^{-1}), μ_{max} is the maximum specific growth rate of microorganisms (h^{-1}), S is the concentration of the limiting substrate for growth, ($g\ L^{-1}$), K_S is the saturation constant ($g\ L^{-1}$).

The specific growth rate of microorganisms (μ) can be expressed mathematically as

$$\mu = \frac{1}{X}\frac{dX}{dt} \tag{4.2}$$

where X is the cell mass concentration at any point in time.,

The Monod equation may be written in the Lineweaver–Burk plot:

$$\frac{1}{\mu} = \frac{K_S}{\mu_{max} S} + \frac{1}{\mu_{max}} \tag{4.3}$$

Regression analysis is used to find the best fit for a straight line of $1/\mu$ versus $1/S$. The values of K_S and μ_{max} are estimated from the intercept and the slope of the straight line, respectively (Fig. 4.1).

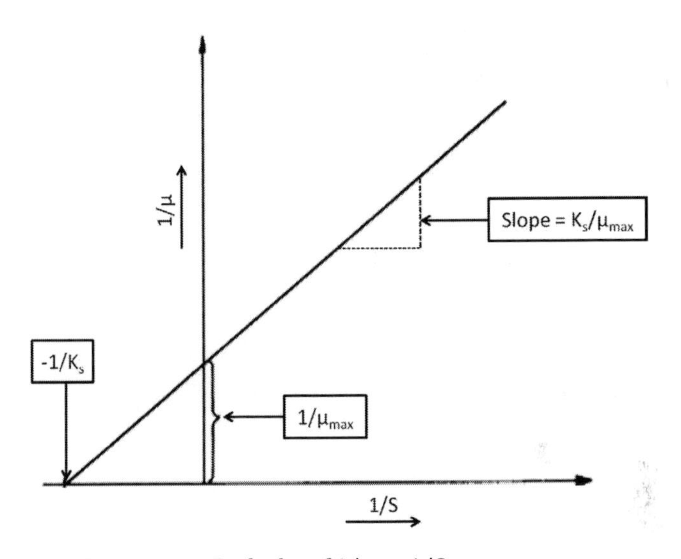

Figure 4.1 Lineweaver–Burk plot of $1/\mu$ vs. $1/S$.

4.2.2 Determination of the Profiles of Cell Mass Concentration and Substrate Concentration

The overall yield coefficient, $Y_{X/S}$, is defined as grams of cell mass generated per gram of substrate degraded and can be expressed as

$$Y_{X/S} = \frac{dX}{dS} \tag{4.4}$$

For experimental data, $Y_{X/S}$ can be calculated by averaging the values obtained using the expression

$$Y_{X/S} = \frac{X - X_o}{S_o - S} \tag{4.5}$$

Again, Eq. 4.1 may be written as

$$\frac{dX}{dT} = \mu X = \frac{\mu_{\max} S}{K_S + S} \cdot X \tag{4.6}$$

$$\frac{dS}{dt} = \frac{-1}{Y_{X/S}} \cdot \frac{\mu_{\max} S}{K_S + S} \cdot X \tag{4.7}$$

X in Eq. 4.7 is replaced by an expression involving S, S_o, X_o, and $Y_{X/S}$ from Eq. 4.5. It can then be integrated to give an expression for simulated values of S as a function of t.

$$\mu_{\max}(X_o + Y_{X/S} S_o)t = [X_o + Y_{X/S}(S_o + K_o)] \ln\left(\frac{X_o + Y_{X/S}(S_o - S)}{X_o}\right) - K_S Y_{X/S} \ln \frac{S}{S_o} \tag{4.8}$$

Now, X as a function of time can be obtained by substituting the value of S from Eq. 4.5 in Eq. 4.8:

$$\mu_{\max}(X_o + Y_{X/S} S_o)t = X_o + Y_{X/S}(S_o + K_s)\ln\left(\frac{X}{X_o}\right) - K_s Y_{X/S} \ln\left(1 - \frac{X - X_o}{S_o Y_{X/S}}\right) \tag{4.9}$$

4.2.3 Modeling and Simulation of the Fermentation Process

The experimental data of the batch biohydrogen production process reported by Nath et al. (2008) (Table 4.1) were used to simulate the substrate and cell mass concentration profiles. These are shown in Figs. 4.2 and 4.3.

Experimental conditions: batch operation; volume of medium, 250 mL; temperature, 37°C; initial medium pH, 6.0 ± 0.2 (Nath et al., 2008).

Table 4.1 Modeling of biomass growth kinetics based on the Monod model for hydrogen production using *Enterobacter cloacae* DM11 (presently known as *K. pneumonia*) in stage 1 (dark fermentation)

Time (h)	Substrate concentration (g L^{-1})	Biomass concentration (g L^{-1})
0	10.0	0.05
2	8.7	0.14
4	7.9	0.26
6	7.6	0.41
8	6.7	0.49
10	5.1	0.51
12	3.9	0.53
14	3.5	0.57
16	3.3	0.60
18	2.9	0.62

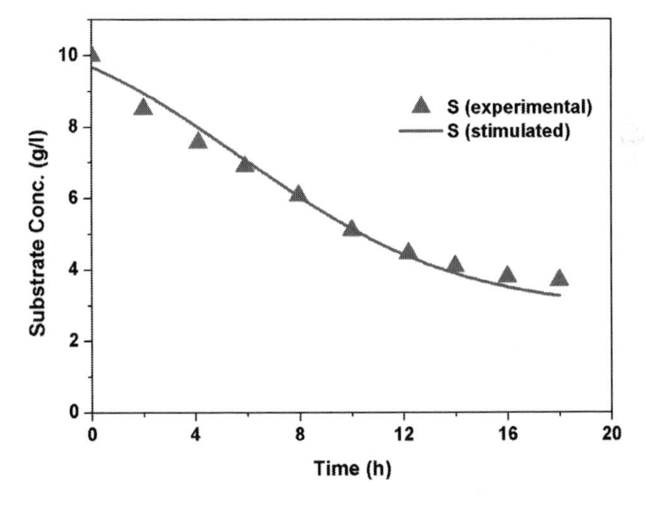

Figure 4.2 Experimental and simulated substrate concentration profiles using the Monod model.

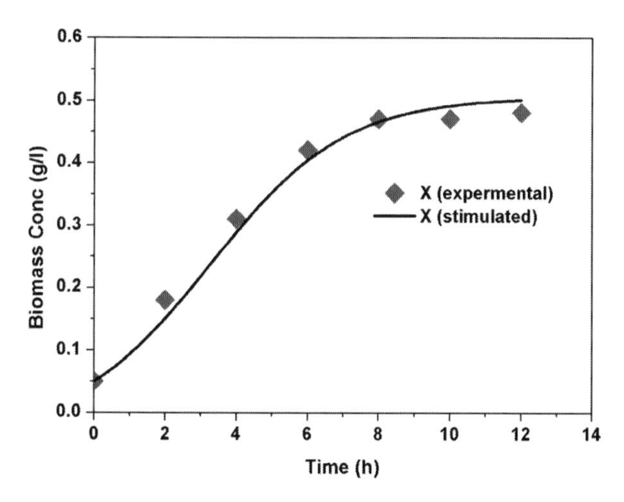

Figure 4.3 Experimental and simulated biomass concentration profiles using the Monod model.

4.2.4 Regression Analysis of Simulated Values Obtained from the Monod Model and Experimentally Obtained Values

The coefficient of determination (R^2) is a statistic that indicates how well data points fit a line or curve. It provides a measure of how well-observed outcomes are replicated by the model as a proportion of the total variation of outcomes explained by the model.

R^2 is mathematically expressed as

$$R^2 = \frac{\text{Total sum of squares due to regression (SSR)}}{\text{Total sum of squares (SST)}} \tag{4.10}$$

$$\text{SSR} = \sum (\hat{y}_i - \bar{y})^2 \tag{4.11}$$

$$\text{SST} = \sum (y_i - \bar{y})^2 \tag{4.12}$$

where $\hat{y}_i$ represents the simulated values, y_i represents the experimental value, and $\bar{y}$ represents the mean of the experimental data.

The closer the value of R^2 is to 1, the better the fit of the model is. A model having R^2 equal to 1 is a perfect fit. However, a value of R^2 greater than 1 indicates the absence of regression.

Table 4.2 Table for calculation of SST and SSR from the experimental values of substrate concentration y_i and simulated values of substrate concentration ($\hat{y}_i$)

S. no.	y_i	$\hat{y}_i$	T	SSR
1	10.0	10.0	16.32	16.32
2	8.7	8.9	7.51	8.64
3	7.9	8.1	3.76	4.58
4	7.6	7.7	2.69	3.03
5	6.7	6.3	0.55	0.12
6	5.1	5.5	0.74	0.21
7	3.9	4.2	4.24	3.10
8	3.5	3.7	6.05	5.11
9	3.3	3.3	7.08	7.08
10	2.9	2.9	9.36	9.36
$\bar{y}$	6.0	Σ	**58.30**	**57.54**

From the values of SSR and SST, as calculated in Table 4.2, we have $R^2 = 0.987$.

Table 4.3 Table for calculation of SST and SSR from the experimental values of biomass concentration y_i and simulated values of biomass concentration ($\hat{y}_i$)

S. no.	y_i	$\hat{y}_i$	T	SSR
1	0.050	0.050	0.135	0.135
2	0.140	0.120	0.077	0.089
3	0.260	0.250	0.025	0.028
4	0.410	0.400	0.000	0.000
5	0.490	0.500	0.005	0.007
6	0.510	0.520	0.008	0.010
7	0.530	0.550	0.013	0.017
8	0.570	0.560	0.023	0.020
9	0.600	0.570	0.033	0.023
10	0.620	0.570	0.041	0.023
$\bar{y}$	0.418	Σ	**0.361**	**0.354**

From the values of SSR and SST, as calculated in Table 4.3, we have $R^2 = 0.979$.

The values of R^2 obtained indicate that the model is a good fit.

4.2.5 Other Monod-Type Models

4.2.5.1 Monod-type model including the pH inhibition term

The Monod-type kinetic expression, incorporating the empirical lower pH inhibition term I_{pH}, was put forward by a number of authors (Lin et al., 2007; Ntaikou et al., 2008, 2009).

$$X = \frac{q_{Glu}^{max} S_{Glu} X}{K_{Glu} + S_{Glu}} \times I_{pH} \tag{4.13}$$

where X denotes biomass concentration, S_{Glu} denotes residual glucose concentration, q_{Glu}^{max} denotes the maximum specific glucose consumption rate, and K_{Glu} represents the saturation constant.

4.2.5.2 Monod-type model with biomass decay constant

Together with the pH inhibition coefficient (I_{pH}), Ntaikou et al. (2008) modified the Monod equation with one additional term, biomass decay constant K_d:

$$\frac{dx}{dt} = \frac{\mu_{max} S}{K_s + S} \times I_{pH} - K_d X \tag{4.14}$$

4.2.5.3 Monod- type model including the pH inhibition term and substrate inhibition factors

The growth of fermentative hydrogen producing *Ruminococcus albus* on glucose was modeled by a modified Monod equation, including both the pH inhibition coefficient (I_{pH}) and substrate inhibition (I_S) factors.

$$\frac{ds}{dt} = -K_m \frac{S}{K_S + S} \times I_{pH} \times I_s \tag{4.15}$$

4.3 Substrate Inhibition Model

4.3.1 Modified Andrew's Model

In the biohydrogen production process, substrate inhibition plays an important role because the product is a gas (hydrogen). The Monod model is found unsuitable for the substrate inhibition process. Andrew proposed a mathematical model for substrate inhibition in microbial fermentation, as given below:

$$\mu = \frac{\mu_{max} S}{K_s + S + K_i S^2} \tag{4.16}$$

Kumar et al. (2000) proposed a modified Andrew's model for the biohydrogen production process with substrate inhibition:

$$\mu = \frac{\mu_{max} S}{K_s + S - K_i S^2} \tag{4.17}$$

It is a modification that suggests a nonlinear relationship between the specific growth rate μ and substrate concentration S.

4.3.1.1 Simulation of cell mass concentration and substrate concentration profiles

$$\text{Now, } \frac{dX}{dT} = \mu X = \frac{\mu_{max} S}{K_s + S - K_i S^2} \cdot X \tag{4.18}$$

$$\text{Thus, } \frac{dS}{dt} = \frac{-1}{Y_{X/S}} \cdot \frac{\mu_{max} S}{K_s + S - K_i S^2} \cdot X \tag{4.19}$$

X in Eq. 4.7 is replaced by an expression involving S, S_0, X_0, and $Y_{X/S}$ (Eq. 4.5).

It can then be integrated to give an expression for simulated values of S as a function of t.

$$\begin{aligned} &\mu_{max}(X_0 + Y_{X/S} S_0)t \\ &= [X_0 + Y_{X/S}(S_0 + K_S)]\ln\left(\frac{X_0 + Y_{X/S}(S_0 - S)}{X_0}\right) - K_S Y_{X/S} \ln\frac{S}{S_0} \\ &- K_i(X_0 + Y_{X/S} S_0)]\left[S - S_0 + \frac{(X_0 + Y_{X/S} S_0)}{Y_{X/S}}\ln\left(\frac{X_0 + Y_{X/S}(S_0 - S)}{X_0}\right)\right] \end{aligned} \tag{4.20}$$

Now, X as a function of time can be obtained by substituting the value of S from Eq. 4.5 in Eq. 4.19:

$$\mu_{\max}(X_0 + Y_{X/S})t = [X_0 + Y_{X/S}(S_0 + K_s)]\ln\left(\frac{X}{X_0}\right) - K_s Y_{X/S}$$

$$\ln\left(1 - \frac{X - X_0}{S_0 Y_{X/S}}\right) - K_i(X_0 + Y_{X/S}S_0)\left[\frac{X_0 - X}{Y_{X/S}} + \frac{(X_0 + Y_{X/S}S_0)}{Y_{X/S}}\ln\left(\frac{X}{X_0}\right)\right]. \tag{4.21}$$

4.3.2 Simulation of the Biohydrogen Production Process with Substrate Inhibition

The experimental values reported by Nath et al. (2008, 2011) were simulated using the Andrew model and are shown in Figs. 4.4 and 4.5.

Table 4.4 Modeling of biomass growth kinetics based on the substrate inhibition model for hydrogen production using *E. cloacae* DM11 (presently known as *K. pneumonia*) in stage 1

Time (h)	Substrate concentration ($g\ L^{-1}$)	Biomass concentration ($g\ L^{-1}$)
0	10.2	0.04
1	9.9	0.08
2	9.0	0.14
3	8.6	0.19
4	8.0	0.31
5	7.7	0.39
6	7.3	0.40
7	6.4	0.46
8	5.7	0.52

Experimental conditions: batch operation; volume of medium, 250 mL; temperature, 37°C; initial medium pH, 6.0 ± 0.2 (Nath et al., 2008)

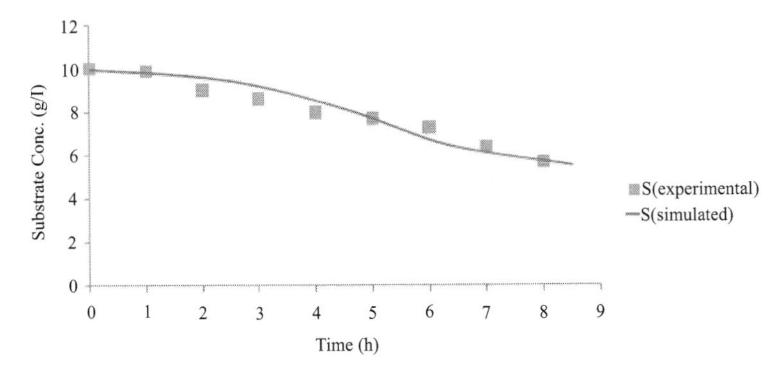

Figure 4.4 Experimental and simulated substrate concentration profiles using the substrate inhibition model.

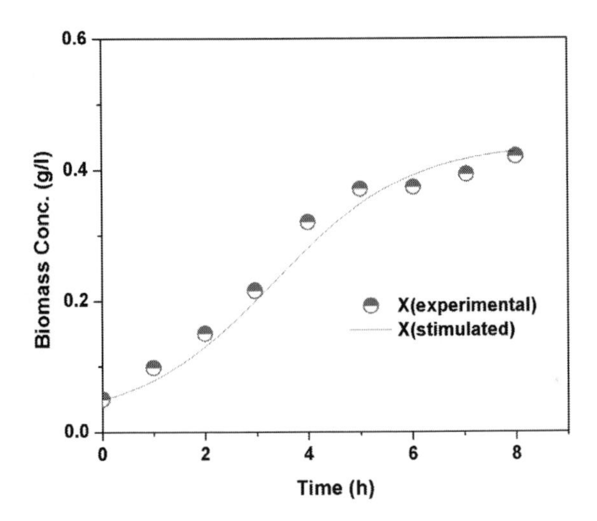

Figure 4.5 Experimental and simulated biomass concentration profiles using the substrate inhibition model.

4.3.3 Regression Analysis of Simulated Values Obtained from the Substrate Inhibition Model and Experimentally Obtained Values

The values of R^2 for the substrate concentration profile and the biomass concentration profile were calculated, as discussed in Section 4.3 for the Andrew model.

For the substrate concentration profile,

$$R^2 = 0.99$$

For the biomass concentration profile,

$$R^2 = 0.97$$

The values of R^2 obtained indicate that the model is a good fit.

4.4 Determination of Cell Growth Kinetic Parameters

4.4.1 Kinetic Parameters and Their Estimation

Kinetic equations, which describe the activity of the enzymes or a microorganism on a particular substrate, are crucial in understanding many phenomena in biotechnological processes. Quantitative experimental data are required for the design and optimization of biological transformation processes. A variety of mathematical models has been proposed to describe the dynamics of metabolism of compounds exposed to pure cultures of microorganisms or microbial populations of the natural environment. The Monod equation has been widely used to describe growth-linked substrate utilization. Characterization of microbe-substrate interactions involves estimation of several parameters in the kinetic models from experimental data. To describe the true behavior of the system, it is important to obtain accurate estimates of the kinetic parameters in these models.

Different approaches have been proposed for estimating the kinetic parameters, but progress curve analysis is the most popular because substrate depletion or product formation data from a single experiment are enough for parameter estimation. In this approach, substrate profile or product profile is used in the integrated form of the kinetic model for parameter estimation. It is important to note that most kinetic models and their integrated forms are nonlinear. This makes parameter estimation relatively difficult. However, some of these models can be linearized. Various linearized forms of the integrated expressions have been used for parameter estimation. However, the use of linearized expression is limited because it transforms the error associated with the dependent variable, making it not to be normally distributed, thus inaccurate parameter estimates. Therefore, nonlinear least-squares regression is often used to estimate kinetic parameters from nonlinear expressions. However, application of nonlinear least-squares regression to the integrated forms of the kinetic expressions is complicated. The parameter estimates obtained from the linearized kinetic expressions can be used as initial estimates in the iterative nonlinear least-squares regression. The kinetic parameters of the Andrew's equation (μ_{max}, q_{max}, K_s, or K_i) can be estimated by the application of the reduced form of the generalized substrate inhibition model reduced to the form of Andrew's equation. The linearized expression of this model was used to obtain initial parameter estimates for use in nonlinear regression.

4.4.2 Calculation of Kinetic Parameters Using the Method of Least Squares

For calculating the kinetic parameter values from Eq. 4.17, a numerical method approach, namely *method of least squares* is used, where the following substitutions are made.

$$S \equiv p \tag{4.22}$$

$$\frac{1}{\mu} \equiv q \tag{4.23}$$

$$\frac{-K_i}{\mu_{\max}} \equiv a \tag{4.24}$$

$$\frac{1}{\mu_{\max}} \equiv b \tag{4.25}$$

$$\frac{K_S}{\mu_{\max}} \equiv c \tag{4.26}$$

Using the above relations, Eq. 4.21 can be rewritten in the form:

$$q = ap + b + \frac{c}{p} \tag{4.27}$$

Now, when a data set of p and q is available, the coefficients a, b, and c can be calculated by solving the following three equations simultaneously:

$$an + b\Sigma\frac{1}{p_i} + c\Sigma\frac{1}{p_i^2} = \Sigma\frac{q_i}{p_i} \quad (4.28)$$

$$a\Sigma p_i + bn + c\Sigma\frac{1}{p_i} = \Sigma q_i \quad (4.29)$$

$$a\Sigma p_i^2 + b\Sigma p_i + cn = \Sigma q_i p_i \quad (4.30)$$

Here, n is the number of data for p and q, and i refers to the ith data of p or q.

When values for a, b, and c are obtained, values of kinetic parameters K_S, μ_{max}, and K_i can be calculated by using equations 4.24, 4.25, and 4.26.

Using the method of least squares, as given in Table 4.5, K_S, μ_{max}, and K_i are estimated from data given in Table 4.4.

Table 4.5 Summation of p_i, q_i, $1/p_i$, $1/p_i^2$, q_i/p_i, p_i^2, and p_iq_i for calculating the values of a, b, and c

T	$S(exp) = p_i$	$1/\mu = q_i$	$1/p_i$	$1/p_i^2$	q_i/p_i	p_i^2	p_iq_i
1	9.9	5.33	0.10	0.01	0.54	98.01	52.80
2	9	4.67	0.11	0.01	0.52	81.00	42.00
3	8.6	7.60	0.12	0.01	0.88	73.96	65.36
4	8	5.17	0.13	0.02	0.65	64.00	41.33
5	7.7	10.86	0.13	0.02	1.41	59.29	83.60
6	7.3	27.33	0.14	0.02	3.74	53.29	199.53
7	6.4	18.40	0.16	0.02	2.88	40.96	117.76
8	5.7	17.33	0.18	0.03	3.04	32.49	98.80
Σ	62.6	96.69	1.05	0.14	13.66	503.00	701.19

Substituting the values in Eqs. 4.28, 4.29, and 4.30, we get

$8a + 1.05b + .14c = 13.66$

$62.6a + 8b + 1.05c = 96.69$

$503a + 62.6b + 8c = 701.19$

On solving for a, b, and c we get

$a = -2.78$, $b = 23.18$, $c = 18.16$.

Now, from Eqs. 4.24, 4.25, and 4.26

$$K_S = 3.49 \text{ g/L}$$

$$\mu_{max} = 0.043 \text{ h}^{-1}$$

$$K_i = 0.12 \text{ g/L}$$

4.5 Cumulative Hydrogen Production by the Modified Gompertz Equation

The products from fermentative biohydrogen processes are broadly grouped into two major categories. The first category is gaseous products (primarily hydrogen and CO_2), while the second group includes volatile fatty acids and solvents. During hydrogen production by anaerobic fermentation the distribution of acidogenic products varies substantially. Moreover, some acidogenic products, for example, acetate, butyrate, hydrogen, and carbon dioxide, may form more complex long-chain fatty acids or alcohols as hydrogen is consumed.

4.5.1 Modified Gompertz Equation for Modeling Hydrogen, Butyrate, and Acetate Production

Mu et al. (2006) used the modified Gompertz equation to model the production of hydrogen as well as butyrate and acetate, during a batch hydrogen production process by mixed anaerobic culture.

$$P_i = P_{\max,i} \exp\left\{-\exp\left[\frac{R_{\max,i}\, e}{P_{\max,i}}(\lambda - t) + 1\right]\right\} \tag{4.31}$$

where i represents hydrogen, butyrate, and acetate, respectively; P_i is the product i formed per liter of the reactor volume at fermentation time t; $P_{\max,i}$ is the potential maximum product formed per liter of the reactor volume; and $R_{\max,i}$ is the maximum rate of the product formed (Table 4.6).

A summary of recent studies where the modified Gompertz equation has been used to model the cumulative production of biohydrogen is shown in Table 4.7.

4.5.2 Product Formation Kinetics by the Luedeking–Piret Model

The relationship between biomass and the products for the anaerobic hydrogen production by mixed anaerobic cultures can be modeled by the Luedeking–Piret model (Mu et al., 2006; Lo et al., 2008; Obeid et al., 2009):

Table 4.6 Kinetic parameters of cumulative hydrogen production for different initial glucose concentrations calculated from nonlinear regression of the Gompertz equation (Kumar and Das, 2000; Nath at al., 2008)

Glucose concentration (%w/v)	P (mL)	R_m (mL h^{-1})
0.2	440	9.7
0.4	450	8.5
0.6	560	10.0
0.8	750	2.3
1.0	1000	13.9
1.2	920	17.6
1.4	800	14.2

$$\frac{dP_i}{dt} = \alpha_i \frac{dx}{dt} + \beta_i X \tag{4.32}$$

where α_i is the growth-associated formation coefficient of the product i and β_i is the non-growth-associated formation coefficient of product i.

$$\frac{1}{x}\frac{dC_{H_2}}{dt} = \alpha\mu = \alpha\frac{1}{X}\frac{dX}{dt} \tag{4.33}$$

Here C_{H_2} represents the hydrogen concentration (mol), x represents the cell concentration (g VSS L^{-1}), and μ represents the specific growth rate (h^{-1}).

On dividing Eq. 4.32 by x (cell concentration) for a specific product, we have

$$\frac{1}{x}\frac{dP}{dt} = \alpha\frac{1}{x}\frac{dx}{dt} + \beta \tag{4.34}$$

Equation 4.35 can be written as

$$v = \alpha\mu + \beta \tag{4.35}$$

where, $v = \frac{1}{x}\frac{dP}{dt}$, is the specific product formation rate and μ is the specific growth rate of the microorganism, as defined in Eq. 4.2.

Growth-associated product formation ($\beta = 0$):

Table 4.7 Comparative studies on the reported values of R_m using the modified Gompertz equation

Substrate	Mode of operation	Microorganism	Substrate concentration	Max. rate of hydrogen production (R_m)	References
Glucose	Batch	*Enterobacter cloacae*	1.4 g L^{-1}	14.14 mL h^{-1}	Nath et al. (2008)
Sucrose	Batch	Mixed microflora	25 g L^{-1}	10.6 mL h^{-1}	Chen et al. (2006)
Lactose	Batch	Anaerobic microflora	15 g L^{-1}	6.8 mL h^{-1}	Davila-Vazquez et al. (2008)
Starch	Batch	Mixed anaerobic culture	20 g COD L^{-1}	37.9 mL h^{-1}	Lin et al. (2008)
Food Waste	Batch	Mixed microflora	25 g COD L^{-1}	32.3 mL h^{-1}	Chen et al. (2006)
Wastewater	Continuous	Seed sludge	20 g COD L^{-1}	12.03 mL h^{-1}	Jung et al. (2010)
Distillery effluent	Batch	Thermophilic mixed anaerobic culture	60 g COD L^{-1}	115 mL h^{-1}	Roy et al. (2012)
Starchy wastewater	Batch	Thermophilic mixed anaerobic culture	30 g COD L^{-1}	200 mL h^{-1}	Roy et al. (2015)

In this case the specific product formation rate is directly proportional to the specific growth rate of the microorganism (Figs. 4.6, 4.7, and 4.8).

Mixed growth-associated product formation ($\alpha \neq 0, \beta \neq 0$):

In this case the specific product formation rate depends not only on the specific growth rate of the microorganism but also on a constant (β).

Non-growth-associated product formation ($\alpha = 0$):

In this case the specific product formation rate is independent of the specific growth rate of the microorganism.

Comparing the above plot with Fig. 4.7, it can be established that biohydrogen production is a growth-associated process.

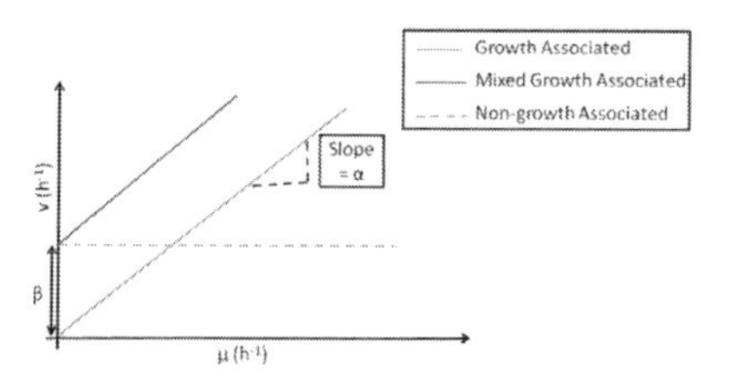

Figure 4.6 Plot of v vs. μ.

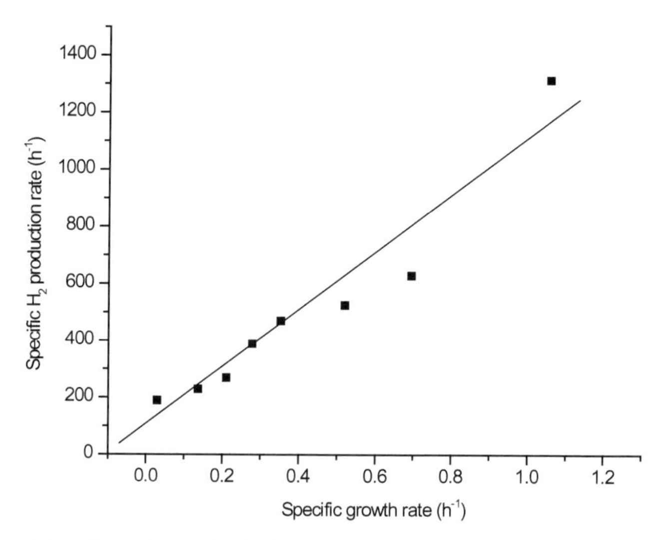

Figure 4.7 Plot of specific hydrogen production rate vs. specific growth rate using the Leudeking–Piret model.

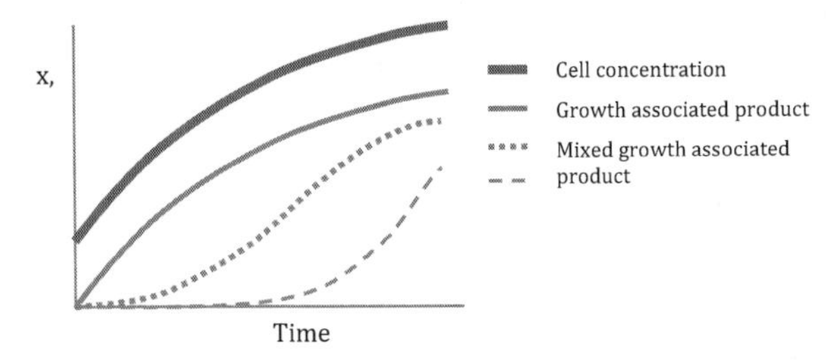

Figure 4.8 Plot of cell concentration and product formation vs. time.

4.6 Development of Mathematical Models for Cell Growth Kinetics in a Packed-Bed Reactor

Continuous hydrogen production in a packed-bed reactor (PBR) is distinctly different from a suspended culture system. In a PBR, substrate diffusivity, pH variance, partial pressure, hydrogen dissolvity, and suspended biomass concentration are the functions of length of the column and the flow rate of feed. It is assumed that the PBR follows a plug flow-type model with axial or radial diffusion. On the basis of these assumptions, the substrate degradation rate was mathematically modeled.

4.6.1 Substrate Degradation Rate in an Ideal Plug Flow Reactor and Related Rate Constant

The mass balance for the substrate in a PBR, assuming ideal plug flow and no axial dispersion, is given by

$$\frac{HQ}{W}\left(\frac{dc}{dt}\right)=r \tag{4.36}$$

where H is the height of the column, Q is the volumetric flow rate, W is the mass of the substrate, and r is the rare constant. Steady-state plug flow, no axial dispersion, and spherical equivalent of a cylindrical immobilized solid matrix are considered for the development of the equation. Generally at lower substrate concentrations and in a PBR, substrate degradation will follow first-order kinetics,

$$r = K_p \times c \tag{4.37}$$

where K_p is the observed first-order substrate degradation constant

$$\frac{HQ}{W}\left(\frac{dc}{dt}\right) = -K_p \times c \tag{4.38}$$

Integrating at boundary conditions, $z = 0$, $c = c_o$ to $z = H$, $c = c$

$$\ln\left(\frac{c}{c_0}\right) = \frac{W}{Q} \times K_p \tag{4.39}$$

The observed first-order substrate degradation constant can be calculated from the above equation at different substrate flow rates for a given quantity of cell loading.

4.6.2 Diffusion Model for a Tubular Packed-Bed Reactor

The concentration gradient along the height of the PBR is given by

$$D \times \frac{d^2c}{dz^2} - u\left(\frac{dc}{dz}\right) = \left(\frac{dc}{dt}\right) \tag{4.40}$$

Assuming the first-order substrate utilization at a lower substrate concentration,

$$\frac{dc}{dt} = K_P \times c \tag{4.41}$$

$$D \times \frac{d^2c}{dz^2} - u\left(\frac{dc}{dz}\right) = K_p \times c \tag{4.42}$$

The solution to this equation will be of the form

$$-(c_z) = c_1 \times e^{\alpha z} + c_2 \times e^{\beta z} \tag{4.43}$$

where c_1 and c_2 are constants.

$$\alpha = \left(\frac{u}{2D}\right) + \left[\left(\frac{u}{2D}\right)^2 + \frac{K_P}{D}\right]^{1/2} \tag{4.44}$$

$$\beta = \left(\frac{u}{2D}\right) - \left[\left(\frac{u}{2D}\right)^2 + \frac{K_P}{D}\right]^{1/2} \tag{4.45}$$

β and α are the roots of the equation

$$m^2 - \left(\frac{u}{D}\right)m - \frac{K_P}{D} = 0 \tag{4.46}$$

where $m = \frac{dc}{dz}$

Thus, on applying boundary conditions

$$c_0 = c_1 + c_2 \tag{4.47}$$

$$c_H = c_1 \times e^{\alpha H} + c_2 \times e^{\beta H} \tag{4.48}$$

$$\left(\frac{dc}{dz}\right) = 0 \text{ at } z = H$$

Therefore,

$$\left(\frac{c_0}{c_H}\right) = \frac{(\alpha \times e^{\beta H} - b \times e^{\alpha H})}{(\alpha - \beta)} \tag{4.49}$$

From the above equation we can calculate the value of K_p and compare this value with its observed values.

4.6.3 Combined Bulk Mass Transfer and Substrate Degradation Modeling for a Packed-Bed Reactor

Assuming that the rate of substrate diffusion from the bulk liquid to the surface of the cells is equal to the rate of substrate transfer from the surface of the cell to bulk liquid,

$$r_m = K_1 \times a_m(c - c_s) \tag{4.50}$$

At a steady state, the rate of mass transfer is equal to the rate of substrate degradation. The rate of substrate degradation can be considered as first-order kinetics.

$$r = K \times c_s \tag{4.51}$$

Then,

$$K_1 a_m(c - c_s) = K \times c_s \tag{4.52}$$

$$c_s = \left(\frac{K_1 \times a_m}{K_1 \times a_m + K}\right) \times c \tag{4.53}$$

$$= K \times c_s = K \times \left(\frac{K_1 \times a_\mathrm{m}}{K_1 \times a_m + K} \right) \times c = K_p \times c \tag{4.54}$$

$$K_p = K \times \left(\frac{K_1 \times a_m}{K_1 \times a_m + K} \right) \tag{4.55}$$

$$\frac{1}{K_\mathrm{p}} = \frac{1}{K} + \frac{1}{K_1 a_\mathrm{m}} \tag{4.56}$$

Parameter K_1 is related to the external diffusion model, which can be represented as $K_1 = A \times G^n$.

$$G = \left(\frac{Q\rho}{A\varepsilon} \right) \tag{4.57}$$

where A is a constant and G is the mass flux, ρ is the density of the fluid, and ε is the void fraction.

$$\frac{1}{K_\mathrm{p}} = \frac{1}{K} + \frac{1}{AG^n a_\mathrm{m}} \tag{4.58}$$

where the value of n ranges from 0.1 to 1.0. We can simulate the value of K_p by using the MATLAB software (method of least squares).

The normalized deviation of K_p from the experimental one were also calculated using the equation

$$\mathrm{ND} = \sum_{i=1}^{n} \frac{\dfrac{\dfrac{1}{k_{\mathrm{p(Experimental)}}} - \dfrac{1}{k_{\mathrm{p(Simulated)}}}}{\dfrac{1}{k_{\mathrm{p(Experimental)}}}}}{n} \tag{4.59}$$

It was observed that the substrate degradation constant was increasing with the flow rate. This might be due to the reduction in space time. The normal deviation between degradation constants of both the diffusion model and the plug flow model was found to be very nominal. This can infer that the plug flow is present in the reactor in spite of axial diffusion. The effect of external film diffusion on the observed substrate degradation rate was studied using the plots of experimentally measured values of $1/K_\mathrm{p}$ and $1/G^n$ for different values of n, ranging from 0.2 to 1.0 (Fig. 4.9).

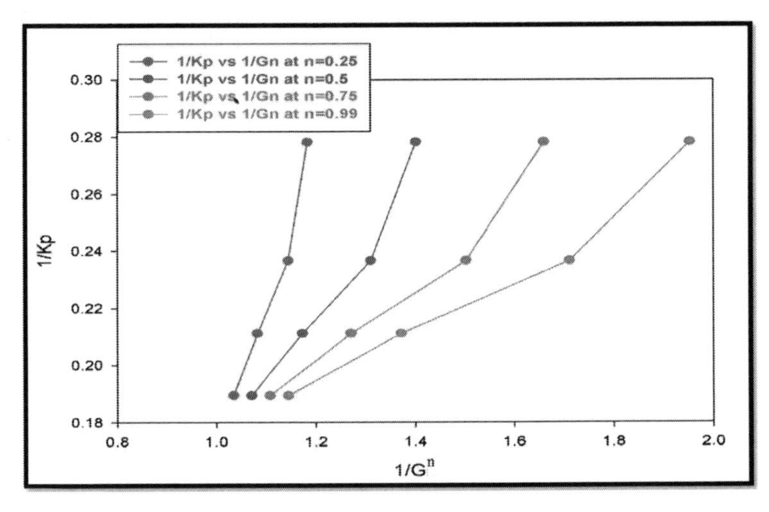

Figure 4.9 Plot of $1/K_{\mathrm{p}}$ vs. $1/G^n$.

It was observed that K values were positive only after $n = 0.75$. It was also found that there was a change from nonlinearity to linearity with an increase in the values from 0.2 to 1.0. From the above plots the values of intrinsic mass transfer constant and Aa_{m} were found out using regression analysis (Table 4.8). By using the method of least squares, normalized deviation of experimental values from the simulated profile was found out. Thus by taking the least error, the generalized equation for the present system was given by

$$\frac{1}{K_p} = 0.079 + \frac{0.096}{G} \tag{4.60}$$

The contribution of the substrate (glucose) degradation rate for the given mass of cells and mass transfer coefficient on the overall observed first-order substrate degradation rate constant is calculated at different flow rates and is given in Table 4.8. The external mass transfer rate coefficient k_1 and the observed first-order reaction rate k_{P} predicted by the model give an idea about the reaction engineering behavior of the system and also help in understanding the controlling step of the process. In the present system both substrate degradation and mass transfer steps are found to contribute almost appreciably to the overall degradation rate, particularly at higher flow rates (Das et al., 2002).

Table 4.8 Comparison of combined effects of the actual substrate degradation rate and mass transfer coefficients on the observed first-order degradation rate constant for $n = 1.00$

Dilution rate ($s^{-1} \times 10^{-4}$)	$1/K_p$ (s) ($\times 10^3$)	$1/k$ (s) ($\times 10^3$)	$1/k_1$ (s) ($\times 10^3$)	Percentage contribution of $1/k_1$	Percentage contribution of $1/k$
0.32	15.64	0.55	15.08	96.45	3.55
0.67	8.26		7.70	93.28	6.71
0.86	6.49		5.94	91.45	8.54
1.12	4.87		4.32	88.62	11.37
1.38	4.04		3.48	86.20	13.70
1.66	3.30		2.74	83.19	16.81
2.01	2.52		1.97	78.03	21.97
2.14	2.23		1.67	75.12	24.87

This gives us an idea about the reaction engineering behavior of the system and helps in understanding the controlling step of the process. In the present system both substrate degradation and mass transfer steps were found to contribute almost appreciably to the overall degradation rate, particularly at high flow rates. In the present system the contribution of the mass transfer coefficient was much higher, even at lower flow rates, proving the advantage of thermophilic systems over the mesophilic process, as reported by Das et al. (2002). Thus it can be inferred that at thermophilic temperatures, enhanced mass transfer could be achieved compared to mesophilic temperatures. The probable reason could be higher kinetic mobility of molecules at elevated temperatures. It was also observed that there was a considerable increase in the rate of substrate degradation with an increase in the flow rate.

4.7 Conclusion

Substrate consumption, increase in biomass and product (biohydrogen) formation in the dark, and photofermentation processes can be predicted with a significant level of accuracy using mathematical models. For the simulation of a process using a mathematical model kinetic parameters must be estimated. Kinetic parameters K_S, μ_{max}, and K_i were estimated from the substrate inhibition model using the method of least squares. Regression analysis was carried out on experimental data and simulated values for Monod's and Andrew's substrate inhibition models. R^2 values obtained for Monod's model were 0.987 and 0.79, and for the substrate inhibition model were 0.99 and 0.97 (substrate concentration profile and biomass concentration profile, respectively). Most of the models discussed in this chapter have reported values of R^2 greater than 0.95, indicating that these models are very reliable tools for process simulation. While studying the Luedeking–Piret model for product formation kinetics it was established that biohydrogen production is a growth-associated process, indicating that the rate of biohydrogen production will increase with the increase in the growth rate of microorganisms. The modified Gompertz equation was found to be a very accurate mathematical model for estimating the cumulative amount of biohydrogen produced. Several researchers have used it to model their processes and reported R^2 values around 0.99. Continuous hydrogen production in the PBR, assuming first-order kinetics, the observed first-order degradation rate constants (K_p) have been calculated at different flow rates. The contribution of the mass transfer coefficient was much higher at lower flow rates.

References

Andrews JF (1968). A mathematical model for continuous culture of microorganisms using inhibitory substrates, *Biotechnol Bioeng,* **10**, 707–723.

Chen WH, Chen SY, Khanal SK, Sung S (2006). Kinetic study of biological hydrogen production by anaerobic fermentation, *Int J Hydrogen Energy,* **31**, 2170–2178.

Davila-Vazquez G, Alatriste-Mondragón F, de León-Rodríguez A, Razo-Flores E (2008). Fermentative hydrogen production in batch experiments using lactose, cheese whey and glucose: influence of initial substrate concentration and pH, *Int J Hydrogen Energy,* **33**, 4989–4997.

Das D, Badri PK, Kumar N, Bhattacharya P (2002). Simulation and modeling of continuous H_2 production process by *Enterobacter cloacae* IIT-BT 08 using different bioreactor configuration, *Enzyme Microbiol Technol,* **31**, 867–875.

Das D, Khanna N, Veziroglu TN (2008). Recent developments in biological hydrogen production processes, *Chem Ind Chem Eng Quart,* **14**, 57–67.

Kumar N, Das D (2000). Enhancement of hydrogen production by *Enterobacter cloacae* IIT-BT 08, *Proc Biochem*, **35**, 589–593.

Kumar N, Monga PS, Biswas AK, Das D (2000). Modeling and simulation of clean fuel production by *Enterobacter cloacae* IIT-BT 08, *Int J Hydrogen Energy*, **25**, 945–952.

Han K, Levenspiel O (1988). Extended Monod kinetics for substrate, product and cell inhibition, *Biotechnol Bioeng*, **5**, 430–447.

Jung KW, Kim DH, Shin HS (2010). Continuous fermentative hydrogen production from coffee drink manufacturing wastewater by applying UASB reactor, *Int J Hydrogen Energy*, **35**, 13370–13378.

Lee KS, Hsu YF, Lo YC, Lin PJ, Lin CY, Chang JS (2008). Exploring optimal environmental factors for fermentative hydrogen production from starch using mixed anaerobic microflora, *Int J Hydrogen Energy*, **33**, 1565–1572.

Lin CY, Chang CC, Hung CH (2008). Fermentative hydrogen production from starch using natural mixed cultures, *Int J Hydrogen Energy*, **33**, 2445–2453.

Monod J (1949). The growth of bacterial cultures, *Ann Rev Microbiol*, **3**, 371–394.

Mu Y, Wang G, Yu HQ (2006). Kinetic modeling of batch hydrogen production process by mixed anaerobic cultures, *Bioresour Technol*, **97**, 1302–1307.

Nath K, Muthukumar M, Kumar A, Das D (2008). Kinetics of two-stage fermentation process for the production of hydrogen, *Int J Hydrogen Energy*, **33**, 1195–11203.

Nath K, Das D (2011). Modeling and optimization of fermentative hydrogen production, *Bioresour Technol*, **102**, 8569–8581.

Ntaikou I, Gavala HN, Kornaros M, Lyberatos G (2008). Hydrogen production from sugars and sweet sorghum biomass using *Ruminococcus albus*, *Int J Hydrogen Energy*, **33**, 1153–1163.

Obeid J, Magnin JP, Flaus JM, Adrot O, Willison JC, Zlatev R (2009). Modeling of hydrogen production in batch cultures of the photosynthetic bacterium *Rhodobacter capsulatus*, *Int J Hydrogen Energy*, **34**, 180–185.

Wang JL, Wan W (2008). The effect of substrate concentration on biohydrogen production by using kinetic models, *Sci China Ser B Chem*, **51**, 1110–1117.

Wang JL, Wan W (2009). Factors influencing fermentative hydrogen production: a review, *Int J Hydrogen Energy*, **34**, 799–811.

Chapter 5

Modeling and Simulation of the Biomethanation Process Using Organic Wastes

5.1 Introduction

Biomethanation refers to a broad spectrum of biochemical processes by which organic materials are upgraded and transformed into higher-grade fuels and other products. Methane and carbon dioxide are the major end products in this process, along with small quantities of nitrogen, hydrogen, ammonia, and hydrogen sulfide (usually less than 1% of the total volume). The organic materials remain mostly either in solid or in liquid form. The chemical compositions of these materials differ from each other. In the last couple of decades, emphasis has been given to study the suitability of different raw materials individually and in combination for the biomethanation process.

5.2 Microbiology

Extensive research work has been done by several workers on the bacteria involved in the biomethanation or anaerobic digestion process in the last several decades (Barker, 1956; Smith and

Biohythane: Fuel for the Future
Debabrata Das and Shantonu Roy

ISBN 978-981-4745-29-1 (Hardcover), 978-981-4745-30-7 (eBook)
www.panstanford.com

Hungate, 1958; Hungate, 1950; Zeikus, 1979; Gunsalus and Wolfe, 1977). Barker was first able to obtain highly enriched cultures of several methane-producing strains which were not an axenic culture (Barker, 1956). It was further inferred that the methane bacteria could utilize a specific number of low-molecular-weight end products of metabolism, such as formate, ethanol, acetate, propionate, and butyrate, obtained by fermentation, besides hydrogen (Gunsalus and Wolfe, 1977). However, the isolation of a pure strain of methane producers was reported by Smith and Hungate (1958).

Pohland and Ghosh (1974) demonstrated anaerobic digestion as a two-stage process, namely acid formation and methane generation. Two physiologically different digesting organisms, acid formers (acidogens) and methane formers (methanogens), are involved in this process (Pohland and Ghosh, 1970; 1971; 1974). Acidogens consist of a number of organisms which can convert proteins, carbohydrates, and lipids mainly to a volatile fatty acid (VFA) by hydrolysis and fermentation (Pohland and Ghosh, 1974). These bacteria are *Ruminococcus albus, R. flavefaciens, Butvrivibrio fibrisolvens, Bacteriodes succinogens, Cellobacterium cellulosolvens, Eubacterium ruminantium, B. amylophilus, B. ruminicola, Anaerovibrio lipolytica*, etc. (Smith and Hungate, 1958; Hungate, 1950; Zeikus, 1979; Gunsalus and Wolfe, 1977). Populations of 10^8–10^9 hydrolytic bacteria per milliliter of mesophilic sewage sludge have been reported by several investigators (Gunsalus and Wolfe, 1977). Studies on hydrolytic bacteria that utilize specific carbon substrates demonstrated 10^7 proteolytic and 10^5 celluloytic bacteria per milliliter of sewage sludge (Das, 1985).

The end products of the metabolism of acidogens are converted to methane and carbon dioxide in the presence of methanogens, which are obligate anaerobes. Different species of methane-producing bacteria were identified. They consisted of *Methanobacterium omelianskii, Methanobacterium bryanti, Methanobacterium formicicum, Methanobacterium thermoautotrophicum, Methanobrevibacter arboriphilus, Methanobrevibacter ruminantium, Methanobrevibacter smithi, Methanobrevibacter vannielli, Methanobacterium mobile, Methanoqenium cariaci, Methanospirillum hungatei, Methanosarcina barkeri, Metnanococcus mazei, Methanobacterium sohuqenii, Methanobacterium suboxydans, Methanobacterium propionicum, Methanococcus thermolithotrophicus*, etc. (Hungate, 1950; Zeikus, 1979; Gunsalus and Wolfe, 1977). Various studies indicate that methanogenic bacteria are present in anaerobic digestion in a number of 10^6 to 10^8 per milliliter (Das, 1985).

The characteristics of methanogenic bacteria are given in Table 5.1. *Methanosarcina barkeri* and *Methanococcus mazei* are of special

Table 5.1 Characteristics of methanogenic species in pure culture

Species	Morphology	Substrate	Cell wall composition
Methanobacterium formicicum thermoautotrophicum	Long rods to filaments	H_2, formate H_2	Pseudomurein
Methanobrevibacter *ruminantium* *smithii* arboriphilus	Lancet-shaped cocci, short rods	H_2, formate H_2, formate H_2	Pseudomurein
Methanococcus vanniellii voltae thermolithotrophics mazei	Motile, irregular, small cocci pseudosarcina	H_2, formate H_2, formate H_2, formate H_2, methanol, methylamines, acetate	Polypeptide subunits
Methanomicrobium mobile	Motile short rods	H_2, formate	Polypeptide subunits
Methanobacterium cariaci	Motile, irregular, small cocci	H_2, formate	Polypeptide subunits
Methanospirillum hungatei	Motile, regular, curved rods	H_2, formate	Polypeptide
Methanosarcina barkeri	Irregular cocci as single cell methanol packets, pseudoparenchyma	H_2, acetate	Heteropoly- saccharide
Methanothrix soehngenii	Rods to long filaments	Acetate	No muramic acid
Methanothermus fervidus	Nonmotile	H_2	Pseudomurein

Source: Das (1985)

interest due to their versatility with respect to the utilization of various substrates (Das, 1985; Sahm, 1984). Methanogenic bacteria are strict anaerobes. They require a lower redox potential (–330 mV, which corresponds to a concentration of 1 molecule of oxygen in about 10^{56} L of water) for growth compared to most other anaerobic bacteria. Oxygen is a potent inhibitor of methanogenesis. These bacteria can be gram-positive or gram-negative, and they have quite different cell shapes (Table 5.1). Balch et al. proposed a new taxonomy for 13 species of methanogens (Balch et al., 1979). Later several new strains were isolated. The more interesting thermophilic strains are *Methanosarcina* species, *Methanobacterium soehngenii, Methanococcus mazei,* and *Methanothermus fervidus* (Bryant, 1963; Dehority, 1971). Among these methanogens, *Methanothermus fervidus* is able to grow near the boiling point of water. Methanogenesis also occurs in nature at 0°C, but most pure strains of methanogens have optimum growth around 40°C (mesophiles) and at 65° to 75°C (thermophiles) (Colleran et al., 1982).

5.3 Biochemistry

The biochemistry of anaerobic digestion has been studied on sewage, farm animal wastes, or small-scale digestions using artificial mixtures of feed constituents, for example, cellulose, hemicellulose, protein, lipid, and lignin (Hungate and Stack, 1982). Carbohydrates play a major role in digester reactions and their breakdown. It is an important rate-controlling step in the digestion (Das, 1985). These carbohydrates are mainly cellulose and hemicellulose. Cellulose degradation takes place in the presence of extracellular enzymes produced by cellulolytic bacteria. These bacteria differ from each other in the relative amounts of cellulases (endoglucanase and exoglucanase) and the ability to attack different forms of cellulose (Hobson, 1976). The absolute rate at which cellulose is attacked depends on its physical form, for example, the degradation of domestic sewage is more compared to agricultural residues (Zeikus, 1980). Lignin, which is in complex association with the cellulosic structure, acts as a barrier to bacterial attack on the cellulose molecule (Stadtman, 1967). It prevents cellulolytic bacteria from adhering to the plant fibers. This is a prerequisite for optimum bacterial attack. Lignin was also shown to be almost entirely undegraded in anaerobic digesters (Bryant et al., 1967). However, a number of compounds of the same types which make up the lignin polymer are themselves degraded to methane and carbon dioxide by the mixture of bacteria from a domestic sewage sludge digester (Kluyver and Schnellen, 1947). The hemicellulose in the digester does not require the complex enzyme system of the cellulases for its hydrolysis. Hemicellulase was found to degrade hemicellulose (Das, 1985). The primary metabolic products of cellulolytic bacteria include aliphatic acids (formic, acetic, propionic, butyric, valeric, etc.), lactic acid, succinic acid, ethanol, carbon dioxide, and hydrogen (Das, 1985). Lactic and succinic acids are rapidly fermented to acetic and propionic acids. The primary breakdown of sugars in fermentation is usually to pyruvic acid with the simultaneous liberation of hydrogen in the form of hydrogen carrier complex.

The degradation of proteins proceeds via extracellular hydrolysis of proteins into peptides and amino acids by protease. Several mechanisms are involved in the degradation of amino acids in different bacteria (Hobson et al., 1981). Organic acids and ammonia are the end products of protein hydrolysis. The presence of extracellular proteases was found by several researchers (Hungate and Stack, 1982).

The degradation of lipids in anaerobic digestion proceeds through the initial breakdown of fats by lipase. The long fatty acids are degraded by β-oxidation, as shown with the ^{14}C tracer, using octanoic and palmitic acids (Das, 1985).

The biochemical qualities of methanogenic bacteria are different from other bacteria in the following manner:

- A very restricted range of oxidizable substrates coupled to the biosynthesis of methane
- Synthesis of an unusual range of cell wall components
- Synthesis of biphytanyl glycerol ethers as well as high amounts of squalene
- Synthesis of unusual coenzymes and growth factors

- Synthesis of rRNA that is distantly related to that of typical bacteria
- Possession of a genome size (DNA) approaching 1/3 that of *Escherichia coli*

Barker and coworkers (Barker, 1956; Das, 1985; Hobson et al., 1981) established the mechanism of methane formation. They showed that carbon dioxide, methanol, or acetate can be used as a precursor of methane by methanogenic bacteria (Fig. 5.1). It was further proved that the methyl group of acetate or methanol was transferred intact and was reduced, accepting one proton from the medium (Wolfe, 1982). The acetate system has been studied extensively as a major substrate of methanogens compared to hydrogen and carbon dioxide (and formate) (Wolfe, 1982).

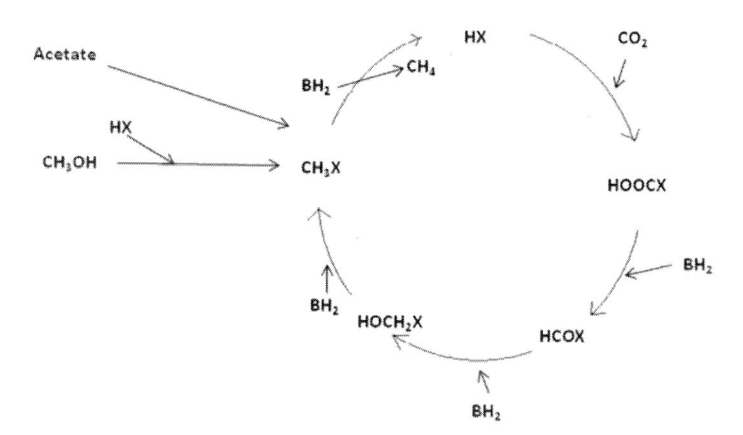

Figure 5.1 Barker's scheme for methanogenesis.

McBride fractionated the acidic molecule formed during the reduction of carbon dioxide to methane. He named this compound coenzyme M since it was involved in methyl transfer. Later, coenzyme M was identified as 2-mercaptoethane sulfonic acid. Coenzyme M was required as a growth factor by *Methanobrevibacter ruminanthium* and actively transported into the organism (Wolfe, 1982). Methanogenesis by *Methanobrevibacter ruminanthium* and *Methanosarcina barkeri* was specifically inhibited by the structural analogs of coenzyme M, for example, 2-bromoethane sulfonic acid and chloroethane sulfonic acid. Significant progress has been made in the study of methyl reductase after the discovery of methyl–coenzyme M. When methyl–coenzyme M was added as a substrate to cell extracts in the presence of excess hydrogen and ATP, methane was formed in stoichiometric quantity, 1 mol of methane being formed from 1 mol of methyl–coenzyme M added. However, if the same experiment was carried out in the presence of a mixture of hydrogen and carbon dioxide (1:1), the rate of methane formation increases 30-fold with a 12-fold increase in the amount of methane formed. On each new addition of substrate, the same effect was seen again and again. This effect is called the RPG effect (Prakash et al., 2014). Each mole of methyl–coenzyme M generated an active complex through which 11 mol of carbon dioxide was activated and reduced to methane. Recently, much research studies are in progress to find out the factor required for carbon dioxide reduction to methane (Wolfe, 1982; Prakash et al., 2014).

In addition to coenzyme M, coenzyme F_{420} has been detected in all methanogens which participate as electron carriers in the NADP-linked hydrogenase and formate dehydrogenase system in methanogens (Hungate, 1950; Das, 1985). Two additional cofactors (F_{430} and F_{342}) were found, in which cofactor F430 from *Methanosarcina barkeri* contained nickel (Thauer, 1982). The stabilization of different acids is reported by several researchers (Hungate, 1950; Das, 1985; Wolfe, 1982; Thauer, 1982).

Acetic acid

$$CH_3COOH + H_2O \rightarrow CH_4 + H_2CO_3 \quad (5.1)$$

Propionic acid

First step, $CH_3CH_2COOH + ½H_2O \rightarrow CH_3COOH + 3/4CH_4 + ¼CO_2$ (5.2)

Second step, $CH_3COOH + H_2O \rightarrow CH_4 + H_2CO_3$ (5.3)

Overall, $CH_3CH_2COOH + 3/2H_2O$ —-$7/4CH_4 + 1/4CO_2 + H_2CO_3$ (5.4)

Butyric acid

First step, $CH_3CH_2CH_2COOH + H_2CO_3 \rightarrow 2\ CH_3COOH$
$+ 1/2CH_4 + 1/2CO_2$ (5.5)

Second step, $CH_3COOH + 2H_2O \rightarrow 2CH_4 + 2H_2CO_3$ (5.6)

Overall, $CH_3CH_2CH_2COOH + 2H_2O \rightarrow 5/2CH_4 + 1/2CO_2 + H_2CO_3$ (5.7)

5.4 Thermodynamics and Kinetics

5.4.1 Thermodynamics

In the biomethanation process, 1 mol of glucose produces 3 mol of acetic acid (Ljungdahl and Andreesen, 1976). The change of free energy is –74.3 kcal according to the equation

$$C_6H_{12}O_6 + 4H_2O \rightarrow 2CH_3COO^- + HCO_3^- + 4H_2 + 4H^+, \quad \Delta F^o = -49.2 \text{ kcal} \quad (5.8)$$

$$2HCO_3 + 4H_2 + H^+ \rightarrow CH_3COO^- + 4H_2, \quad \Delta F^o = -25.1 \text{ kcal} \quad (5.9)$$

$$C_6H_{12}O_6 \rightarrow 3CH_3COO^-, \quad \Delta F^o = -74.3 \text{ kcal} \quad (5.10)$$

In this process, the formation of acetic acid from carbon dioxide is also favorable because the standard free-energy change is found to be negative (Lozano et al., 1980).

$$2\,CO_2 + 4H_2 \rightarrow CH_3COOH + 2H_2O, \quad \Delta F^o = -18.7 \text{ kcal} \quad (5.11)$$

However, this reaction is not favorable in a process involving methanogens due to the lower standard free-energy change in the following reaction compared to the above reaction:

$$CO_2 + 4H_2 \rightarrow CH_4 + 2H_2O, \quad \Delta F^o = -33.2 \text{ kcal} \quad (5.12)$$

The degradation of VFAs to methane, as occurred in this process, has a different change of standard free energy.

$$CH_3COO^- + H_2O \rightarrow CH_4 + HCO_3^-, \quad \Delta F^o = -7.4 \text{ kcal} \quad (5.13)$$

$$CH_3CH_2COO^- + 2H_2O \rightarrow CH_3COO^- + 3H_2 + CO_2, \quad \Delta F^o = +19.5 \text{ kcal} \quad (5.14)$$

$$CH_3CH_2CH_2COO^- + 2H_2O \rightarrow 2CH_3COO^- + 2H_2 + H^+, \quad \Delta F^o = +9.95 \text{ kcal} \quad (5.15)$$

The degradation of butyric and propionic acids is not favored in a thermodynamic sense. The free-energy change involved in the reduction of carbon dioxide by NAD(P)H is shown below.

$$4NAD(P)H + 4H^+ \rightarrow 4NAD(P)^+ + 4H_2, \quad \Delta F^\circ = +18.4 \text{ kcal} \quad (5.16)$$

$$4H_2 + CO_2 \rightarrow CH_4 + H_2O, \quad \Delta F^\circ = -33.2 \text{ kcal} \quad (5.17)$$

$$\text{SUM: } 4H^+ + 4NAD(P)H + CO_2 \rightarrow 4NAD(P)^+ + CH_4 + 2H_2O, \quad \Delta F^\circ = -14.8 \text{ kcal} \quad (5.18)$$

The change of standard free energy is very much negative in the case of the reduction of carbon dioxide to methane as compared to other methanation reactions mentioned earlier.

5.4.2 Kinetics

The degradation of solid biomass containing cellulose, hemicellulose, proteins, and lipids in anaerobic digestion involves the sequential step of hydrolysis, acidification, and methanation. Only the last two steps are involved in the digestion of a simple compound such as monosaccharides or organic acids. Study of the kinetics of these three digestion steps requires detailed information. Kinetic information, however, is available for these steps. This is mainly applicable for the hydrolysis and acidification.

Pretorius reviewed the published kinetic data applicable to the overall anaerobic digestion process (Pretorius, 1969). Lawrence and McCarty (1970) developed the rate constants for biomethanation of different VFAs, for example, acetic, propionic, and butyric acids. It was suggested that methanation was the rate-limiting step. This was still questionable because faster rates of reactions were assumed for hydrolysis and acidification compared to the methanation reaction. However, several researchers reported some indirect evidence that hydrolysis could be the rate-limiting step in the overall digestion of cellulosic feeds (Das, 1985). In the case of biomethanation of simple carbohydrates, for example, glucose, acetic acid degradation to methane was the rate-limiting step (Van den Berg, 1981).

The factors influencing the design and operation of a digester are (a) rate of product formation, (b) acid spectrum, and (c) rate of substrate utilization. Different operating conditions are important, for example, retention time, loading rate, temperature, pH, and biokinetic parameters like μ, μ_{max}, and K_S. Characteristics of acid formers were studied by Ghosh and Pohland (1974) using wastewater sludge, and Lozano et al. studied with a pure substrate such as sucrose (Lozano et al., 1980). Andrews and Pearson also showed the kinetics and characteristics of VFA formation in this process (Andrews and Pearson, 1965). Lozano et al. showed the correlation between specific rate of acid formation and cell growth rate using sucrose as a substrate (Lozano et al., 1980).

$$(1/x)r_{\mathrm{p}} = 1.7\mu + 1.6 \quad (5.19)$$

where x is the cell mass concentration (g L^{-1}), μ is the specific growth rate, (h^{-1}), and r_{p} is the rate of acid formation (g L^{-1} h^{-1}).

The maximum VFA productivity reported was 1.7 g L^{-1} h^{-1} as acetic acid (Lozano et al., 1980).

The kinetic constants for methanogenic bacteria were reported by several researchers (Andrews and Graff, 1971; Ghosh and Klass, 1977). At mesophilic temperature, Ghosh et al. (1980) reported the maximum specific growth rate (μ_{max}) of 0.49 d^{-1} and a saturation constant (K_{s}) of 4.2 g L^{-1} in a batch system of mixed methanogenic population, with acetic acid as the substrate. Chen and Hashimoto studied different kinetic parameters in the case of biomethanation of different wastes, for example, sewage sludge, municipal refuse, and livestock residue (Chen and Hashimoto, 1978). Ghosh et al. considered a kinetic relationship for anaerobic digestion of kelp and sewage sludge at the steady state without recycling (Ghosh et al., 1980).

$$\theta = \frac{k}{\mu_{\mathrm{m}}} \frac{1}{(1 - VS_{\mathrm{R}})L\theta} + \frac{1}{\mu_{\mathrm{m}}} \quad (5.20)$$

where θ is the hydraulic detention time (d), k is the half velocity constant (g L^{-1} d^{-1}), μ_{m} is the maximum specific growth rate constant (d^{-1}), VSR is the volatile solid reduction (%), and L is the loading rate (g L^{-1} d^{-1}). μ_{m} and k values using kelp and sewage sludge were found to be equal to 0.09 d^{-1}, 0.38 d^{-1} and 6.8 g $L^{-1}d^{-1}$, 63 g $L^{-1}d^{-1}$, respectively (Chen and Hashimoto, 1978).

Buswell (1939) studied the anaerobic digestion of different organic residues and presented a general formula for the conversion of complex material to carbon dioxide and methane, neglecting the fraction of substrate converted to microorganisms.

$$\mathrm{C}_a\mathrm{H}_b\mathrm{C}_c + \left(a - \frac{b}{4} - \frac{c}{2}\right)\mathrm{H_2O} \rightarrow \left(\frac{a}{2} - \frac{b}{8} + \frac{c}{4}\right)\mathrm{CO_2} + \left(\frac{a}{2} + \frac{b}{8} - \frac{c}{4}\right)\mathrm{CH_4} \quad (5.21)$$

Klass and Ghosh (1977) reported an empirical formula for kelp as

$C_{2.61}H_{4.63}O_{2.23}N_{0.1}S_{0.01}Ash_{26.7}$. Theoretical methane yield, neglecting the effect of nitrogen and sulfur, can be written as follows:

$$C_{2.61}H_{4.63}O_{2.23} + 0.34H_2O \rightarrow 1.285CO_2 + 1.32CH_4 \qquad (5.22)$$

From the above formula, it is found that 1 kg of kelp is considered with a mineral content of 267 g. The maximum methane and carbon dioxide yields are about 297 L and 288 L, respectively

5.5 Mathematical Modeling of the Biomethanation Process

5.5.1 Kinetics of Methanogenesis of Volatile Fatty Acids

Different intermediary VFAs such as acetic, propionic, and butyric acids were found to be the main substrates for methanogens. To evaluate the order of reactions and rate constants for the individual components, the overall reaction steps are considered.

$$\text{VFA (Si)} \xrightarrow{K_i} \text{Methane} + \text{Carbon dioxide} + \text{Cell mass}$$

The rates of degradation of these acids can be written as

$$-(ds_i/dt) = K_iS_i^{ni} \qquad (5.23)$$

where $i = 1, 2, 3$, and 4 for acetic, propionic, butyric, and mixed acids, respectively; S is the substrate concentration (g L^{-1}), K is the rate constant, and n is the order of the reaction.

These kinetic parameters are calculated for the composite reaction scheme. For estimation of rate constants and order of reactions, the logarithmic form of Eq. 5.23 is taken, which can be written as

$$\log(-dS_i/dt) = n_i \log S_i + \log K_i \qquad (5.24)$$

A plot of $\log(-dS_i/dt)$ versus $\log S_i$ should give a straight line whose slope gives the order of the reaction (n_i) and from whose intercept the kinetic rate constant (K_i) can be found out (Das, 1985).

5.5.2 Development of Overall Kinetic Model for Organic Wastes

In biological systems, the substrate balance equation can be written as follows:

Rate of substrate utilization	=	Rate of substrate utilization for cell growth	+	Rate of substrate utilization for cell maintenance	+	Rate of substrate utilization for gaseous product formation

$$\frac{d}{dt}(S_{\mathrm{T}})=\frac{d}{dt}(S_{\mathrm{g}})+\frac{d}{dt}(S_{\mathrm{m}})+\frac{d}{dt}(S_{\mathrm{G}}) \tag{5.25}$$

where S_{T} is the total biodegradable substrate concentration (g L^{-1}), S_{g} is the substrate utilized for cell growth (g L^{-1}), S_{m} is the substrate utilized for cell maintenance (g L^{-1}), and S_{G} is the substrate converted to gaseous form (g L^{-1}).

If substrate concentration is considered equivalent to the amount of biodegradable carbon available and product concentration is considered equivalent to the gaseous carbon available, the above equation may be presented as

$$\frac{d}{dt}(\mathrm{C_T})=\frac{d}{dt}(\mathrm{C_g})+\frac{d}{dt}(\mathrm{C_m})+\frac{d}{dt}(\mathrm{C_G}) \tag{5.26}$$

where $\mathrm{C_T}$ is the total biodegradable carbon input (g), $\mathrm{C_g}$ is the carbon utilized for cell growth (g), $\mathrm{C_m}$ is the carbon utilized for cell maintenance (g), and $\mathrm{C_G}$ is the carbon present in gaseous form (g).

It has already been reported that the carbon utilized for cell growth and maintenance was much less compared to the carbon converted to gaseous form (Ghosh and Pohland, 1974). The stoichiometry of the biodegradation of mixed biomass (A) can be written as

$$\mathrm{C_T'} = \alpha_{\mathrm{g,m}}\,(\mathrm{C_T'}) + \alpha_{\mathrm{G}}\,(\mathrm{C_T'}) \tag{5.27}$$

where $\alpha_{\mathrm{g,m}}$ is the fraction of carbon utilized for cell growth and maintenance, α_{G} is the fraction of carbon converted to gaseous form, and $\mathrm{C_T'}$ is the amount of carbon utilized for gas formation, cell growth, and maintenance.

The modeling of the reaction scheme has been made on the basis of major acid formation and gas generation steps. The conversion of substrate material for cell growth and maintenance has been taken into consideration in determining the acidification rate constant. The assumption has been made that the rate of consumption of substrate for cell mass growth and maintenance follows the similar pattern as that of an acidification step. However, the contribution of cell mass growth and maintenance for the methanation step has been neglected due to its reported low value (Ghosh and Pohland, 1974).

The following major reaction steps involved in the anaerobic digestion of mixed biomass (A) to methane can then be considered:

$$\text{Mixed biomass (A)} \xrightarrow{K_1'} \text{VFAs (B)} \xrightarrow{K_2'} \text{(Methane + Carbon dioxide) (G)}$$

From the above relationships, the following equation can be written:

$$\frac{dC_A}{dt} = -K_1' C_A^{n'} \tag{5.28}$$

$$\frac{dC_B}{dt} = \alpha_G K_1' C_A^{n_1'} - K_2' C_B^{n_2'} \tag{5.29}$$

$$\frac{dC_G}{dt} = K_2' C_B^{n_2'} \tag{5.30}$$

where C_A, C_B, and C_G are the amount of carbon present in solid biomass, VFAs, and gas, respectively; n_1' and n_2' are the order of reactions; and K_1' and K_2' are rate constants.

However, experimental determination of the carbon equivalent of the substrate (CS) (net balance of original total carbon minus the amount of carbon present in VFAs and gas) also includes the carbon converted into cell mass,

Equation 5.28 can be modified as

$$\frac{dC_A}{dt} = \frac{d}{dt}(C_s - C_g) = \frac{d}{dt}[C_s - \alpha_g(C_T - C_A)] \tag{5.31}$$

where C_g is the amount of carbon present in the cell,

$$\frac{dC_A}{dt} = \frac{1}{(1-\alpha_g)}\frac{dC_s}{dt} \tag{5.32}$$

since $\dfrac{dC_T}{dt} = 0.$

Again the combination of Eq. 5.28 and Eq. 5.32 gives

$$\frac{dC_s}{dt} = K_1' (1-\alpha_g)[C_s - \alpha_g(C_T - C_A)]^{n_1'} \tag{5.33}$$

For estimation of rate constants and order of reactions (K_1', n_1' and K_2', n_2'), logarithmic forms of Eq. 5.33 and Eq. 5.30 are taken, respectively, which gives

$$\log\left(-\frac{dC_G}{dt}\right) = \log[(1-\alpha_g)K_1'] + n_1' \log[C_s - \alpha_g(C_B + C_G)] \quad (5.34)$$

$$\log\left(\frac{dC_G}{dt}\right) = \log K_2' + n_2' \log C_B \quad (5.35)$$

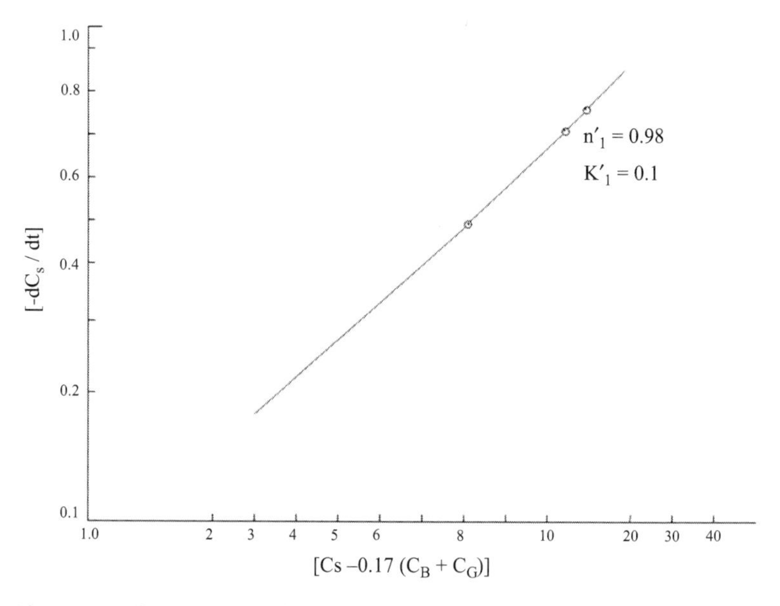

Figure 5.2 Computation of order of reaction and reaction rate constant for the degradation of mixed biomass (A) to VFAs.

From the slopes of the plots, $\log\left(-\frac{dC_G}{dt}\right)$ versus $\log [C_s - \alpha_g (C_B + C_G)]$ and $\log\left(\frac{dC_G}{dt}\right)$ versus $\log C_B$, we get the values of n_1' and n_2'. From the intercepts, the values of K_1' and K_2' can be estimated. It has been found experimentally that the value of n_1' = 0.98 (Fig. 5.2). So for the simplicity of Eq. 5.29 it is assumed that n_1' = 1. Equation 5.8 when combined with Eq. 5.29 may be written as

$$\frac{dC_B}{dt} = \alpha_G K_1' C_T e^{-K_1' t} - K_2' C_B^{n_2'} \quad (5.36)$$

Equation 5.36 could be used to predict the concentration of acid at any time of the reaction. From this profile of concentration, the time for the maximum acid concentration can also be determined.

5.6 Biomethanation of Volatile Fatty Acids

The experimental setup consisted of a specially designed 2 L reactor. Several batch experiments were carried out in this regard. Temperature was maintained at 37°C. VFA concentration is generally expressed in terms of acetic acid equivalents. Sanderson et al. reported a few disproportionation reactions (Sanderson and Wise, 1978), which are given below.

$$7CH_3COOH \rightarrow 4CH_3CH_2COOH + 2CO_2 + 2H_2O \quad (5.37)$$

$$5CH_3COOH \rightarrow 2CH_3CH_2COOH + 2CO_2 + 2H_2O \quad (5.38)$$

Hence each mole of propionic acid is equivalent to 1.75 mol of acetic acid and 1 mol of butyric acid is equivalent to 2.5 mol of acetic acid. Such calculation and data analysis are followed in biomethanation. These have already been reported (Sanderson and Wise, 1978).

A specially developed methanogenic culture was used from 15-day-old digested cattle dung through several generations using acetic acid as a substrate. Next, 10 %v/v inoculum was used for seeding the biogas reactor. The intermediate VFAs produced during biomethanation of organic residue contain mostly acetic, propionic, and butyric acids. The maximum acid concentration was found to vary from 2.5 to 5.5 g L^{-1} in terms of acetic acid. So we considered the acid concentration of individual fatty acids as 3000 mg L^{-1}. The degradation profile of several VFAs was studied and then compared to the methanogenesis of mixed acids.

5.6.1 Methanogenesis from Acetic Acid

Acetic acid addition results in a drastic change in pH of the solution. It has already been reported that pH has a great influence on the biomethanation process (Cohen, 1980). Different alkaline solutions, for example, $Ca(OH)_2$ and NaOH, can be used to control the pH. Effects of different minerals like Ca^{2+} and Na^+ on the biomethanation process were already studied (Das, 1985). The sodium ion has more inhibitory effect when compared to the calcium ion. So the initial pH was adjusted to 7.3 by using $Ca(OH)_2$. Figure 5.3 provides the gas production and acid concentration profiles. The total carbon balance of the system in terms of acetic acid and gaseous products was monitored, and 20% carbon appeared to be utilized for cell formation, metabolism, and maintenance. The maximum rates of degradation of acetic acid and gas production were 0.83 g L^{-1} d^{-1} and 0.7 L g^{-1}, respectively (Table 5.2).

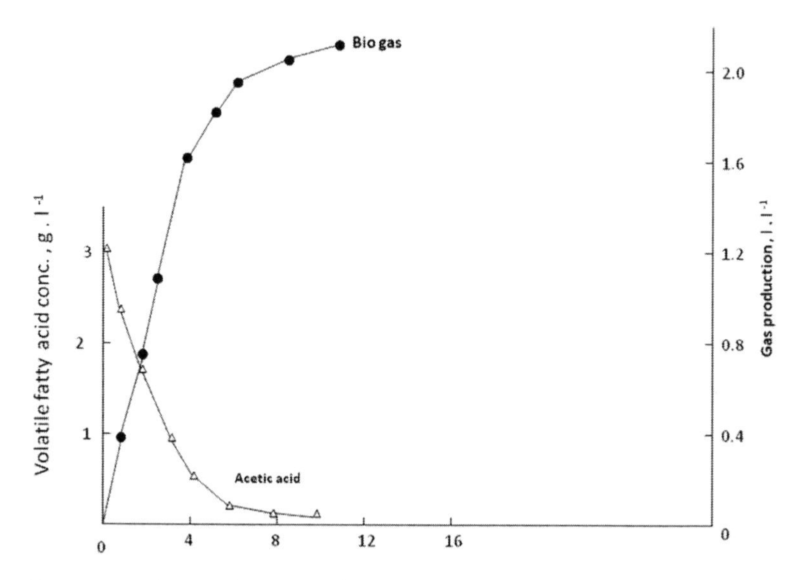

Figure 5.3 Gas production and acetic acid degradation profiles in a batch system.

Table 5.2 Comparative studies on the rate of degradation of different VFAs and methane yield

Volatile fatty acid	Maximum acid degradation rate (g L^{-1} d^{-1})	Methane yield (L g^{-1})	Per cent input carbon in gas form
Acetic acid	0.83	0.50	80.0
Propionic acid	0.40	0.28	72.5
Butyric acid	0.583	0.25	79.0
Mixed acids	0.416	0.44	76.0

Source: Das (1985)

5.6.2 Methanogenesis from Propionic Acid

Propionic acid is an important intermediary VFA during anaerobic digestion. It has been reported that propionic acid possesses some inhibitory effect on methanogens at a higher concentration level (>6000 mg L^{-1}). Extensive studies on the inhibition of propionic acid were reported by Kugelman and Chin, 1971. In the present study, the rate of degradation of propionic acid in relation to gas production was monitored at a concentration of 3000 mg L^{-1}. Acetic acid was found as an intermediary VFA during methanogenesis from propionic acid. Propionic acid degradation and gas formation profiles are shown in Fig. 5.4. The maximum rate of degradation of propionic acid was 0.4 g L^{-1} d^{-1}, and gas yield was 0.28 L g^{-1} of acid in terms of acetic acid.

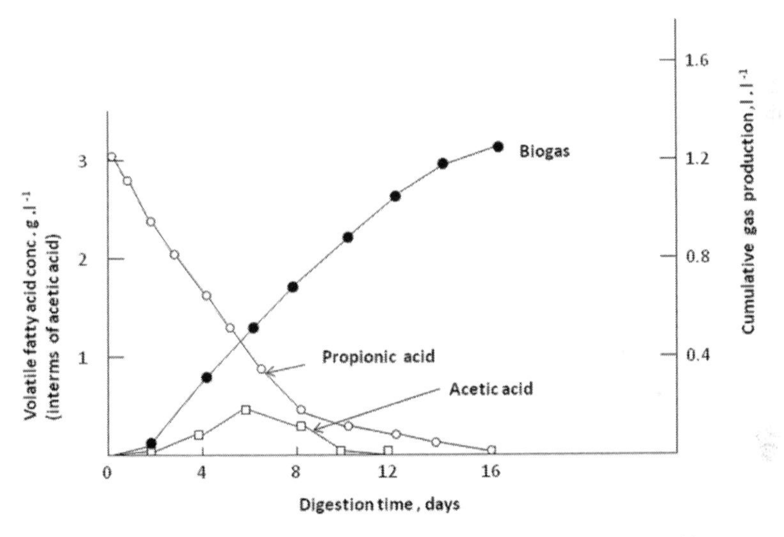

Figure 5.4 Gas production and propionic acid degradation profiles in a batch system.

5.6.3 Methanogenesis from Butyric Acid

Acetic acid is an intermediate VFA in butyric acid digestion. Butyric acid degradation in relation to gas production was monitored. The maximum rate of degradation of butyric acid was 0.583 g L^{-1} d^{-1}, and gas yield 0.25 L g^{-1}.

The kinetic constants were computed by using Eq. 5.23 (Fig. 5.5). Different kinetic parameters, such as reaction rate constants (K_1) and order of reactions (n_1), using acetic, propionic, and butyric acids are presented in Table 5.3. From the data shown in the table, it appears that the rates of degradation of these acids are in the following order: acetic > butyric > propionic. It appears that the degradation of acetic acid by the cell mass present in the system occurs at a higher rate mainly due to the higher number of cells (2.5×10^7 mL^{-1}) and reaction mechanisms.

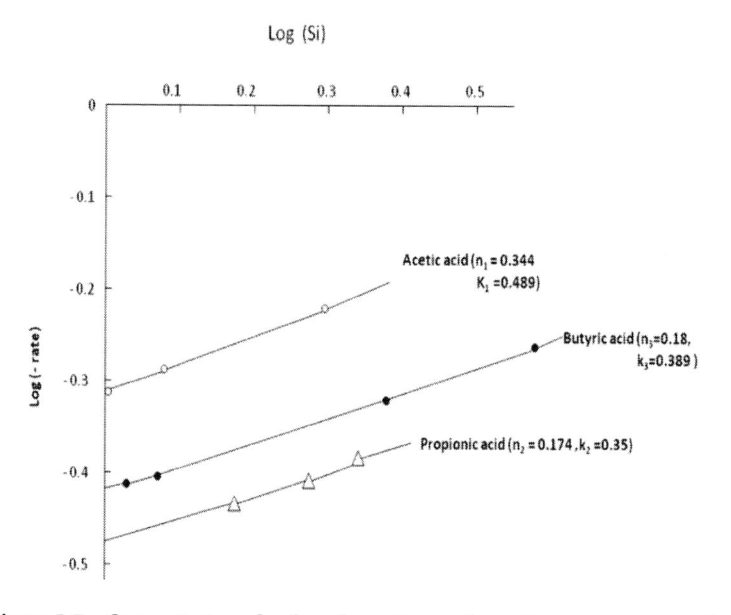

Figure 5.5 Computation of order of reaction and reaction rate constants for methanogenesis of acetic, propionic, and butyric acids.

Table 5.3 Kinetic constants by using different volatile fatty acids

Volatile fatty acid	Order of reaction	Reaction rate constant
Acetic acid	0.344	0.489
Butyric acid	0.18	0.389
Propionic acid	0.174	0.350
Mixed acids	0.419	0.272

5.6.4 Effects of Mixed Acids

The maximum acid produced during anaerobic digestion of mixed biomass (A) was 4.28 g L^{-1} in terms of acetic acid. This acid consisted of 38% acetic, 40% propionic, and 22% butyric acid. Mixed acids containing identical concentration and composition of these acids were considered as substrates for the batch biomethanation process. The main purpose of these experiments was first to find out the time required for mixed acids fermentation and second to correlate this gas production with the gas yield from mixed biomass (A) fermentation. The mixed acid fermentation gave less than 25% methane production compared to mixed biomass (A). Therefore, it appears that the acidogens remained active even after reaching a maximum acid concentration of 4.28 g L^{-1} at about the fourth day of fermentation. The rate constant for the overall acid degradation was found out. The value of the rate constant and the order of the reaction were 0.272 and 0.419, respectively (Fig. 5.6). The order of the overall reaction approaches the order of reaction in the case of acetic acid. From this, it may be concluded that acetic acid is the main reaction-controlling intermediary component in the biomethanation process (Das, 1985).

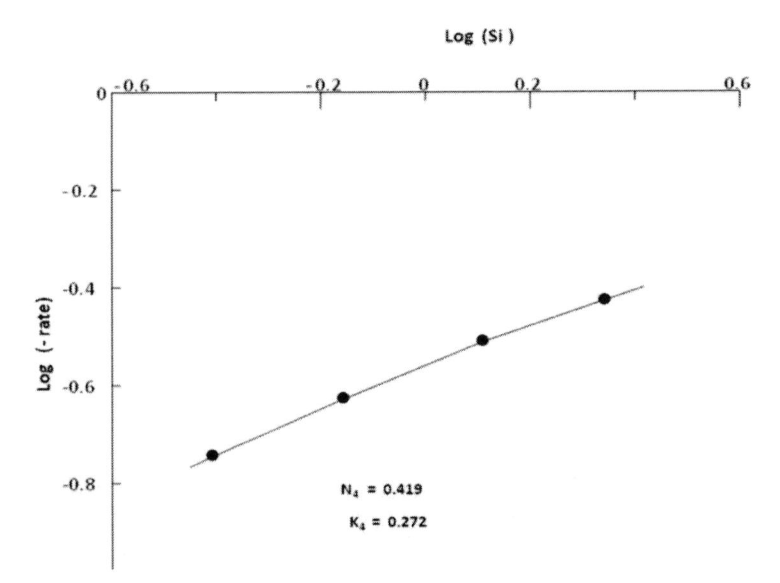

Figure 5.6 Computation of order of reaction and reaction rate constant for methanogenesis of mixed acids.

5.6.5 Computation of Maximum Energy Conversion and Kinetics of the Batch Biomethanation Process

5.6.5.1 Computation of maximum energy conversion

The energy relationship between stored energy in the raw material and gaseous energy recovery was studied in more detail. The enthalpy of the biomethanation process of mixed biomass (A) at 298 K was estimated, as shown in Table 5.4. It was assumed in these calculations that the minerals are carried through the process unchanged and that nitrogen and sulfur can be neglected since their concentrations are small. The calculations, as shown in Table 5.4, indicate that the process is slightly exothermic and the heat of reaction is 38.93 kcal/100 g of mixed biomass (A).

A. *Composition of mixed biomass (A)*

Basis:	100 g total solids
C	31 g
H	6 g
N	2.38 g
S	0.25 g
O	33.67 g
Minerals	26.7 g = M (assumed)
	100.0 g

B. *Empirical formula*: $C_{2.58}H_6O_{2.104}N_{0.17}O_{0.0078}M = 100\text{ g}$ (5.39)

C. *Heat of formation of mixed biomass (A)*

Neglecting S and N in the mixture and assuming the minerals as inert, the mixed residue can be considered to follow the reaction path given below for estimation of the heat of formation:

$$C_{2.58}H_6O_{2.104} + 3.028O_2 \rightarrow 2.58CO_2 + 3H_2O \quad (5.40)$$

$$(\Delta H)_R = \text{Heat of combustion}$$

$$= H'_{\text{Product}} - H'_{\text{Feed}}$$

$$= 2.58H'_{CO_2} + 3H'_{H_2O} - H'_{C_{2.58}}H_6O_{2.104} \quad (5.41)$$

where H' stands for enthalpy.

$\therefore$ Heat of formation of the substrate $(H'_{C_{2.58}}H_6O_{2.104})$

$= 2.58H'_{CO_2} + 3H'_{H_2O} - (\Delta H)_R$

$= -86.23\text{kcal}$ (5.42)

$(\Delta H)_R$ was estimated experimentally.

Table 5.4 Fuel values of different biomasses (basis: 1 kg dry biomass)

	Energy content of solid biomass (MJ)	
Biomass	**Estimated**	**Calculated**
Algae (mixed)	12.3	12.50
Water hyacinth	13.0	14.96
Cow dung	13.0	15.00
Rice husk	12.0	15.90
Mixed biomass (A)	13.8	14.55
Mixed biomass (B) [Algae (mixed) + Water hyacinth + Cow dung (1:1:1)]	13.6	14.15

D. *Heat of biomethanation of mixed biomass (A)*

To compute the heat of biomethanation of mixed biomass (A), the following reaction path is considered:

$C_{2.58}H_6O_{2.104} + 0.028H_2O \rightarrow 1.514CH_4 + 1.066CO_2$ (5.43)

Heat of reaction, $(\Delta H)_R$

$$= C_{2.58}H_6O_{2.104} + 0.028\,H_2O \rightarrow 1.514H'_{CH_4} + 1.066H'_{CO_2} - H'_{C_{2.58}}H_6O_{2.104} - 0.028H'_{H_2O}$$

$= -38.93$ kcals (5.44)

E. *Thermodynamic energy balance*

Input: 1 kg mixed biomass (A) reacted, 862.3 kcal.

Output: 15.14 g moles of CH_4, 271.197 kcal

10.66 g moles of CO_2, 1002.5 kcal

Heat of reaction, 389.3 kcal

F. *Energy analysis based on heat of combustion*

Basis: 1 kg of mixed biomass (A)

Total energy available in the residue: 3298.2 kcal.

Available gaseous energy in product form = 15.14 g moles of CH_4

Energy output = 15.14 × 19176 kcal = 2903.23 kcal.

Maximum available energy conversion efficiency

$$= \frac{2903.23}{3298.20} \times 100$$

= 88%

5.6.5.2 Overall kinetics for biomethanation of mixed biomass (A)

The kinetics model of the biomethanation process using mixed biomass (A) as the substrate has been considered. The carbon contents have been taken as equivalent to the concentration of substrates, VFAs, and gaseous products because of the mixed nature of the substrates, intermediates, and products (Fig. 5.7).

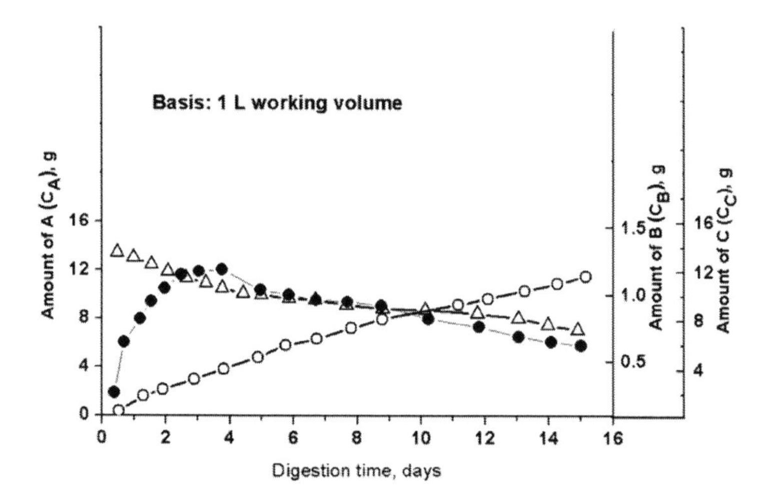

Figure 5.7 Carbon balance of the batch system using mixed biomass (A).

The two-stage reaction mechanism has been well established by many research workers (Das, 1985; Ghosh and Pohland, 1974). The kinetic model assumed in this particular case is based on the homogeneous reaction system. The validity of the model is established from Fig. 5.8. In this plot the initial substrate concentration was determined by multiplying the total initial carbon content with a factor of 0.43, representing the fraction of total mixed biomass (A), which was biodegradable. The fraction was experimentally determined and supported by other researchers (Ashare and Buivid, 1981). It has been experimentally determined that 80% of the biodegradable substrate is converted to a gaseous product. The rest of the substrate (20%) appeared to be utilized for cell growth and maintenance. This has also been reported by several workers (Das, 1985; Ghosh and Pohland, 1974). So the values of the constants $\alpha_{g,m}$ and α_g were assumed to be equal to 0.2 and 0.17, respectively. Figure 5.2 shows the rate of biodegradation of the substrate plotted against the residual substrate carbon content. The linear plot indicates that the reaction followed the general homogeneous mechanism. The order of reaction was determined from the slope of the curve as $n_1' = 0.98$, which was approximately unity. The rate constant was determined from the intercept of the straight line and was found to be $K_1' = 0.1$. Similarly, Fig. 5.9 indicates the logarithmic plot of the rate of gaseous product evolution against the carbon content in the VFAs. From this plot the order and rate of reaction were estimated as $n_2' = 0.7$ and $K_2' = 0.57$, respectively.

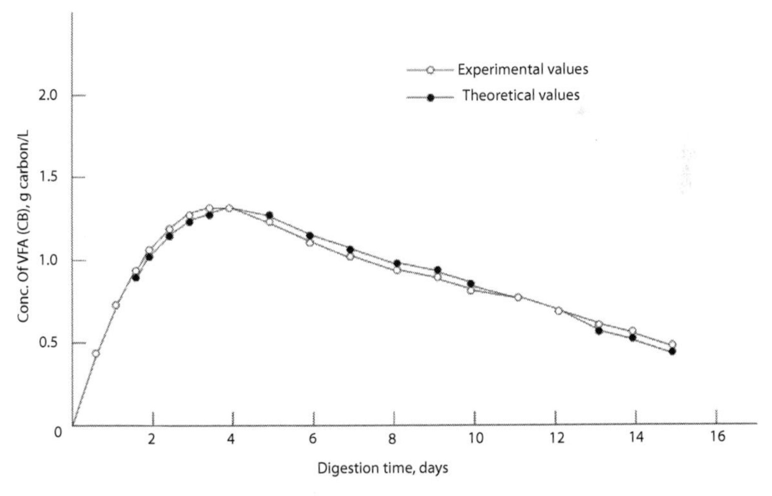

Figure 5.8 Comparative studies between the predicted and observed volatile fatty acid concentration profiles from the mixed residue (A).

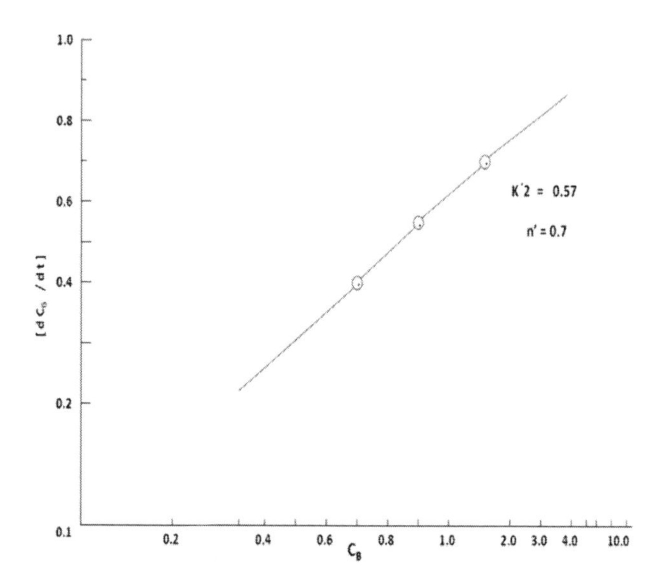

Figure 5.9 Computation of order of reaction and reaction rate constant for methanogenesis of intermediary VFAs.

Using the kinetic parameters established, the equation (Das, 1985) for the balance of intermediate VFAs is modified as follows:

$$\frac{dC_B}{dt} = 1.064e^{-0.1t} - 0.57{C_B}^{0.7} \tag{5.45}$$

This nonlinear first-order differential equation was solved by using the simple R-K method and computing by the ICL-2960 computer. Figure 5.8 shows the mode of production of VFA concentration. The experimental values of acid concentration were compared with the predicted one. The agreement between the two plots is excellent and indicates the validity of the application of the homogeneous consecutive reaction scheme to the biomethanation system. The main feature of the plots is the occurrence of maxima in acid concentration sometime around the fourth day. Experimentally it has been found that the acid concentration reaches a plateau sometime around the third day and stays till the end of the fourth day, indicating that the maximum occurs somewhere in between. This is also confirmed by the theoretical plot, which indicates the maximum to occur somewhere near the fourth day. Statistical analysis of the experimental and calculated values was carried out by using chi-square test. The observed chi-square (X^2) being insignificant at both levels of significance (α' = 0.05, 0.005), the experimental results seem compatible with the calculated values of the reaction kinetic model (Eq. 5.45).

The major significance of the concurrence of the two results is the prediction of the time period at which the maximum of acid concentration occurs at various operating conditions of temperature and biomass concentration. This period of maxima also indicates the prevalence of acidogenic and methanogenic bacteria before and after the occurrence, respectively. The model prediction can also be used for the development of a two-stage system where the first stage, up to the maximum acid concentration, will be containing predominantly acidogens. The second stage will be having mostly methanogens and will be for the period of declining acid concentration.

This two-step process has been successfully operated at the temperature and process conditions maintained in the batch process. In this case the acid formation step was continued for the period to reach a maximum acid concentration and followed up to 15 days, coinciding with the acid declining stage. The successful application of this model highlights the possible modification and control of operating conditions of the two-stage system and maintains the optimum process parameters for development of reactors of different configurations other than the conventional batch or semicontinuous process. Kinetics of the biomethanation of different types of wastes was reported by several researchers. Batch anaerobic digestion of glucose and its mathematical modeling were reported by Kalyuzhnyi (1997). Rodríguez-Martínez et al. (2002) discussed the kinetics of the biomethanation of slaughterhouse wastes.

5.7 Conclusion

Kinetics of the degradation of pure VFAs by methanogens was studied. The order of reactions of acetic, propionic, butyric, and mixed acids was 0.344, 0.18, 0.174, and 0.419, respectively. The order of reaction for acetic and mixed acids is similar; hence it could be concluded that acetic acid is the main reaction-controlling intermediary VFAs. The higher order of reaction in the case of mixed acids is due to the synergistic effect of mixed microflora.

Biomethanation is a two-step process, namely acid formation and methane generation. From the experimental data it has been found that the first reaction is a first-order reaction and the second has a fractional order of reaction of 0.7. The proposed kinetic model is found to give the concentration profile of individual components at any time satisfactorily, conforming to the experimental data. The predicted results showed a maximum acid concentration at the end of the fourth day, corroborating experimental observations. So it is concluded that the activities of acidogens are maximum at the end of the fourth day of fermentation in the batch system.

References

Andrews JF, Pearson EA (1965). Kinetics and characteristics of volatile acid production in anaerobic fermentation processes, *Int J Air Water Pollut*, **9**, 439–461.

Andrews JF, Graef SP (1971). Dynamic modeling and simulation of the anaerobic digestion process. In: *Anaerobic Biological Treatment Processes*, American Chemical Society Advances in Chemistry Series, **105**, 126–162.

Ashare E, Buivid MG (1981). In: Wise WL (ed), *Fuel Gas Production from Biomass*, Vol. I, CRC Press, Florida, 20–46.

Barker HA (1956). *Bacterial Fermentations*, John Wiley, New York.

Balch WE, Fox GE, Magrum LJ, Woese CR, Wolfe RS (1979). Methanogens: reevaluation of a unique biological group, *Micro Rev*, **43**, 260–296.

Bryant MP (1963). Biosynthesis of branched-chain amino acids from branched-chain fatty acids by rumen bacteria, *Arch Biochem Biophys*, **101**, 269–277.

Bryant MP, Wolin EA, Wolin MJ, Wolf RS (1967). Methanobacillus omelianskii, a symbiotic association of two species of bacteria, *Arch Biochem Biophy*, **59**, 20–31.

Buswell AM (1939). Anaerobic fermentation, *Illinois State Water Survey Bulletin*, 32.

Chen YR, Hashimoto AG (1978). Kinetics of methane fermentation, *Biotechnol Bioeng Symp*, **8**, 269–282.

Colleran E, Barry M, Wilkie A (1982). The application of the anaerobic filter to biogas production from solid and liquid agricultural wastes, Symposium papers *Energy from Biomass and Wastes VI*, IGT, Chicago, 443–482.

Das D (1985). PhD dissertation: Optimization of methane production from agricultural residues, BERC, IIT Delhi, India.

Dehority BA (1971). Carbon dioxide requirement of various species of rumen bacteria, *J Bacteriol*, **105**(1), 70–76.

Cohen A, Breure AM, Van Andel JG, Van Dersen A (1980). Influence of phase separation on the anaerobic digestion of glucose - I maximum COD-turnover rate during continuous operation, *Water Res*, **14**, 1439–1448.

Ghosh S, Pohland FG (1974). Kinetics of substrate assimilation and product formation in anaerobic digestion, *J Water Pollut Control Fed*, **46**, 748–759.

Ghosh S, Klass DL (1977). Symposium papers *Clean Fuels from Biomass and Wastes*, IGT, Chicago, 373–415.

Ghosh S, Henry MP, Klass DL (1980). Bioconversion of water hyacinth-coastal Bermuda grass-MSW-Sludge blends to methane, *Biotechnol Bioeng Symp*, **10**, 163–187.

Gunsalus RP, Wolfe RS (1977). Stimulation of CO2 reduction to methane by methylcoenzyme M in extracts Methanobacterium, *Biochem Biophys Res Commun*, **76**, 790–795.

Hobson PN (1976). The microflora of the rumen. In: Cook JG (ed) *Patterns of Progress*. Meadowfield Press, England, 33–36.

Hobson PN, Bousfield S, Summers R (1981). *Methane Production from Agricultural and Domestic Wastes*, ASP, London.

Hungate RE (1950). The anaerobic mesophilic cellulolytic bacteria, *Bac Rev*, **14**, 1–49.

Hungate RE, Stack RJ (1982). Phenylpropanoic acid: growth factor for *Ruminococcus albus*, *Appl Environ Microbiol*, **44**, 79–83.

Kalyuzhnyi SV (1997). Batch anaerobic digestion of glucose and its mathematical modeling II: description, verification and application of model, *Bioresour Technol*, **59**, 249–258.

Klass DL, Ghosh S (1977). Symposium papers *Clean Fuels from Biomass and Wastes*, IGT, Chicago, 323–351.

Kluyver AJ, Schnellen GTP (1947). On the fermentation of carbon monoxide by the pure culture of methane bacteria, *Arch Biochem*, **14**, 57–70.

Kugelman IJ, Chin KK (1971). *Anaerobic Biological Treatment Processes*, American Chemical Society Advances in Chemistry Series, **105**, 55–90.

Lawrence AW, McCarty PL (1970). Unified basis for biological treatment design and operation, *J Sanitary Eng*, **96**, 757.

Ljungdahl LG, Andreesen JR (1976). Reduction of CO_2 to acetate in homoacetate fermenting *Clostridia* and the involvement of tungsten in formate dehydrogenase. In: Schlegel H, Gottschalk G, Pfennig N (eds), *Microbial Production and Utilization of Gasses*. Goltze, Gottingen, 163–172.

Lozano IT, Cornok I, Goma G (1980). Symposium on, *'Bioconversion and Biochemical Engineering II'*, Ghose TK (ed), IIT Delhi, India, 113–132.

Pohland FG, Ghosh S (1970). Developments in anaerobic treatment process, *Biotechnol Bioeng Symp*, **2**, 85–106.

Pohland FG, Ghosh S (1971). Developments in anaerobic stabilization of organic wastes-the two-phase concept, *Environ Lett*, **1**, 255–266.

Prakash D, Wu Y, Suh SJ, Duina EC (2014). Elucidating the process of activation of methyl-coenzyme M reductase, *J Bacteriol*, **196**, 2491–2498.

Pretorius WA (1969). Anaerobic digestion III, kinetics of anaerobic fermentation, *Water Res*, **3**, 545–558.

Jesús R-M, Ivan R-G, Estaban P-F, Nagamani B, Sosa-Santillan Gerardo S-S, Yolanda G-G (2002). Kinetics of anaerobic treatment of slaughterhouse wastewater in batch and upflow anaerobic sludge blanket reactor, *Bioresour Technol*, **85**, 235–241.

Sahm H (1984). Anaerobic wastewater treatment. In: Fiechter A (ed) *Advances in Biochemical Engineering/Biotechnology*, Vol. 29, Springer-Verlag Berlin Heidelberg, New York, 83–115.

Sanderson JE, Wise DL, Augenstein DC (1978). Organic chemicals and liquid fuels from algal biomass, *Biotechnol Bioeng Symp*, **8**, 131–151.

Smith PH, Hungate RE (1958). Isolation and characterization of Methanobacterium ruminantium n. sp, *J Bacteriol*, **75**(6), 713–718.

Stadtman, TC (1967). Methane fermentation. In: Clifton CE, Raffel S, Starr, MP (eds), *Annual Review of Microbiology*, Vol. 21, Annual Reviews Inc, California, USA, 121–142.

Thauer RK (1982). In: Hughes DE, et al. (eds.), *Anaerobic Digestion 1981*, Elsevier Biomedical Press, BV, The Netherlands, 37–44.

Van den Berg C, Lentz CP, Armstrong DW (1981). In: Moo-Young M, Robinson CW (eds), *Advances in Biotechnology*, Vol II, Pergamon Press, Canada, 251.

Wolfe RS (1982). The formation of methane from biomass-ecology, biochemistry and applications, *Experimentia*, **38**, 198–201.

Zeikus JG (1979). Thermophilic bacteria: ecology, physiology and technology, *Enz Microbiol Technol*, **1**, 243–252.

Zeikus JG (1980). Microbial populations in digesters. In: Stafford DA, Wheatley BI, Hughes DE (eds), *Anaerobic Digestion*, Applied Science Publishers, London, UK, 61–89.

Chapter 6

Scale-Up of Biohydrogen Production Processes

6.1 Introduction

It is quite evident that present hydrogen production processes are dependent on utilization of nonrenewable energy sources. Moreover, the conventional processes of hydrogen production consume lot of energy and also have a high carbon footprint. For hydrogen to be considered as a future renewable energy source, it should be produced from renewable feedstock. Biohydrogen may be produced from organic wastes at ambient temperature and atmospheric pressure. So a sustainable process could be developed where energy could be generated from organic wastes, subsequently helping in waste management.

The technologies available for hydrogen production have certain advantages and disadvantages. A rigorous evaluation of these processes in terms of the cost for commercialization is the need of the hour. In the case of biological hydrogen production, few reports dealing with economic analyses of biohydrogen production processes are available.

At present, the biohydrogen production is not cost effective compared to conventional fuels. Therefore, there is immense scope of improvement in terms of development of cost-effective reactors,

Biohythane: Fuel for the Future
Debabrata Das and Shantonu Roy

ISBN 978-981-4745-29-1 (Hardcover), 978-981-4745-30-7 (eBook)
www.panstanford.com

high-yielding strains, and use of cheaper feedstock for sustainable biohydrogen production. A comprehensive technoeconomic analysis is required to compare the prospect of biohydrogen with other conventional fossil fuels. An economic survey based on fuel production cost, transport system, and storage makes the technoeconomic studies complicated. The acceptance of a fuel for anthropogenic activity at relevant costs depends upon many factors such as the emission profile in terms of pollution, short-term/long-term environmental effects, direct/indirect health costs, and socioeconomical acceptance. Since hydrogen is considered a truly carbon-neutral fuel, it is the most logical choice worldwide as a future energy source.

The advent of fuel cell technologies has partially addressed the need of hydrogen storage as hydrogen produced in any system can be stored transiently and can be channelized to fuel cells to convert it to electricity. With low CO_2 emissions, fuel cells have the potential to become major factors in catalyzing the transition to a future sustainable energy system. A large number of countries are now implementing roadmaps for the advancement of fuel cell and hydrogen technologies. The European Union has set up a roadmap which is a potent example of the development and deployment of hydrogen and fuel cell technologies. It set a target of 1 GW of distributed power generation capacity from fuel cells by 2015 and 0.4–1.8 million hydrogen vehicles sold per year by 2020. The three major technological barriers that must be prevailing over for a transition from a carbon-based (fossil fuel) energy system to a hydrogen-based economy are as follows:

- Cheap and sustainable hydrogen production is required.
- New generations of hydrogen storage systems are needed that would be cheap and easy to implement for both vehicular and stationary applications.
- The cost of fuel cells and other hydrogen-based systems must be reduced.

The combination of large and small fuel cells running on local hydrogen supply networks for domestic and decentralized heat and electricity power generation would help in realizing the vision of integrated energy systems. Using hydrogen as a fuel has certain safety issues. This encompasses scientific and technological aspects of the development of a safety protocol and strategies. The psychological and societal acceptance of using hydrogen as a fuel is also critical. Even though hydrogen has been maligned for its safety issues, it has an exemplary safety record when it comes to industrial use. As hydrogen is the lightest element on earth, it has a tendency to escape from small pores and channels, and on mixing with atmospheric oxygen, it becomes inflammable.

There are many technologies available for hydrogen production. It can be produced from fossil fuels such as natural gas, coal, and hydrocarbons, or it can be produced from gasification of biomass, municipal waste, and industrial wastewater. The majority of hydrogen is produced by electrolysis of water. Steam reforming of methane is also a commercial way of producing hydrogen. This process leads to CO_2 emission, but the amount of CO_2 released is far less compared to the combustion of methane. High-temperature pyrolysis was used to convert hydrocarbons, biomass, and municipal solid waste into hydrogen. The by-products of this process are carbon-rich charcoal, CO, and CO_2. At present, the cost of this process is higher than that of steam reforming of natural gas. The production of hydrogen by electrolysis of water has a very high efficiency of 70%–75%, but the cost of production is several times higher than that of production from fossil fuels. With the advent of technology, fuel cells convert hydrogen to electricity. This has infused new life into the implementation of a hydrogen-based economy. A fuel cell is a device that is similar to a continuously recharging battery which generates electricity by the low-temperature electrochemical reaction of hydrogen and oxygen. The contrasting difference with batteries store energy is that a fuel cell can produce electricity continuously as long as hydrogen and oxygen are supplied to it. Hydrogen-powered fuel cells produce water as a by-product, with virtually no pollutant. Fuel cells operate at temperatures much below an internal combustion (IC) engine. Fuel cells are not bound by the limitations of the Carnot cycle; thus they can efficiently convert fuel to electricity as compared to IC engines. The operating temperature, the type of fuel, and a range of applications of fuel cells depend on the electrolyte the fuel cells use. The electrolyte can be an acid, a base, a salt, or a solid ceramic or polymeric membrane that conducts ions.

However, at present, fuel cells cannot compete with conventional energy conversion technologies in terms of cost and reliability. High-temperature solid oxide fuel cells (SOFCs) and molten carbonate fuel cells (MCFCs) are ideal for distributed energy supplies operating today with natural gas, which enables the development and use of this technology independently from the establishment of a hydrogen infrastructure. Indeed, they offer an interesting transition to the hydrogen economy. It has given a fresh breath to biofuel research, where gaseous energy or ethanol can be converted to electricity directly. The types of fuel cell technology available are:

- Low-temperature proton exchange membrane fuel cells (PEMFCs): The highest power density is provided by PEMFCs and alkaline fuel cells (AFCs). The requirement of a costly platinum catalyst and highly pure hydrogen is the major bottleneck in using such fuel cells. PEMFCs are the most favored for mass-market automotive and small-scale combined heat and power (CHP) applications, and there is a massive global effort to develop commercial systems.
- Phosphoric acid fuel cells (PAFCs): These types of fuel cells are more tolerant toward impurities that remain as contaminants with hydrogen. PAFCs could be potentially used for stationary power generation as well as large vehicles. They are commercially available today but have a relatively high cost. Direct methanol fuel cells (DMFCs) are powered by methanol and are considered for a number of applications, particularly those based around replacing batteries in consumer applications such as mobile phones and laptop computers.

Thus a hydrogen economy has great potential for creating employment. Moreover, its environmental benefits also give it an upper hand in considering it as a future fuel. Implementing hydrogen production technologies and its use in fuel cells for electricity generation would require dedicated human resources. Human resources having expertise in various domains of hydrogen production technologies and fuel cells application would be required. Policies and measures are needed to minimize the production cost of such technologies. Moreover, safety issues need serious consideration so that a hydrogen economy booms in a real way. It is essential to change the misconception of hydrogen energy and fuel cell technologies being not commercially viable, and markets do not readily accept them. To promote alternative fuels, a market transformation approach will help in penetration of the product to the masses and services associated with it. It includes promotion of total technology transfers and gradual upgrading rather than one-time transfers. Supportive government policies could help in effective implementation of the market transformation approach, which eventually would help in ending the supremacy of traditional technologies and pave the path for renewable energy-related technologies. Technology adaptation, incentive measures, command, enterprise technological capability, control measures, consumer awareness, and marketing are a few vital points that are related to public policy. In addition to these holistic policy measures, successful deployment requires some targeted technology-/end-use-specific policy prescriptions. The continuous biohydrogen production process with suspended cells and an immobilized whole cell system are shown in Fig. 6.1 and Fig. 6.2.

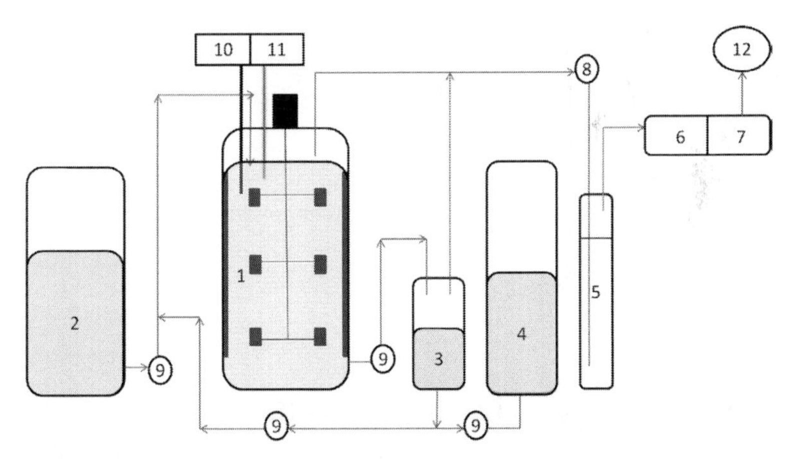

Figure 6.1 Schematic flow diagram of an industrial-scale biohydrogen production in a continuous stirred tank reactor (CSTR): (1) reactor, (2) feed tank, (3) recycle tank, (4) drain tank, (5) CO_2 absorber, (6) flow meter, (7) sensor, (8) pressure control pump, (9) peristaltic pump, (10) pH monitoring system, (11) thermostat, and (12) hydrogen storage section.

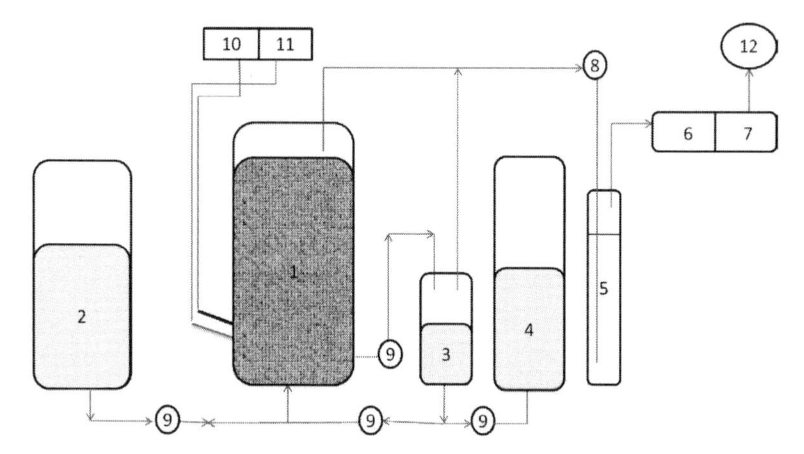

Figure 6.2 Schematic flow diagram of an industrial-scale biohydrogen production in a pack bed reactor (PBR): (1) reactor, (2) feed tank, (3) recycle tank, (4) drain tank, (5) CO_2 absorber, (6) flow meter, (7) sensor, (8) pressure control pump, (9) peristaltic pump, (10) pH monitoring system, (11) thermostat, and (12) hydrogen storage section.

6.2 Determination of Scale-Up Parameters

The suitability of any process lies in its capacity to be scaled up to the industrial level. When it comes to scaling up of dark-fermentative hydrogen production, the following are the parameters that could be considered:

- Geometric similarity in scale-up
- Scale-up based on volumetric power consumption
- Constant impeller tip speed
- Reynolds number
- Constant mixing time

6.2.1 Geometric Similarity in Scale-Up

During fabrication of any bioreactor, the geometry of the reactor is the pivotal point. There is a certain set of ratios that are needed to be maintained during scaling up of the vessels. The following are the vital ratios (Junker, 2004):

- The H/D ratio, that is, the ratio of the height of the reactor to the diameter of the reactor
- The ratio of the impeller diameter to the reactor diameter
- The ratio between the reactor diameter and the reactor volume

The above-mentioned set of ratios is essentially required to be conserved for maintenance of geometric similarity. Though these parameters are of prime importance, it is also unlikely that they can be completely conserved. Thus on improvising the process from a shaken flask to a bioreactor (stirred tank), the geometric similarity cannot be conserved. Optimization of the above-mentioned ratios and improvement of design could help in operation of the process with minimum energy consumption. If a considerable number of changes is made in the volume of the reactor (scale-up), then the aforementioned ratios are also likely to be changed.

6.2.2 Scale-Up Based on Volumetric Power Consumption

In the case of biofuel research, the power consumption during the production process plays a critical role in deciding the relevance of the biofuel produced. The volumetric power consumption (P/V) is one of the important parameters considered for scale-up studies. It's the total energy consumed during mixing of liquid media inside the reactor. Moreover, there are many other power sinks such as power consumed by the gearbox, motor, and frictional losses. Thus, the overall power consumption is not considered during scale-up studies.

Volumetric power consumption is directly affected by the following characteristics:

- Degree of turbulence and media circulation in vessels
- Heat transfer
- Mass transfer
- Mixing or circulation times

For industrial-scale penicillin production that requires low energy input, volumetric power input was considered an important parameter. It was around 1 hp per gallon, equivalent to 1.8 kW/m^3.

While the fermentative process required high energy inputs, such as recombinant *Escherichia coli* cultures, the role of volumetric power input was limited. It could be possibly due to the high cost of operation and high shear stress developed in the larger-scale stirred vessels.

In an anaerobic process, the power input during agitation is denoted by P_o and is defined by the equation

$$P_o = N_p \rho N^3 D^5 \tag{6.1}$$

where N is the stirrer speed, ρ is the density of the fluid, and D is the stirrer diameter (Rushton et al., 1950). The power number itself is dependent on other dimensionless groups such as Reynolds number and Froude number, as well as on the number of agitator turbines.

The following relationship for estimating the volumetric power consumption (P) has been suggested by Hughmark:

$$\frac{P}{P_o} = 0.1\left(\frac{N^2 D^4}{gwv^{2/3}}\right)^{-\frac{1}{6}}\left(\frac{Q}{NV}\right)^{-\frac{1}{4}} \tag{6.2}$$

where P_o is the power consumption of the nonaerated system, W is the width of turbine blades, and Q is the volumetric gas flow rate.

6.2.3 Volumetric Power Consumption in an Agitated System

The power consumption in shaken flasks at high and low medium viscosities was studied by Büchs et al. (2011), where a simple rotary shaking machine fixed to a frame, combined with a torque sensor attached to the powering drive, was used. Torque and shaking speed were measured and correlated with the following equation:

$$\frac{P}{V} = Ne\rho\frac{N^3 d^4}{V^{2/3}} = C_3\rho\frac{N^3 d^4}{V^{2/3}}Re^{-0.2} \tag{6.3}$$

where Ne' is the modified Newton number of shake flasks, d is the impeller diameter, and C_3 is the fitting parameter.

The specific power consumption is found to be a function of the shaking frequency according to $P \approx N^{2.8}$ in nonbaffled agitated tank reactors. The model includes C_3 as the only fitting parameter using least-squares nonlinear fitting for the description of all the experimental results obtained in their study, having a value of 1.94.

Using working volumes of 4% to 20% of the culture flask volume and a shaking diameter of 2.5 to 5 cm, the volumetric power consumption was observed in the range of 0.01 to 0.2 kW with 80 to 380 min^{-1} shaking frequency. Between calculated and measured values of power consumption, a contrasting variation was observed. The probable reason for such deviation in power consumption values could be due to the fact the data were collected at a low shaking frequency (80 to 120 min^{-1}), and thereby, a low Reynolds number (Re ≈ 500 to 5000) was observed, which eventually corresponds to the transition zone between laminar to turbulent flow.

6.2.4 Constant Impeller Tip Speed

The impeller tip speed, v_{tip}, is expressed as

$$v_{tip} = \pi ND \tag{6.4}$$

where N is the number of revolutions per second and D is the impeller diameter.

Tip speed is a vital parameter for scale-up. The shear force generated at the tip of the impeller could lead to changes in cellular integrity. A tip speed above 3.2 m s^{-1} could lead to cell wall distortion. The exact value of tip speed varies with the rheological properties of the fermentation broth, such as viscosity, operation temperature, etc. During scaling-up studies, the volumetric power consumption (P/V) is lowered if the system is operated at a constant tip speed (the geometric consideration kept in mind). Such demerits could be overcome by having more impellers so that both volumetric power consumption and tip speed could be kept constant.

Tip speed is useful for the fermentation process where filamentous microbes such as fungi are used. It helps in hyphae disintegration and thereby alteration of broth characteristics, whereas the tip speed is less constructive for single-cell microbial fermentation processes. If scale-up is performed under a constant tip speed (in accordance with geometric similarity), then the value of P/V tends to get lower. To overcome this problem, it is recommended that more impellers be used for larger vessels, which would eventually lead to maintenance of a constant tip speed and P/V. The tip speed modulates the value of impeller shear. Under a turbulent flow regime, the impeller shear is proportional to the product of impeller tip speed and impeller diameter, ND_i.

6.2.5 Reynolds Number

The agitator Reynolds number is expressed as

$$\mathrm{Re} = \frac{\rho N D^2}{\mu} \tag{6.5}$$

where N is the number of revolutions per second, D is the impeller diameter, μ is the viscosity, and ρ is the density of the fluid.

It is a dimensionless number whose value could help in understanding the rheological property of a fluid in motion, that is, whether the flow of fluid follows a laminar regime or a turbulent one (Table 6.1). The extent of homogeneity in a bioreactor is measured in the scale of turbulence, but the Reynolds number generally is not used for scaling up of the bioreactor as it does not influence the effect of aeration and its value also changes with the reactor configuration (Junker, 2004; Ju and Chase, 1992). During scale-up studies, use of dimensionless groups in conjunction with other process parameters could lead to infeasible operating conditions.

Table 6.1 Variation in Reynolds number with respect to agitation rates

Bioreactor geometry	*N* (rpm)	Re
Microwell (1000 μL)	500	700
	750	1060
	1000	1400
Shake flask (100 mL)	300	106,800
Stirred tank (1.4 L)	700	24,720
	1000	35,320

Source: Micheletti et al. (2006)

6.2.6 Constant Mixing Time

The time required for achieving the desired homogeneity inside the reactor on infusion of a pulse at a single point in the reactor is regarded as the mixing time t_m. It is represented by Eq. 6.6:

$$t_{\mathrm{m}} = \frac{V}{N_{\mathrm{f}} N D^3} \tag{6.6}$$

where N is the number of revolutions per second, D is the impeller diameter, N_f is the pumping number, and V is the volume of the reactor. Recently, a high-speed video technique has been developed for precise determination of the mixing time inside the reactor. To reach 95% homogeneity, the following correlation was developed (Eq. 6.7):

$$t_{95} = \frac{2.6\, h^{0.5} D^{15}}{u_0 d_n} \tag{6.7}$$

where d_n is the nozzle diameter, D is the well or vessel diameter, h is the liquid height, and u_0 is the nozzle velocity.

The mixing time gives an idea about the flow and mixing within the reactor and can be useful for biosynthesis process scale-up. In some reports, a mixing probe has been used to study the fluid mixing in a 1 L culture flask with a working volume of 540 mL at different shaking frequencies (Gerson et al., 2001). It was observed that the mixing time decreased proportionally in concurrence with the shaking frequency in both stirred reactors and culture flasks.

A high-speed video technique has been developed for accurate estimation of jet macromixing times in static microwell plates (Nealon et al., 2006). A general correlation has been established for the minimum time taken to reach 95% homogeneity:

$$t_{95} = \frac{2.6\, h^{1.5} D^{0.5}}{u_0 d_n} \tag{6.8}$$

where d_n is the nozzle diameter, D is the well or vessel diameter, h is the liquid height, and u_0 is the nozzle velocity.

6.3 Scale-Up Methods

Deciphering the scale-up criteria helps in maintaining consistent process parameters from the laboratory-scale to the industrial-scale production process. The basic parameters used for scaling up of the process are pH, temperature, homogeneity, agitation, etc. Manipulation of these process parameters is required as the volume of reactor changes. The magnification in the volume has to be met with appropriate changes in the other parameters.

A case study has been provided below that will help in understanding the basic challenges associated with scaling up of a process from a 500 mL culture flask to a 20 L pilot-scale reactor. Both batch as well as continuous operation was performed at the lab scale, while the pilot scale was a continuous operation.

6.3.1 Laboratory-Scale Study

6.3.1.1 Batch fermentation

A double-jacketed glass reactor of 500 mL working volume has been used to perform batch fermentation (Fig. 6.2). The maximum hydrogen yield of 8.23 mol hydrogen (kg $COD_{removed}$)$^{-1}$ was observed in the batch process. The media consisted of molasses (10 g COD L^{-1}), malt extract (1% w/v), and yeast extract (0.4% w/v). A temperature of 37°C and a pH of 6.5 ± 0.2 were maintained throughout the experiment. The malt and yeast extracts present in the media have been the source of nitrogen and other essential nutrients such as phosphorus, phosphate, and growth factors. The formulated media has an initial COD of 22.4 g COD L^{-1}. The maximum cumulative hydrogen production 1100 mL was observed. The total substrate degradation efficiency in terms of COD and reducing sugar concentration was 53% and 69%, respectively. The amount of hydrogen produced using the formulated complex media showed a 1.5 times higher value compared to the nonsupplemented media.

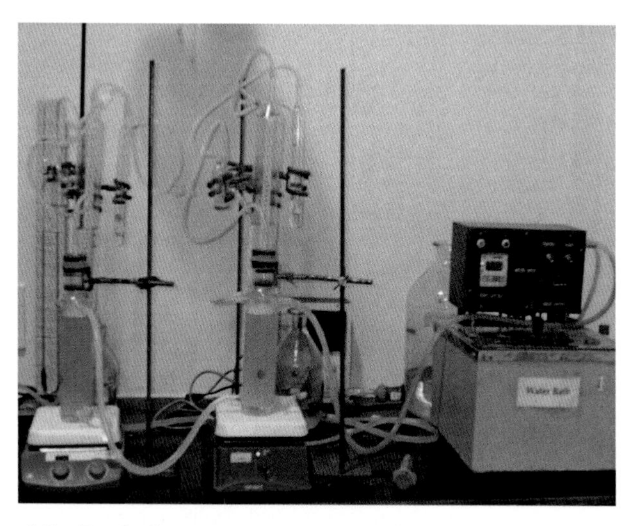

Figure 6.3 Batch fermentation in a double-jacketed reactor (500 mL working volume) for biohydrogen production.

6.3.1.2 Continuous fermentation

The same double-jacketed reactor was used for continuous hydrogen production. The reactor is modified in terms of configuration. The working volume of the reactor has been replaced with a biodegradable, agriculture-based immobilization matrix having whole cells immobilized on its surface. Thereby, a packed-bed reactor is established for continuous hydrogen production.

The actual working volume is reduced to 400 mL (void volume) with a porosity of 0.8. The reactor was packed with coconut coir as a carrier material to a packing density 50 g L^{-1}. For immobilization of whole cells on a solid matrix, 30 g L^{-1} of suspended cells in a 0.1 M phosphate buffer (pH 6) was gradually passed through the packed column at a dilution rate of 0.2 h^{-1} with the help of a peristaltic pump. For effective immobilization, the packed-bed reactor with a cell suspension was incubated for 12 h at 4°C. This protocol is suitable for facultative hydrogen-producing microorganisms. But for the use of obligate anaerobes, the cell suspension has to be made in anaerobically made phosphate buffer. Moreover, prior to charging of the cell suspension inside the packed-bed reactor, the void space of the reactor should be thoroughly flushed with nitrogen gas. Once immobilization is done, nonimmobilized cells are flushed out of the reactor and the production medium is charged inside it. Initially, the immobilized whole cells were allowed to perform batch fermentation until the rate of gas production reached its maximum and became steady. Following this, the system is eventually shifted to a continuous mode of operation. In this, sterile feed is passed to the reactor at a predefined flow rate, and simultaneously, the spent medium is taken out from the reactor with the help of a peristaltic pump. By this process, dilution rates were varied from 0.1 to 0.95 h^{-1}. At a dilution rate of 0.6 h^{-1}, the maximum rate of hydrogen production and a yield of 1250 mL $L^{-1}h^{-1}$ and 11.6 mol hydrogen (kg $COD_{removed})^{-1}$, respectively, were observed. A COD reduction efficiency of 47% was also observed. A 1.4-fold higher hydrogen yield was observed compared to batch fermentation.

6.3.2 Scale-Up Study

6.3.2.1 Experimental setup for the continuous process

Organism: Immobilized *Enterobacter cloacae* IIT-BT 08; bioreactor packed with pretreated lignocellulosic solid matrices

Production media: Malt extract 1% (w/v), yeast extract 0.4% (w/v), and cane molasses 1% (w/v)

Vessel: Total volume 20 L; void volume 14 L

Temperature: 37°C

pH: 6.5 ± 0.2

A 50 g L^{-1} of cell suspension in 0.1 M phosphate buffer (pH 6) was used for the immobilization process. The cell suspension was charged into the reactor with the help of a peristaltic pump (Fig. 6.2). Thereafter, the reactor was kept at rest for 12 h in 4°C. The nonimmobilized cells were flushed out from the reactor and charged with the sterile media. The highest hydrogen yield of 12.7 mol hydrogen (kg $COD_{removed}$)$^{-1}$ was observed at 0.60 h^{-1}. The overall yield improved by 12%.

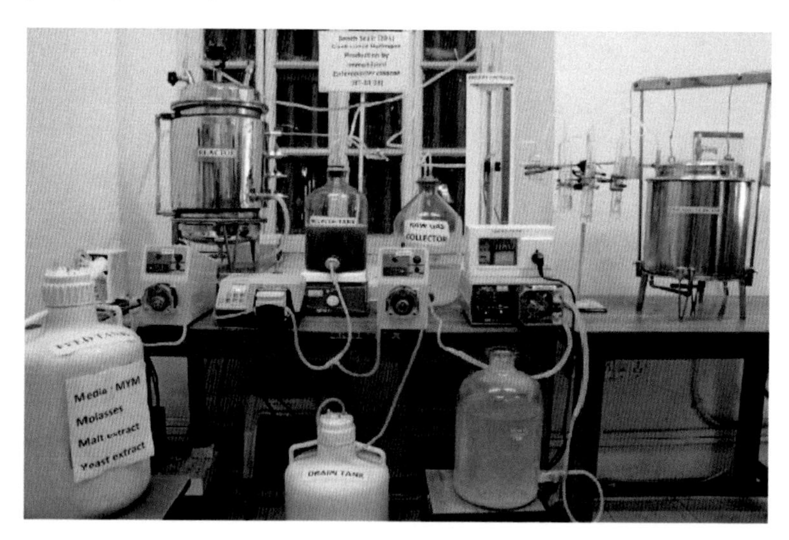

Figure 6.4 Whole-cell immobilized bench-scale packed-bed bioreactor for continuous hydrogen production.

Table 6.2 Yield of hydrogen at different dilution rates in a 20 L bioreactor using 1% w/v cane molasses

Dilution rate (h^{-1})	Rate of hydrogen production (mmol hydrogen $L^{-1}h^{-1}$)	COD reduction efficiency (%)	Hydrogen yield [mol (kg $COD_{reduced})^{-1}$]
0.2	33	53.7	2.54
0.3	82.3	49.3	7
0.5	119.2	46.3	10.66
0.6	123	39.4	12.7
0.7	110	39	11.5
0.8	98	37	10.8
0.9	88	35	10.2

6.4 Case Studies on Pilot-Scale Plants

6.4.1 Case I: Pilot-Scale Plant Using Mixed Microflora, Feng Chia University, Taiwan (Lin et al., 2011)

Characteristic features of a pilot-scale plant:

- Feedstock storage tanks (2) having a carrying capacity of 0.75 m^3
- Feedstock: Synthetic wastewater containing pure carbon sources (glucose, xylose, and sucrose)
- Nutrient storage tank having a carrying capacity of 0.75 m^3
- Mixing tank having a carrying capacity of 0.6 m^3
- Agitated granular sludge bed reactor with a working volume of 0.4 m^3
- Gas-liquid-solid separator with a volume of 0.4 m^3

Process conditions:

- Duration = 67 days
- Temperature = 35°C
- Organic loading rate (OLR) = 40–240 kg COD/m^3 d
- Influent sucrose concentration = 20 and 40 kg COD/m^3
- Fermentor agitation speed = 10–15 rpm

Microbial characteristics:

Seed culture: *Clostridium pasteurianum*, *Bifidobacteria* sp., and *Clostridium tyrobutyricum*

Operation flowchart for hydrogen production:

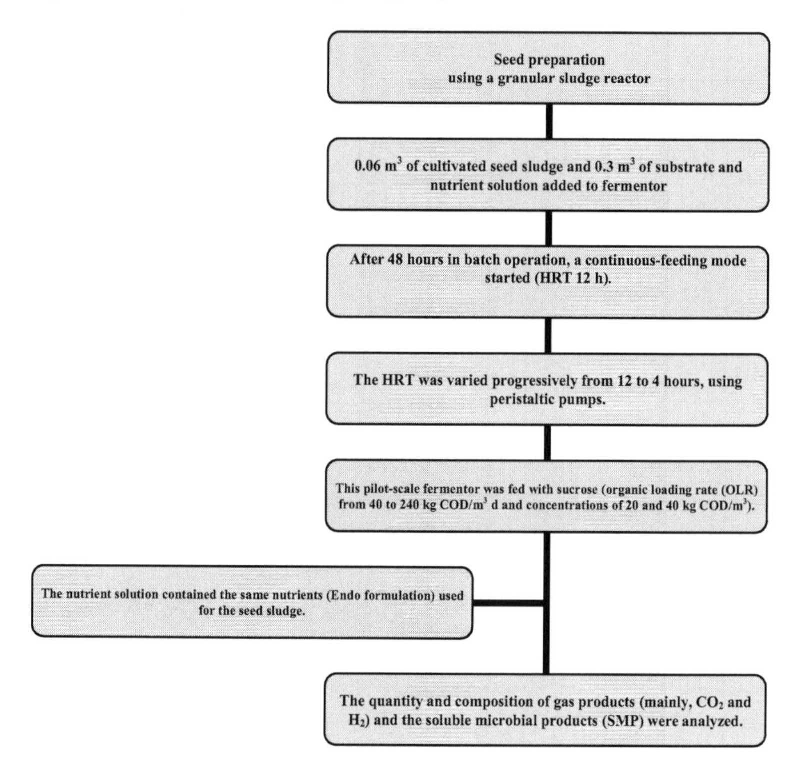

Seed preparation
using a granular sludge reactor
0.06 m^3 of cultivated seed sludge and 0.3 m^3 of substrate and nutrient solution added to fermentor
After 48 hours in batch operation, a continuous-feeding mode started (HRT 12 h).
The HRT was varied progressively from 12 to 4 hours, using peristaltic pumps.
This pilot-scale fermentor was fed with sucrose (organic loading rate (OLR) from 40 to 240 kg COD/m^3 d and concentrations of 20 and 40 kg COD/m^3).
The nutrient solution contained the same nutrients (Endo formulation) used for the seed sludge.
The quantity and composition of gas products (mainly, CO_2 and H_2) and the soluble microbial products (SMP) were analyzed.

Hydrogen production rate and yield:

- The highest hydrogen yield observed was 1.04 mol of hydrogen per mol of sucrose.
- The highest hydrogen production rate was 15.59 m^3/m^3 d.

Study of energy efficiency of the pilot plant:

- The energy sinks present in the process:
    - Motors used for agitation in substrate tank, nutrient tank, and fermentor
    - Feeding pump and drainage pump
    - Temperature control system

  - Air compressor
  - Electricity needed for the control center panel
- The heating value of hydrogen is 285.8 kJ mol^{-1} at the normal condition (temperature 20°C and pressure 1 atm). The energy output was obtained by detecting the hydrogen production rate and using the following relation:

  $$\text{Energy output} = \text{HPR(L/h L)} \times 389\text{(L)}/24.5\text{(L/mol)} \times 285.8\text{(kJ)} \quad (6.9)$$

- The energy factor (E_f) was calculated using

  $$E_f = \frac{\text{Energy Output}}{\text{Energy Input}} \quad (6.10)$$

  E_f was found to lie in the range of 13.65–28.68.

6.4.2 Case II: Pilot-Scale Plant Using Distillery Effluent to Produce Biohydrogen (Vatsala et al., 2008)

Characteristic features of a pilot-scale plant:

- Reactor volume: 100 m^3
- Hydrogen production scaled up from 1 L to 100,000 L
- A sequence of bioreactors used for inoculum preparation for the 100 m^3 reactor
- Different reactor volumes considered: 0.125, 1.25, 12.5, and 125 m^3
- Working volumes set from 0.1 to 100 m^3
- Height-to-diameter (H/D) ratio maintained at 1.28
- Feedstock for biohydrogen production: Distillery effluent

Process conditions:

- Time of fermentation: 40 h
- Temperature: 37°C
- pH: Not controlled
- Initial effluent temperature: 37°C
- Initial COD of the feedstock: 101.2 g/L
- Initial BOD of the feedstock: 58.8 g/L

Microbial characteristics:

Seed culture: *Citrobacter freundii* C01, *Enterobacter aerogenes* E10, and *Rhodopseudomonas palustris* P2

Heterotrophic bacterial strains (HTBs) were isolated from the effluent treatment plant, EID Parry (I) Ltd., Nellikuppam, Tamil Nadu, India. Photoheterotrophic bacterial strains (PTBs) were isolated from pond soil, Chengalpet, Tamil Nadu, India.

Performance of the reactor:

- 21.38 kg of hydrogen corresponding to 10692.6 mol obtained through the batch method from reducing sugar (3862.3 mol) as glucose
- Average yield of hydrogen = 2.76 mol/mol glucose
- Rate of hydrogen production estimated to be 0.53 kg/100 m^3h

6.4.3 Case III: Biohydrogen Production from Molasses by Anaerobic Fermentation with a Pilot-Scale Bioreactor System (Ren et al., 2006)

Characteristic features of a pilot-scale plant:

- Volume of the reactor: 2 m^3 anaerobic hydrogen bioproducing reactor (working volume of 1.48 m^3)
- Feedstock: Cane molasses (53% w/w reducing sugars)
- OLR varied in the range of 3.11–85.57 kg COD/m^3d

Process conditions:

- Time of fermentation: 200 days
- Temperature: 35°C ± 1°C
- pH: Influent pH kept above 7 by adding $CaCO_3$ to prevent low pH (<4)
- Start-up OLR: 6.32 kg COD/$m^3_{reactor}$ d with an HRT of 11.4 h and a substrate concentration of 3000 mg COD/L

Microbial characteristics:

Seed culture: *Clostridium* sp., *Enterobacter aerogenes,* and *Ectothiorhodospira vacuolata*

The seed sludge in the HBR was collected from a local municipal wastewater treatment plant.

Performance of the reactor:

- Maximum hydrogen production of 5.57 m^3 hydrogen m^{-3}
- Specific hydrogen production rate of 0.75 m^3 hydrogen kg^{-1} MLVSS d^{-1}
- Hydrogen yield 26.13 mol kg^{-1} $COD_{removed}$ within the OLR range (35–55 kg COD $m^{-3}{}_{reactor}$ d^{-1})
- Hydrogen production decreased when the OLR more over 68.21 kg COD $m^{-3}{}_{reactor}$ d^{-1} due to the accumulation of VFAs in the HBR system
- Hydrogen production rate influencedby accumulation of ethanol, acetate, and butyrate
- Hydrogen yield improved to a considerable degree by high concentration of biomass in the pilot-scale HBR system due to good settling ability and compactness of anaerobic activated sludge
- Hydrogen production rate of 0.75 m^3 kg^{-1} MLVSS d^{-1} observed

6.4.4 Comparative Study among Different Pilot-Scale Plants

Comparative studies on the above three pilot plants are tabulated in Table 6.3.

Table 6.3 Comparison of features of pilot-scale plants for biohydrogen production

Features/ parameters	Case I	Case II	Case III
Reactor volume (m^3)	0.4	100	1.48
Feed	Synthetic wastewater containing pure carbon sources	Distillery effluent	Normal molasses containing about 53% sugars

(*Continued*)

Table 6.3 (*Continued*)

Features/ parameters	Case I	Case II	Case III
Duration of process	67 d	40 h	>200 d
Mode of operation	Continuous	Batch	Continuous
Temperature	35°C	Not controlled	35°C ± 1
Initial COD	20 and 40 kg COD m^{-3}	Initial COD: 101.2 g L^{-1}	3.11–85.57 kg COD $m^{-3}d^{-1}$
OLR	40–240 kg COD $m^{-3}d^{-1}$	N/A	>3 kg COD $m^{-3}d^{-1}$
Microbial community	*Clostridium pasteurianum*, *Bifidobacteria* sp., and *Clostridium tyrobutyricum*	*Citrobacter freundii*, *Enterobacter aerogenes*, and *Rhodopseudomonas palustris*	*Clostridium* sp., *Enterobacter aerogenes*, and *Ectothiorhodospira vacuolata*
Yield	1.04 mol hydrogen mol^{-1} sucrose	2.76 mol mol^{-1} glucose	26.13 mol kg^{-1} $COD_{removed}$
HPR	15.59 m^3 $m^{-3}d^{-1}$	0.53 kg 100 $m^{-3}h^{-1}$	0.75 m^3 kg^{-1} MLVSS d^{-1}

6.5 Energy Analysis of the Biohydrogen Production Process

The data related to energy released as product and energy provided as feedstock during biohydrogen production in a batch process are presented in Table 6.4, and the gross energy requirement (GER) and net utilizable energy product (NUEP) for the process have been calculated.

The NUEP is expressed as

$$\text{NUEP} = \frac{\text{Externally utilizable energy} - \text{Secondary energy energy input}}{\text{Primary energy available (raw material)}} \tag{6.11}$$

Table 6.4 Energy production and requirement for biohydrogen production (basis: 100 m^3 biohydrogen production d^{-1})

Parameters	Energy calculations
Gross energy of product as biohydrogen (GJ)	1.277
Energy provided by substrate (GJ)	1.3
Total process energy required (GJ)	0.875
Gross energy requirement (GER)	0.685
Net utilizable energy product (NUEP) (%)	30.92

Souce: Nayak et al. (2013)

From Table 6.4, it has been found that the GER is equal to 68.5% of the energy recovered as hydrogen. This comprises the energy required for mixing, heating, and pumping. The NUEP was estimated to be 30.92%.

6.5.1 Biological Route vs. Chemical Route

The input energy required for production of hydrogen via biological and physicochemical routes (steam reforming) is compared in Table 6.5. The majority of commercially produced hydrogen comes from the steam reforming process, where it uses hydrocarbons as feedstock. The hydrocarbons are subjected to high-pressure steam under a process known as reformation in order to produce hydrogen. The chemistry behind steam reforming of methane to hydrogen is shown in the equation below:

$$CH_4 + H_2O \xrightarrow{1000°C} CO + 3H_2 \quad (6.12)$$

$$CO + 3H_2 + H_2O \xrightarrow{700°C} 4H_2 + CO_2 \quad (6.13)$$

The biological hydrogen production process is more promising compared to the chemical process (Table 6.5).

6.5.2 Electrolysis of Water vs. Biological Route for Hydrogen Production

The input energy required for production of hydrogen via biological and electrochemical routes (electrolysis) was also studied. On electrolysis, water (H_2O) is split into oxygen (O_2) and hydrogen gas (H_2). Electrolysis of water into hydrogen and oxygen at standard temperature and pressure is not thermodynamically favorable.

Anode (oxidation):

$$2H_2O\ (l) \rightarrow O_2\ (g) + 4H^+\ (aq) + 4e^- \quad Eo_{ox} = -1.23\ V \tag{6.14}$$

Cathode (reduction):

$$2H^+\ (aq) + 2e^- \rightarrow H_2\ (g) \quad Eo_{red} = 0.00\ V \tag{6.15}$$

A higher electrical input in terms of voltage is required for electrolysis of water to provide the required enthalpy of 286 kJ/mol to produce hydrogen and oxygen as products. Thus to generate 100 m^3 of hydrogen, 2.5 GJ of energy would be required if the electrolytic pathway is implemented, whereas the biological route would require only 0.875 GJ of energy, which is 2.9 times less compared to the former. Moreover, the overall energy efficiency of electrochemical hydrogen production process is low since half the electrical energy supplied to the cell is dissipated as heat.

Table 6.5 Comparison of energy production and requirement during production of hydrogen via the chemical and biological routes (basis: 100 m^3 hydrogen production per day)

Parameter	Chemical route	Biological route
Total energy available from product (GJ)	1.717	1.277
Energy content in substrate (GJ)	1.623	1.3
Total process energy (GJ)	1.3185	0.875
GER	0.768	0.685
NUEP (%)	24.65	30.92

Source: Das (1985)

6.6 Cost Analysis of the Process

6.6.1 Hydrogen as a Commercial Fuel

The suitability of any process lies in its capacity to be scaled up to the industrial level. Many industrial-scale hydrogen production processes are in existence, but hydrogen has not gained much attention as an alternative fuel due to its higher cost of production and naive technological advancement in terms of hydrogen-based IC engines and storage. The list of technologies which are available for large-scale production of hydrogen and their unit costs is shown in Table 6.6.

Table 6.6 Various methods of hydrogen production and their costs

Source and process (large-scale technology)	Cost of hydrogen
Natural gas (produced via steam reforming at a fueling station)	\$4–\$5/kg
Wind (via electrolysis)	\$8–\$10/kg
Nuclear (via electrolysis)	\$7.50–\$9.50/kg
Nuclear (via thermochemical cycles)	\$6.50–\$8.50/kg
Solar (via electrolysis)	\$10–\$12/kg
Solar (thermochemical cycles)	\$7.50–\$9.50/kg
Wastewater (dark fermentation)	\$1.3/MBTU
Gasoline	\$23.5/MBTU
Natural gas	\$7/MBTU

Source: Das et al. (2008)

6.6.2 Cost Calculation of Continuous Biohydrogen Production Process Using Cane Molasses

A 20 L (pilot-scale) immobilized anaerobic bioreactor was used to study suitability of cane molasses for continuous hydrogen production (Das, 2009). An indigenously developed automated logic control system was used to maintain reduced partial pressure in the range of 380–760 mm Hg. The rate of hydrogen production varied in response to the variation in OLRs. At an OLR of 0.03 kg COD $L^{-1}h^{-1}$, the highest hydrogen production rate of 8.41 L hydrogen $L^{-1}h^{-1}$ was observed. The purity of hydrogen gas was in the range of 58%–66% v/v. On operating with a recycle ratio of 6.4 and a dilution rate of 0.068 h^{-1}, the maximum specific hydrogen production rate was 0.093 L hydrogen (g VSS)$^{-1}h^{-1}$. This rate, however, was observed to be inversely proportional to a specific OLR when the dilution rate was more than 0.07 h^{-1}. At an OLR of 0.02 kg COD $L^{-1}h^{-1}$, the maximum hydrogen yield of 17.94 mol hydrogen (kg $COD_{removed}$)$^{-1}$ was observed.

Table 6.7 Comparative analysis on the bench-scale or pilot-scale operations reported for biohydrogen production

Substrate	Organism	Reactor capacity	Days of operation	Yield/HPR	References
Wastewater from citric acid	*Clostridium pasteurianum*	50 m^3 UASB	10–15	0.7 m^3 hydrogen $m^{-3}{}_{reactor}$ d^{-1}	Yanga et al., 2006
Molasses	Anaerobic activated sludge	1.48 m^3	200	26.1 mol (kg $COD_{removed}$)$^{-1}$	Ren et al., 1997
Cane molasses	*Enterobacter cloacae DM 11*	20 L immobilized packed-bed reactor	20	17.9 mol hydrogen (kg $COD_{removed}$)$^{-1}$	Kotay, 2008

Table 6.8 Cost analysis of the biohydrogen production process

Highest hydrogen yield	45.4 mL (g $COD_{reduced}$)$^{-1}$
Average COD value of the sludge	88 g L^{-1}
COD reduction efficiency	46%
Heating value of hydrogen	142,000 kJ (kg of hydrogen)$^{-1}$
Heating value of sewage sludge	18,640 kJ kg^{-1}
Hydrogen yield	0.0019 kg hydrogen (kg sludge)$^{-1}$
Energy recovery from the substrate = [(Heating value of hydrogen × Yeld of hydrogen)/(Heating value of sewage sludge)] × 100	{(142,000 × 0.0019)/18,640} × 100 = 1.44%
Amount of gaseous energy generated as hydrogen	142,000 × 0.0019 = 270 kJ (kg sludge)$^{-1}$

Total volume of sludge produced in India	~3 million L d^{-1}
Total amount of sludge produced per day	$3 \times 10^6 \times 0.088$ kg d^{-1}
Therefore, total amount of energy produced as hydrogen per day	$3 \times 10^6 \times 0.088 \times 270$ KJ = 1200 MBTU (provided all the activated sludge is devoted for hydrogen biogeneration)
Dilution rate	1 h^{-1}
HRT of the process	1 h
Volume of the reactor	150 m^3 (considering 20% head space)
Cost of the reactor and accessories	150 × \$1000 (at the rate of \$1000 m^{-3}) = \$150,000
Life of the reactor	10 years
Assuming the cost of sludge	0
Money to be returned/day	\$45 (capital + interest)
Labor/recurring expenditure	\$1500 d^{-1}
Total expenditure	\$1545 d^{-1}
Cost of hydrogen	\$1545/1200 MBTU = \$1.3 $(MBTU)^{-1}$
Cost of natural gas (2013, India)	\$8.4 $(MBTU)^{-1}$
Cost of gasoline 2013	\$3.57 $gallon^{-1}$
Heat of combustion of gasoline	47,000 KJ/kg = 44,407 BTU/ kg = 0.17 MBTU $gallon^{-1}$
Cost of gasoline	\$20.9 $(MBTU)^{-1}$

The above-mentioned values could become more promising if the availability of sludge from sludge treatment plants (STPs) increases significantly. For economical production, a biohydrogen plant could be installed near to an STP, which could eventually help in decreasing the cost related to transport and storage of sludge. The establishments of STPs are increasing exponentially with every coming year. This would ensure a steady supply of sludge (feedstock) for sustainable biohydrogen production.

In the near future, sugarcane juice, molasses, biomass, or distillery effluents can be considered as feedstock due to their high content of fermentable sugars. At present a comprehensive technoeconomic analysis of the biohydrogen production process needs greater emphasis. This would help in analyzing and comparing the cost-effectiveness of biological hydrogen production and other conventional ways of hydrogen production. As of now, the fuel cost estimation of the biohydrogen production process on the basis of an economic survey is complicated and no definite methodologies have been developed for accessing its potentiality as a fuel. This is largely due to the involvement of many intervening technoeconomic parameters in terms of inclusiveness, complexity, impurities, life cycle analysis, and cost of installing all the commercially undemonstrated equipment as a percentage of the total installed equipment cost.

At present, when pure sugar is used as a substrate I hydrogen production, the unit cost is approximately three times higher than that of the gasoline unit cost of production. In the near future the cost of production of biohydrogen can be brought down to a competitive level with gasoline with the advent of genetically modified robust strains, improved reactor configurations, and cheap feedstock. Moreover, considering the unconstructive social and environmental impact of fossil fuels, the use of hydrogen as a fuel would always be preferred. The socioeconomic factors related to the introduction of any new fuel in the market generally emphasizes the emission profile, short-/long-term environmental costs, and direct or indirect impact on health. The revenues generated from a large-scale (100 m^3) wastewater-to-biohydrogen plant have been shown in Table 6.9.

Table 6.9 Revenue from the biohydrogen production process using wastewater in a 100 m^3 reactor

Sales	Revenue (1000 USD)
Revenue from hydrogen sales	160
Revenue from CO_2 sales	24
Annual total revenue	184
Annual total cost for the system	100
Annual administration costs	3
Annual total profit	81
Return on investment (%)	81

Source: Ya-Chieh Li et al. (2012)

6.7 Conclusion

A clean, sustainable renewable energy supply is the ultimate goal for a hydrogen-based economy. Reduction in greenhouse gases and a stable supply of energy can lead to sustainable growth. Research on biohydrogen has given a lot of promise in terms of energy generation and waste management. Focused research on commercialization of the biohydrogen production process is under realization. With the advent of newer technologies for reactor designing, development of improved strains could be exploited in the near future for cost-effective hydrogen production.

References

Büchs L, Zoels B (2011). Evaluation of maximum to specific power consumption ratio in shaking bioreactors, *J Chem Eng Jpn*, **34**, 647–653.

Das D (1985). *Biomethanation of Mixed Agricultural Residues*, PhD thesis, IIT Delhi.

Das D (2009). Advances in biohydrogen production processes: an approach towards commercialization, *Int J Hydrogen Energy*, **34**, 7349–7357.

Das D, Khanna N, Veziroğlu NT (2008). Recent developments in biological hydrogen production processes, *Chem Ind Chem Eng Q*, **14**, 57–67.

Gerson DF, et al. (2001). Measuring bioreactor mixing with mixmeter, *Genet Eng News*, **21**, 68–69.

Hughmark GA (1980). Power requirements and interfacial area in gas-liquid turbine agitated systems, *Ind Eng Chem Proc Des Dev*, **19**, 638–641.

Ju LK, Chase GG (1992). Improved scale-up strategies of bioreactors, *Bioproc Eng*, **8**, 49–53.

Junker, B (2004). Scaleup methodologies for *Escherichia coli* and yeast fermentation processes, *J Biosci Bioeng*, **97**, 347– 364.

Kotay SM (2008). *Microbial Hydrogen Production from Sewage Sludge*, PhD thesis, IIT, Kharagpur.

Li YC, Liu YF, Chu CY, Chang PL, Hsu CW, Lin PJ, Wu SY (2012). Techno-economic evaluation of biohydrogen production from wastewater and agricultural waste, *Int J Hydrogen Energy*, **37**, 15704–15710.

Lin CY, et al. (2011). A pilot-scale high-rate biohydrogen production system with mixed microflora, *Int J Hydrogen Energy*, **36**, 8758–8764.

Morimoto M, Atsuko M, Atif AAY, Ngan MA, Fakhru'l-Razi A, Iyuke SE (2004). Biological production of hydrogen from glucose by natural anaerobic microflora, *Int J Hydrogen Energy*, **29**, 709–713.

Nath, K, Das D (2004). Biohydrogen production as a potential energy resource: present state-of-art, *J Sci Ind Res*, **63**, 729–738.

Nayak BK, Pandit S, Das D (2013). Biohydrogen. In: Kennes C, Veiga Ría C (eds), *Air Pollution Prevention and Control*, John Wiley & Sons, Chichester, UK, 345–381.

Nealon AJ, O'Kennedy RD, Titchener Hooker NJ, Lye GJ (2006). Quantification and prediction of jet macro-mixing times in static microwell plates, *Chem Eng Sci*, **61**, 4860–4870.

Ren N, Li J, Li B, Wang Y, Liu S (2006). Biohydrogen production from molasses by anaerobic fermentation with a pilot-scale bioreactor system, *Int J Hydrogen Energy*, **31**, 2147–2157.

Ren N, Wang B, Huang JC (1997). Ethanol-type fermentation from carbohydrate in high rate acidogenic reactor, *Biotechnol Bioeng*, **54**, 429–433.

Rushton JH, Costich EW, Everett HJ (1950). Power characteristics of mixing impellers: part I, *Chem Eng Prog*, **46**, 395–404.

Rushton JH, Costich EW, Everett HJ (1950). Power characteristics of mixing impellers: part II, *Chem Eng Prog*, **46**, 467–476.

Vatsala TM, Mohan Raj S, Manimaran A (2008). A pilot-scale study of biohydrogen production from distillery effluent using defined bacterial co-culture, *Int J Hydrogen Energy*, **33**, 5404–5415

Yanga H, Shaoc P, Lub T, Shena J, Wangb D, Xub Z (2006). Continuous bio-hydrogen production from citric acid wastewater via facultative anaerobic bacteria, *Int J Hydrogen Energy*, **31**, 1306–1313.

Chapter 7

Two-Stage Biomethanation Process for the Stabilization of Distillery Effluents Using an Immobilized Whole-Cell System: A Case Study

7.1 Introduction

Biological treatment processes for alternative fuel production have been receiving great attention recently. Considering the increasing cost of nonrenewable fuels, intensive attempts have been made to achieve independence from imported fossil fuels, not only in the advanced countries, but in developing countries also. This has led to an appreciation of the potential of solar-derived synfuels. The substitution of petrofuels by renewable, nonfossil biomass assures us a perpetual, indigenously produced energy supply which could be available now and in the future long after fossil fuel resources are exhausted. In 2012, the International Energy Agency (IEA) estimated that the world energy consumption was 155,505 terawatt-hour (TWh), or 5.598 × 1020 joules (Statistical Review of World Energy, 2013). This works out to 17.7 TW or a bit less than the estimated 20 TW produced by radioactive decay on earth. From 2000 to 2012 coal was the source of energy with the largest growth. The use of oil and natural gas also had considerable growth, followed by

Biohythane: Fuel for the Future
Debabrata Das and Shantonu Roy

ISBN 978-981-4745-29-1 (Hardcover), 978-981-4745-30-7 (eBook)
www.panstanford.com

hydropower and renewable energy. Renewable energy grew at a rate faster than at any other time in history during this period, which can possibly be explained by an increase in international investment in renewable energy. Most developing countries consume energy from noncommercial resources, whose share is about 45%. The developed countries consumed 83% of the world's commercial energy (fossil) consumed in 1974. The per capita consumption of 187 billion joules in developed countries in 1974 was about 13 times that of developing countries (14 billion joules). These data stress the importance of properly utilizing integrated renewable resource–recycling systems. Processes for the utilization of solid organic wastes like animal manure and agricultural residues for gaseous energy recovery are well established (Ghose and Bhadra, 1981; Ghose and Das, 1982; Klass, 1981; Hobson, 1982). Research work on the treatment of industrial effluents by anaerobic digestion process is encouraging (Das, 1983; Ghosh et al., 1982; Donnelly, 1978; Kwietniewska and Tys, 2014). The distillery waste treatment by anaerobic digestion is also reported (Scammel, 1975; Skogman, 1979; Pipyn and Verstraete, 1979; Braun and Hass, 1982; Sankaran et al., 2014).

Brazil, India, and Cuba are the major countries producing alcohol from cane molasses. The water pollution due to the distillery effluents in these countries is severe. In India the total alcohol production from molasses is about 400×10^6 L per annum, resulting in more than 400×10^7 L of distillery effluents with high organic matter which directly reach and pollute the water streams (Monteiro, 1975). Though there are about 200 distilleries in India, little attention has been paid to commercially use these effluents as an energy source (Roychowdhury, 1981; Sen and Bhaskaran, 1962).

Anaerobic digestion of organic residues to methane has several advantages over thermal gasification procedures since it is applicable to more types of high-moisture organic feeds and is operated at moderate temperatures and atmospheric pressures but with relatively high overall thermal efficiencies (Radhakrishnan et al., 1969).

The anaerobic system is capable of treating medium- and high-strength effluents, and yet only minimal operating skills are needed. The aerobic system on the other hand is merely a conversion of soluble organic disposal material to solid organic disposal material, resulting in high sludge accumulation (Jones et al., 1980).

Several industries have used this process and have noted several disadvantages such as:

- High energy consumption
- Dilution before treatment
- Large sludge production, with its inherent problems of disposal
- Odor and air pollution

The lagoon treatment of wastes has the following drawbacks:

- Seepage of the distillery wastes through the porous soil, polluting the water sources
- Large land requirement (5 to 10 acres), depending on the size of the distillery
- Odor nuisance in the surrounding area

The advantages of the anaerobic system over activated sludge process are:

- Gaseous energy recovered as methane
- Low nutrient requirement and low cell synthesis
- No oxygen transfer limitation
- Low power requirement
- No odor as it is a closed system

The disadvantages of the anaerobic process are longer contact time and oxygen toxicity to methanogens. However, these can be minimized by using perhaps immobilized whole cells (Gorenszy, 1978; Messing, 1982). With the advancement of science and technology, different industries are growing at a rapid rate throughout the world. It has been found by the Central Pollution Control Board, India, that the total industrial wastewater generated in the country constitutes only 10% by the volume of the total wastewater generated by all-urban settlements. One industrial survey report on large and medium sector industries responsible for water pollution was prepared by the central and 14 state water pollution boards in 1981. Out of a total of 27,000 large and medium industries, 1700 water pollution industries were identified. Distillery industries are included in this category. There are about 200 distilleries in India, producing more than 400×10^7 L of distillery effluents per annum. Again, the quantity of distillery effluent production can be reduced with the advancement of the ethanol fermentation technology (Fig. 7.1). In the conventional alcohol fermentation process, 7%–8% (v/v) ethanol is usually present in the fermentation broth. This is corresponding to 13 L effluent generation per liter of rectified spirit production. The raw materials used in the distilleries are mostly cane molasses and beet molasses. In India, cane molasses is usually considered as a raw material for the alcohol fermentation processes. The characteristics of the distillery effluents collected from several locations in India is shown in Table 7.1. The biochemical oxygen demand (BOD_5) of the effluent is extremely high, that is, from 40,000 to 78,000 mg/L.

The objectives of this chapter are:

- To demonstrate how a modified two-phase anaerobic digestion process can provide superior performance in terms of
  - Energy recovery
  - Waste stabilization (i.e., significant reduction of BOD of the effluent)

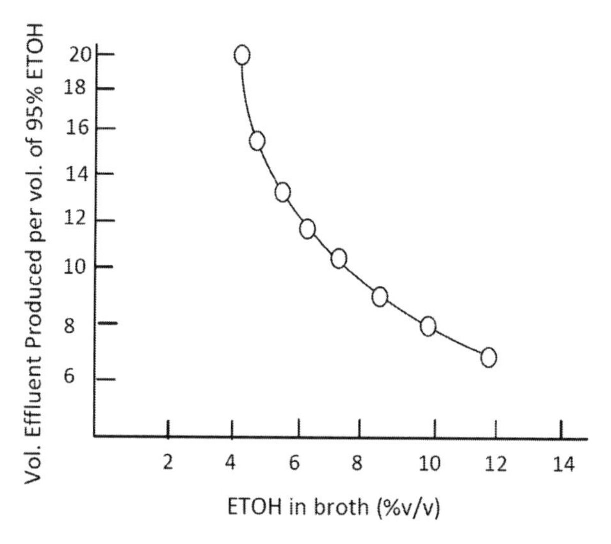

Figure 7.1 Variation of distillery effluent production with respect to ethanol concentration.

- To present the successful scale-up of the laboratory data to an on-site distillery unit in India

- To present a total energy analysis of the two-stage process
- To examine the reasons for the unpopularity of the anaerobic process (biomethanation process) for treating distillery wastes

Table 7.1 Analysis of effluents collected from several distilleries in India

Characteristics of distillery effluent	North-central India	South-central India	South India
pH	4.5–4.7	4.5–5.0	4.0–4.5
T.S. (% w/v)	12.0–13.00	11.0–12.0	5.5–7.0
V.S. (% w/v)	9.0–1.0	8.0–9.0	4.5–5.0
C (% w/v)	3.03	2.90	1.41
N			
A. Free ammonia (% w/w)	0.02	0.01	0.01–0.02
B. Total nitrogen (%w/w)	0.218	0.20	0.1–0.14
COD (g L^{-1})	90–140	90–120	60–70
BOD_5 (g L^{-1})	55–80	55–78	40–50
P (g L^{-1})	0.13	0.13	–
Sucrose (% w/v)	0.5	0.55	0.27
Volatile fatty acids (g L^{-1})	2–4	2–4	1.0–1.4

7.2 Wastewater Treatment Processes

Two types of processes are available for the treatment of distillery effluents:

- Incineration processes
- Biological waste treatment processes

The incineration process is energy intensive as well as costly. The total solid (TS) content of the distillery effluent is to be increased to 60% before it goes to the combustion process. This also causes environmental pollution due to the presence of dust particles (containing mostly K_2O) in the air. However, biological waste treatment processes are commonly used for the stabilization of high-moisture-content wastes, such as distillery effluents.

The advantages of the anaerobic system over the activated sludge process have already been mentioned before. The main disadvantages of the anaerobic processes, however, are a higher hydraulic residence time (HRT) and oxygen toxicity to methanogens. The anaerobic waste treatment processes for alternative fuel production has been receiving great attention recently. The anaerobic biological waste treatment processes, which are also known as biomethanation processes, are capable of producing gaseous energy as methane to a great extent.

7.2.1 MINAS

The Minimum National Standards (MINAS) for the disposal of distillery effluents were prescribed by the Central Pollution Control Board, India: BOD_5 3000 mg/L to be achieved after anaerobic treatment and the resultant effluent to be diluted with water or low BOD_5 effluents to bring the BOD_5 to 500 mg/L and used on land for irrigation. Secondary two-stage aerobic treatment may bring down the BOD_5 to 100 mg/L for application on land and to 30 mg/L for discharge into water courses.

7.3 Materials and Methods

7.3.1 Process Description

The process employed for bench-scale studies is photographically displayed in Fig. 7.2a and Fig. 7.2b. The process essentially consists of two reactors, two pumps, and suitable connections and valves between the reactors. In the bench-scale studies, the reactor temperature was controlled by a heating tape connected through a variac. In scale-up studies the temperature of the first reactor was controlled at 35°C ± 3°C by using insulation to the reactor. In the second reactor, wet steam was circulated through a heat exchanger and the resulting hot water (70°C–80°C) passed through a copper coil inside the reactor. The temperature was maintained at 45°C ± 2°C. Peristaltic and centrifugal pumps were used for the bench and the pilot study, respectively. While glass reactors were used in the bench studies, scale-up operations were conducted in mild steel (MS) reactors with epoxy coating.

A: Peristaltic pump
B: Magnetic stirrer
C: Variac
D: Heating tape
E: Setting tank
R-I: Acidogenic phase
R-II: Methanogenic phase

75% v/v CH_4
95% v/v CO_2
Inflow
Mixer
Outflow
Temp- 37° C
RT- 1.5 d, 2 L
Temp- 45° C
RT- 6 d, 10 L

(a)

(b)

Figure 7.2 (a) Flow diagram of the two-stage biomethanation process (bench scale). (b) Experimental setup of the two-stage biomethanation process at IIT, Delhi.

7.3.2 Input to the Reactors

The studies were conducted at different loading rates varying from 30 to 80 kg m^{-3} d. Experiments were carried out in duplicate to confirm reproducibility of the reactor performance. The initial pH was adjusted to 6.5 by using sodium hydroxide or caustic soda.

7.3.3 Reactor Capacities

Different sizes of reactors were used in these studies. The capacity of the reactors varied from 2 L to 1000 L.

7.3.3.1 Feeding

The feeding was done in several ways:

- Once a day
- Several times a day
- Recycling of the effluent

The distillery wastes were collected from three different distilleries:

- D.C.M. Sugar Division (North-central India)
- E.T.D. Parry Ltd. (South India)
- Sakthi Sugars Ltd. (South-central India)

Table 7.1 illustrates the composition of these wastes. Most analyses were done in duplicate and a few in triplicates according to the American Public Health Association (APHA) standard methods (Riera et al., 1982). Sucrose was estimated by the dinitrosalicylic acid (DNS) method following acid hydrolysis of the sample. Carbon was estimated by Wakley and Blank's rapid titration method (Bernfeld, 1955). The heating value of the distillery waste was estimated by a bomb calorimeter (MATREX). Gas and volatile fatty acid (VFA) analyses were done by gas chromatography (GC) (AMIL) using Porapak Q and Chromosorb 101 columns by thermal conductivity detector (TCD) and flame ionization detector (FID) controllers, respectively.

7.3.4 Inoculum and Start-Up

The enriched culture used in these studies was especially developed over several generations from fresh cattle dung as the original source.

Initial experimental work was done in 2 L and 10 L reactors and then transferred to 40 L reactors. The 20 L reactors were sometimes used to study the different phases individually. The scale-up studies were carried out in 215 L and 1000 L reactors.

7.3.5 Operations

Digester performance was monitored by measuring pH, VFAs, ammonia nitrogen and total nitrogen, and the quality and quantity of gas. In addition, at the steady state (defined as digester operation for a period of two or more dilution rates during which time the variation in the volume of the gas yield and gas production rates did not exceed 5%), the effluent slurry was analyzed for the chemical oxygen demand (COD), BOD, TS, volatile solids (VS), ash, sucrose, VFAs, etc. The gas was collected in a collector by acidified displacing water.

7.3.6 Two-Stage Biomethanation Process

Anaerobic digestion of organic wastes is usually carried out by two groups of microflora which are different from each other with respect to temperature, substrate requirement, process optima, etc. These microflora are comprised of acidogens and methanogens. It has been found that the overall efficiency of the biomethanation process can be drastically increased by allowing acidogens and methanogens to grow and multiply in separate reactors.

7.3.7 Whole-Cell Immobilization

Both acidogens and methanogens are usually immobilized by using adsorption techniques. Polypropylene (PP) was used as a solid matrix. The shape of the matrices was cylindrical (Fig. 7.3) (size: D:H = 0.5:1.0). The PP materials were treated with acetic acid, which was followed by washing with water. Then cells were immobilized on the surface of the solid matrices. The surface of PP was embedded with the cells (Fig. 7.3). It has been observed that the pore distribution at the surface is more uniform after the acid treatment. The microbial cells usually formed a slim layer at the surface of the solid matrices.

Figure 7.3 Immobilized solid matrices before and after immobilization.

Both bench-scale and pilot-scale studies were carried out by using this whole-cell immobilization system, as shown in Fig. 7.2 and Fig. 7.4. At the bench scale, the temperature was controlled by using heating tape and variac. Similarly, at the pilot scale, the temperature was controlled by circulation of water through the jacketed reactors. This whole-cell immobilization system is also known as fixed-film reactors.

7.4 Results and Discussions

The conventional single-reactor batch system for generating methane using distillery wastes was unsuccessful due to accumulation of volatile organic acids (greater than 15 g L^{-1}), presumably because the acid concentration was more than 3.5 g L^{-1} which is known to inhibit the methanogens (Huang et al., 2013). Therefore, a two-phase system—a modification of the one proposed by several workers (Ghosh et al., 1977)—was designed. The flow diagram of the two-phase system for the bench and scale-up studies is shown in Figs. 7.2 and 7.4.

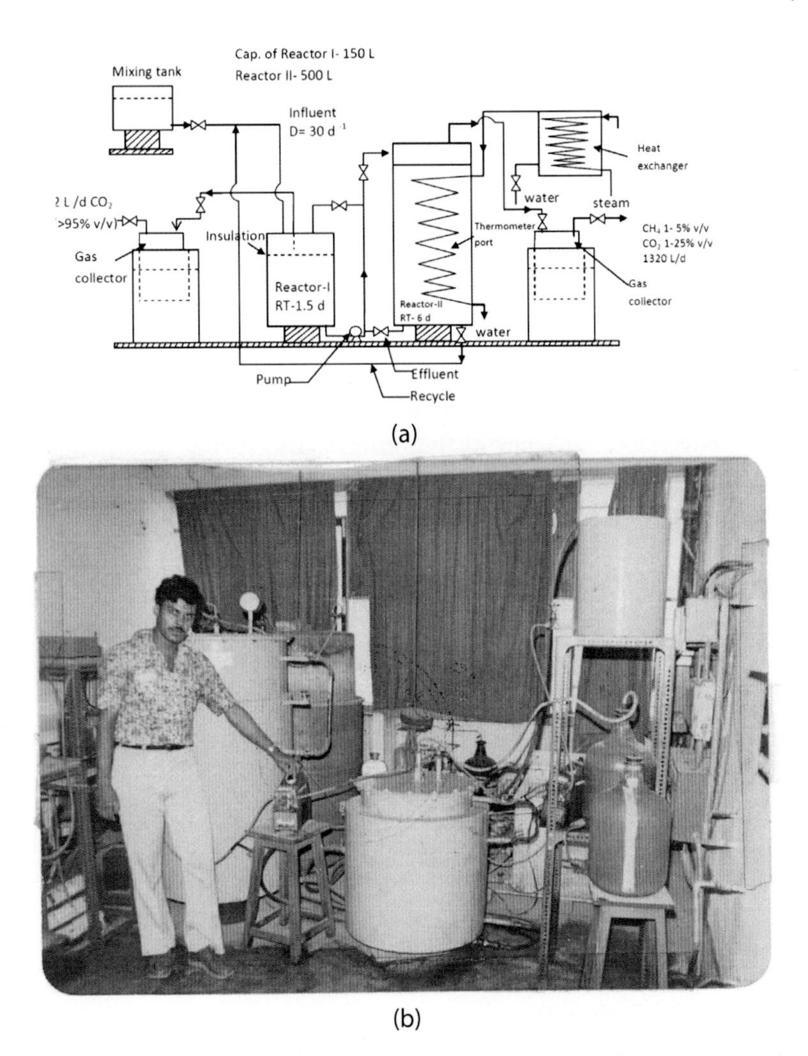

Figure 7.4 (a) Flow diagram of a scale-up unit for anaerobic treatment of distillery wastes. (b) Two-stage biomethanation process at IIT, Delhi, using a distillery effluent as the substrate.

7.4.1 Waste Characteristics

The distillery wastes, obtained from different locations (north-central, south-central, and south) varied in strength, suspended solid content, and other constituents present (Table 7.1). Early studies were carried out using distillery wastes from the northcentral unit.

7.4.2 Quality and Quantity of Gas

The amount of gas produced was calculated on the basis of organic matter available in the waste. The maximum gas production was 47 L/L of distillery wastes with an average gas composition of 75% v/v CH_4 and 24% v/v CO_2. The optimum loading rate was 35 kg m^{-3} d^{-1}. The effect of recycling of effluents from Reactor-II (R-II) in different proportions with distillery effluents was also determined (Table 7.2). In our experience a mixture of distillery waste and effluents (either supernatant or total mixture) in a 1:1 proportion results in higher gas production. The effluent recycling is thus seen to improve the conversion efficiency of the process. The sucrose degradation profile is illustrated in Fig. 7.5. It is evident that 90% of sucrose was degraded.

Table 7.2 Performance of the fixed-film reactors

Reactor parameters	Reactor-I	Reactor-II
Loading (kg COD $m^{-3}d^{-1}$)	24.26	7.28
HRT (d)	1.5	6.0
Biogas productivity (m^3 m^{-3} d^{-1})	1.33	6.66
Biogas (m^3 m^{-3} distillery effluents)	2	44
Gasification rate (L methane kg^{-1} COD)	–	329
COD reduction	6%	81%
Acid concentration (g L^{-1} in terms of acetic acid)	24	7

The spectrum of gases obtained from the two reactors is shown in Fig. 7.6. It is clear from the GC analysis that the gas obtained from Reactor-I (R-I) contains mainly CO_2, whereas in R-II the gas contains 75% v/v CH_4 and 24% v/c CO_2 and traces of nitrogen. From the gas spectrum and acid analysis of R-I, it may be concluded that the system was dominated by acidogens. Acid and gas spectra also tell us that in R-II, the population of mixed flora is dominated by methanogens.

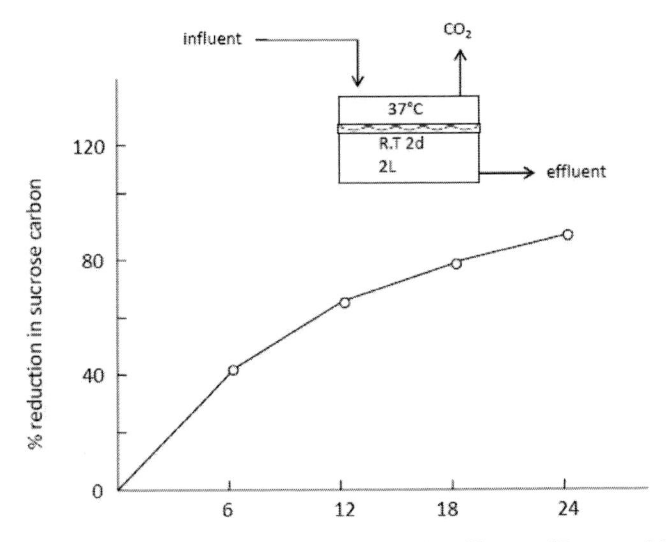

Figure 7.5 Degradation of sucrose carbon in the effluent of Reactor-I (R-I).

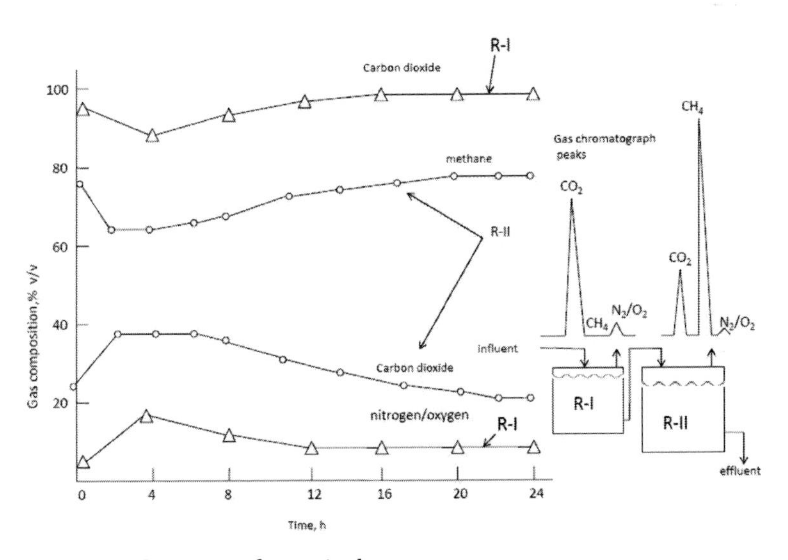

Figure 7.6 Spectrum of gases in the reactors.

7.4.3 Dilution Rates

To optimize the process, the dilution rates were considered important (which influences the retention time). The effect of the dilution rate on the acid-producing phase (R-I) and methane-producing phase (R-II) was studied. Different dilution rates were used to determine the most suitable one for the process. Optimum dilution rates for R-I and R-II were found to be 0.5 d^{-1} and 0.125 d^{-1}, respectively. With the higher dilution rate in R-I, the reactions remain incomplete, resulting in lower acid concentration. It is also seen that higher dilution rates cause an increase in acid concentration in R-II. This is inhibitory to methanogenesis.

7.4.4 COD, BOD, and Gas Production Profile

When the system reached steady-state conditions, studies were carried out to estimate COD and BOD removal profiles in the process as correlated to gas production (Fig. 7.7). In our observation with effluent recycling, the BOD and COD reduction is about 87% and 73%, respectively. In the scale-up unit initially recycling could not be done due to technical problems. Later we observed that the recycling significantly improves the total conversion efficiency of the process.

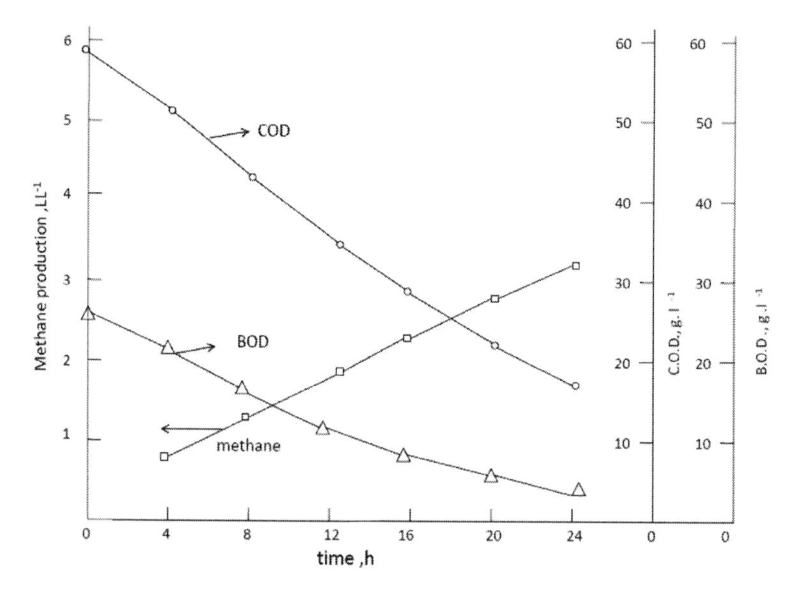

Figure 7.7 Methane production and COD and BOD reduction in the two-phase system in 24 h (with recycling).

A semicontinuous process with its obvious advantage of feeding was done once and several times a day. Feeding several times a day had only a marginal advantage, and therefore, it was decided that feeding would be done once every 24 h for the sake of economy and simplicity of the process. In R-I, mainly, acetic, propionic, and butyric acids were present. Propionic acid concentration decreased after 12 h. According to Barker (1956), 4 mol of propionic acid may be converted to 4 mol of acetic acid, 3 mol of methane, and 1 mol of carbon dioxide. Degradation of propionic acid perhaps results in an increase in the acetic acid concentration in the system. Later this was confirmed in pure substrate studies. There is no significant variation in pH (Fig. 7.8a). The mode of degradation of these precursor acids in the methanogenic phase (in R-II) is depicted in Fig. 7.8b. The rate of degradation of these higher acids was lower than acetic acid. The pH of R-II was varied from 7.0 to 7.5. In the scale-up studies, some unidentified fatty acids (C_4) were detected after running the system for 3 months continuously. However, the concentration of these acids was less than 5% of the total.

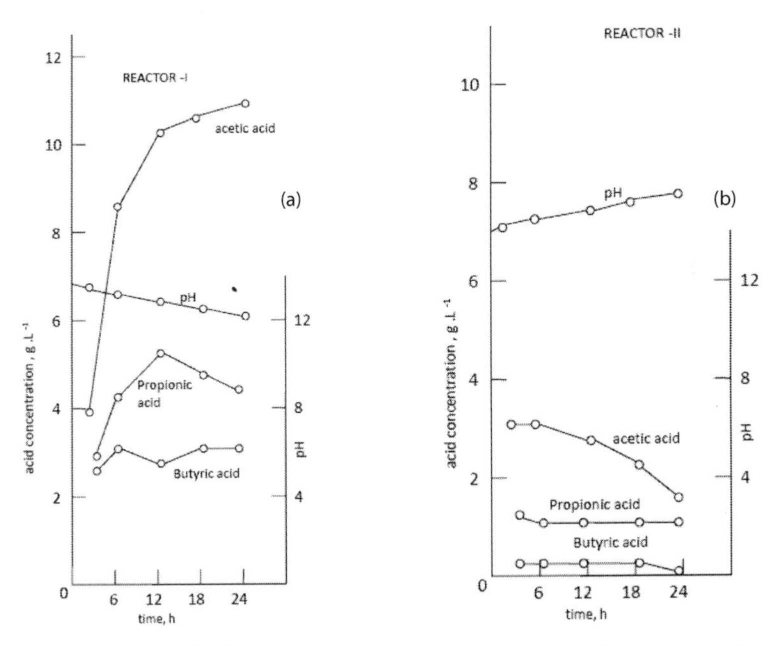

Figure 7.8 (a) Volatile fatty acids and pH status in R-I and (b) volatile fatty acids and pH status in R-II.

7.4.5 Scale-Up Studies

The flow diagram of the scale-up study is already presented. Both reactors were of MS construction. R-I was insulated to reduce the heat loss and fed with diluted distillery wastes at 55°C. Figure 7.2b illustrates details of the total setup. The performance of the reactor compared to bench-scale studies is illustrated in Fig. 7.9. The BOD reduction varied from 70% to 75%, which is a good correlation with the bench-scale results. The waste stabilization and gaseous fuel production efficiency with recycling of the effluent is shown in Fig. 7.10. The data indicate that recycling improves the total conversion efficiency of the process. Figure 7.11 illustrates the correlation between the percentage energy recovery and the percentage feed carbon consumption. The percentage energy recovery is, favorable considering the feed carbon consumption, confirming that distillery waste is a suitable substrate for the application of the anaerobic process. A drastic change in the yeast population in the process samples is interesting. Many researchers have reported (Kobayashi et al., 1981; Kobayashi et al., 1982) that glucanase can degrade the yeast cell wall. Whether this enzyme is present and is responsible for reduction in the yeast cell number/mL is being determined.

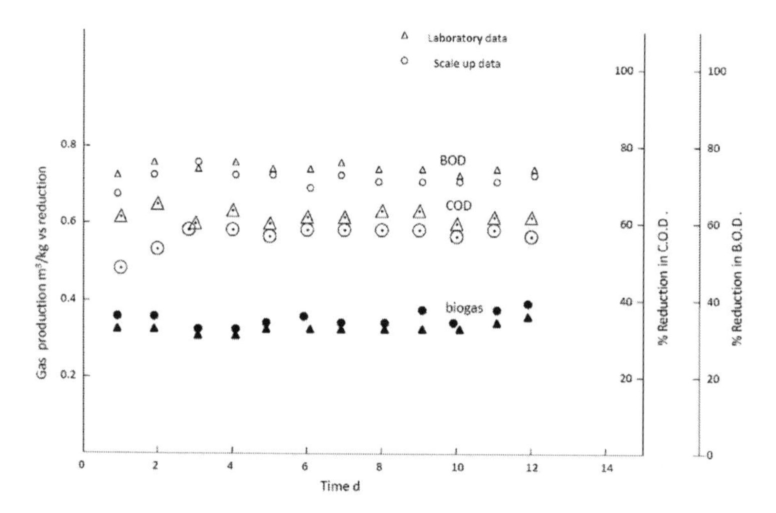

Figure 7.9 Operation performance of the two-stage system (without recycling).

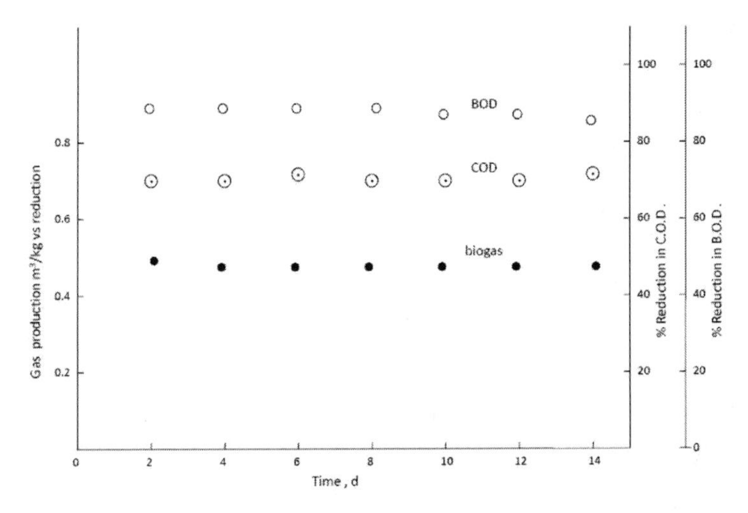

Figure 7.10 Operation performance of the two-stage system (with recycling).

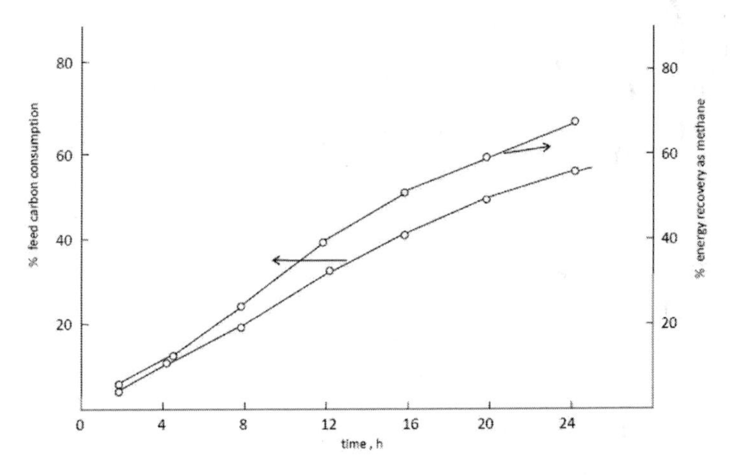

Figure 7.11 Feed carbon consumption and energy recovery.

Kinetic studies of the biomethanation process are usually carried out through the observation of increase in biomass (Monod, 1949; Jennelle et al., 1970). As the present system concerns gas production, studies were carried out on the basis of the exponential gas production rate (k, k'). This is illustrated in Fig. 7.12. As the gas production rate in the acidogenic phase is low, attempts were made to determine the rate during the methanogenic phase.

7.4.6 Materials and Energy Analysis

These were computed on the basis of volatile substrates. The net energy recovery would be nearly 65.5% of the energy potentially available in the distillery waste. Because the organic matter is converted to gas, the relative proportion of inorganic nutrients in the residue increases considerably, rendering it useful in agriculture. The material and energy calculation was made both for the inflow and for the outflow of the system.

7.4.6.1 Energy analysis

The materials and energy analyses (Fig. 7.13 and Fig. 7.14) were computed on the basis of 10 $m^3\ d^{-1}$ alcohol production in a distillery unit. Assuming that 1 L ethanol provides 10 L of distillery wastes, the total wastes available will be 100 $m^3\ d^{-1}$. The energy required for distillation has been reported by several workers. For the calculation it was assumed that 21,260 Btu/gal will be required to convert 10% w/w to 92.4% w/w alcohol. The energy required to maintain the temperature, pumping and mixing, etc., for the two-stage digestion process has also been computed. It is apparent that gaseous energy available is higher compared to energy required for distillation as well as waste stabilization.

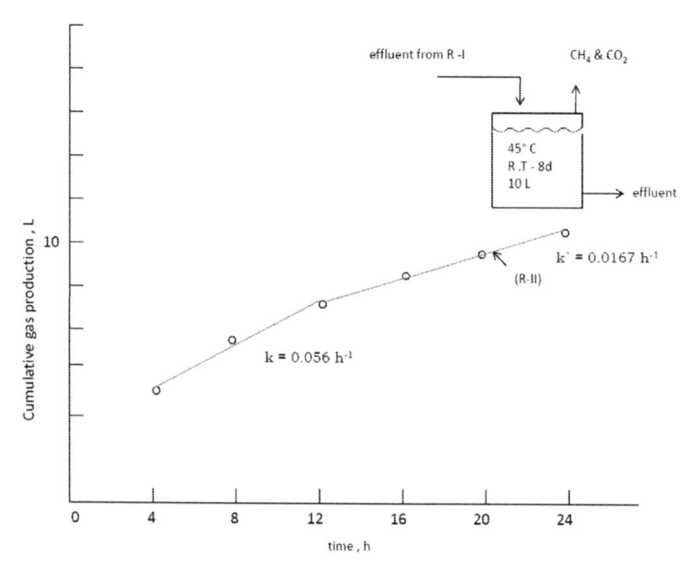

Figure 7.12 Exponential gas production rate of the two-stage system.

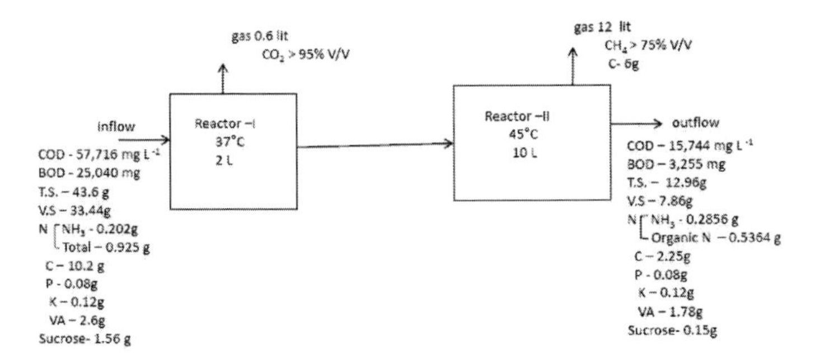

Figure 7.13 Materials balance of the two-stage biomethanation process (with recycling).

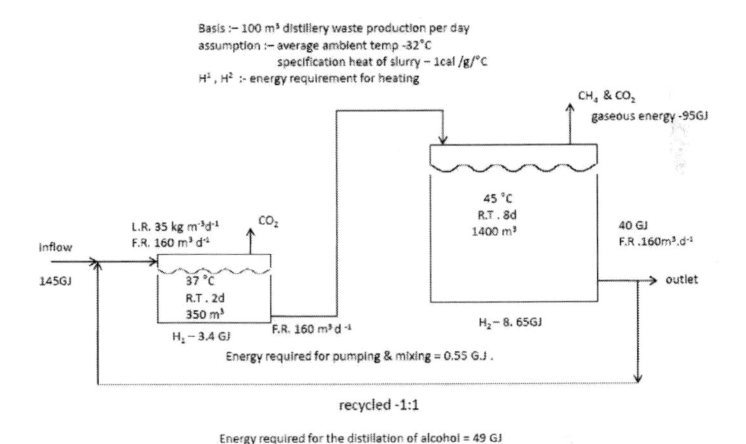

Figure 7.14 Energy balance of the two-stage biomethanation process (with recycling).

7.4.6.2 Material balance

Figure 7.13 illustrates the degradation of different components present in the wastes. The percentage carbon degraded was 78%. The BOD and COD removal was 87% and 73%, respectively. Anaerobic digestion is a natural two-step process: the first is the acidogenic phase, leading to the conversion of complex organic molecules to predominantly VFAs, and the second is the methanogenic phase, involving reduction of intermediates into gaseous products, viz., methane and carbon dioxide. The two groups of microflora involved in this biomethanation process differ in physiology, nutritional requirements, and growth optima. The physical phase separation and maintenance of segregation of cultures of acidogens and methanogens should be possible in a continuous culture by controlling dilution rates. The two-stage anaerobic system as reported here is capable of successfully treating heavily loaded distillery wastes of variable composition and ensures fair COD correction with minimum sludge production. It also withstands uncontrollable stress conditions like organic overloading, variable pH, and sudden temperature variations. This is in sharp contrast to the mixed residue system (Ghose and Das, 1982).

The performance of the reactors is given below:

- The VFA accumulation in R-I was 24 g L^{-1}.
- 30 L of methane was produced per liter of effluent.
- The methane content in the biogas was 75% (v/v).
- The immobilized whole-cell reactor system reduced the retention time to 7.5 days in comparison to the free-cell system (10 days).
- COD reduction was 71%.
- BOD reduction was 87%.

The efficiency of the two reactors is summarized in Table 7.2.

The main problem of this immobilization system is the gas hold-up. The system was operated for about one and a half years successfully. Deposition of solid particles on the solid matrices was observed. This may be due to the formation of $CaCO_3$, Fe_2S_3, etc. (Hung et al., 2015). The biofilm formation taking place on the surface of PP solid matrices was analyzed, which is tabulated in Table 7.3.

Table 7.3 Analysis of the biofilm on a plastic support

Parameters	Content
Dry matter	0.322–0.89 g/g support
Solid volatile matter	65% (on the basis of dry matter)
Nitrogen Kjeldahl (N)	3.85 g/g
Calcium (Ca)	5.57 g/g
Potassium (K)	2.78 g/g
Magnesium (Mg)	1.05 g/g
Sodium (Na)	0.40 g/g

7.4.7 Comparison with the Chemical Process

Energy analysis of the two-stage biomethanation process is given in Fig. 7.14. It is possible to recover 79% of the energy in the form of a gas. Energy required for maintaining the temperature was about 5% of the energy recovered as methane.

Methane can also be produced by thermochemical process using coal as a raw material. A comparison is made to find out the suitability of the biomethanation process compared to the chemical routes (Das, 1985) (Table 7.4). The gross energy requirement was comparatively low in the case of the biomethanation process.

Table 7.4 Energy analysis of different routes for the production of 100 m^3 of methane

Parameters	Chemical routes	Biomethanation process
Energy content in substrate (GJ)	3.999	4.3
External energy required for the process (GJ)	1.205	0.204
Net energy required (GJ)	5.204	4.504
GER (GJ/GJ product)	0.35	0.06
NUEP (%)	54.9	74.3

7.4.8 Bottlenecks in the Whole-Cell Immobilization System

The major problem of these immobilization systems is the clogging of the filter. So, it is desirable to open the reactor after one and a half years and refill it with new packing materials immobilized with microbial cells. The shape and size of the packing materials can be changed to avoid difficulties to some extent. For example, corrugated sheet has been used as packing material in one Bacardi rum distillery in the United States.

7.4.8.1 Other immobilized systems

Microbial cells can be immobilized in fine particles such as sand, poly(vinyl chloride) (PVC), porous glass, etc. (Ramm et al., 2014).

The reactors may be of different types: fluidized bed and expanded bed. These processes are energy intensive (mainly due to recycling) compared to fixed-film reactors. However, the performance of these reactors is better compared to fixed-film reactors (Table 7.5).

Table 7.5 Comparison of different immobilized whole-cell reactors

Parameters	Units	Fixed-film reactors	Expanded-bed and fluidized-bed reactors
Typical COD loading rate	kg m^{-3} d^{-1}	Up to 25	19
Typical COD removal	%	>70	>70
Typical methane	m^3 kg^{-1}	0.35	0.32
Productivity	COD destroyed		
Hydraulic retention time (HRT)	h	4.5–96	34
Solid retention time (SRT)	H	Very high	C.360
Typical biomass concentration	kg m^{-3}	15	25
Recycling ratio	%	2–100	5–500

7.4.9 Anaerobic-Aerobic System

It has been found that it is possible to reduce 87% BOD by using a two-stage biomethanation system. So, it is difficult to achieve the required disposal standard of the treated effluent in the case of a distillery effluent. The aerobic processes may be of two types: free cells (e.g., activated sludge) and immobilized whole-cell systems (e.g., trickling filter and rotary disc biological contractors). In the case of a trickling filter, the microorganism usually forms a slime layer on the surface of gravel, plastics, etc. In a rotary disc biological contractor, the microbial cells are immobilized on the surface of the rotating disc. The flow diagram of the anaerobic-aerobic system is shown in Fig. 7.15.

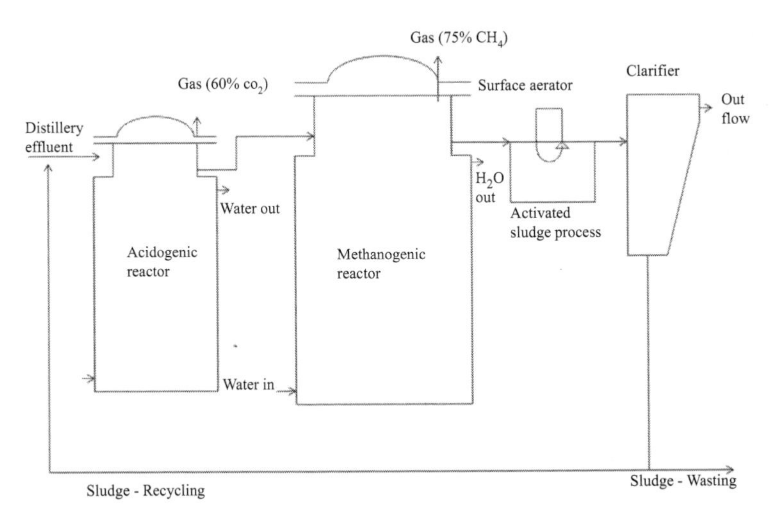

Figure 7.15 Anaerobic-aerobic stabilization processes for distillery effluents.

7.5 Conclusion

The findings of this study indicate that the amount of energy recovery as methane is higher compared to the energy required for the distillation of alcohol. The simplicity of operation of this system is attractive.

The most important points identified in this study producing methane from distillery wastes as a source of energy are:

- Though the characteristics of wastes differ from location to location, they have little influence on the efficiency of the process.
- The maximum VFA concentration obtained is 20 g L^{-1} in R-I. Higher fatty acid accumulation takes place after running the system for a longer time.
- 1 L of distillery waste produces 30 L of methane. The net energy surplus is 82.4 GJ, and the methane content in the gas is 75% v/v.
- The yeast population in R-I decreases. BOD and COD reduction was 87% and 73%, respectively. The percentage energy recovery is favorable, considering the feed carbon consumption.

- Scale-up was successfully carried out, indicating the full-scale industrial feasibility of this process.

Acknowledgments

The authors acknowledge the Indian Institute of Technology (IIT), Delhi, for the research facilities and also thank Dr. T. K. Ghose, Dr. K. S. Gopalkrishnan, Dr. B. K. Guha, and Dr. A. P. Joshi for their valuable suggestions.

References

Barker HA (1956). *Bacterial Fermentations*, John Wiley & Sons, New York.

Bernfeld P (1955). Amylase α and β. In Colowic SP, Kaplan NO (eds), *Methods in Enzymology*, Academic Press, New York, **1**, 149–158.

Braun R, Hass S (1982). Anaerobic digestion of distillery effluent, *Proc Biochem*, **17**, 25–27.

Chopra SL, Kanwar JS (1976). *Analytical Agricultural Chemistry*, Kalyani, Ludhiana.

Das D (1985). *Optimization of Methane Production from Agricultural Residues, BERC*, PhD dissertation IIT Delhi, India.

Das D, et al. (1983). Treatment of distillery wastes by a two phase biomethanation process, Symposium papers *Energy from Biomass and Wastes VII*, Florida, USA, 601–626.

Donnelly T (1978). Industrial effluent treatment with bioenergy process, *Proc Biochem*, **13**, 14–16.

Ghose TK, Bhadra A (1981). Maximization of energy recovery in biomethanation process, part 1: use of cowdung as substrate in multireactor system, *Proc Biochem*, **16**, 23–25.

Ghose TK, Das D (1982). Maximization of energy recovery in the biomethanation process, part 2 - use of mixed residue in batch system, *Proc Biochem*, **17**, 39–42.

Ghosh S, Klass DL (1977). Symposium papers *Clean Fuels from Biomass and Wastes*, IGT, Florida, USA.

Ghosh S, Ombregt JP, De Proost VH, Pipyn P (1982). Methane production from industrial wastes by two phase anaerobic digestion, Symposium papers *Energy from Biomass and Wastes VI*, IGT, Florida, 323–340.

Gorenszy MC (1978). Single vessel activated sludge treatment for small systems, *Proc Biochem*, **13**, 19–25.

Hobson PN (1982). Biogas production from agricultural wastes, *Experientia*, **38**, 206–209.

Huang BS, Chu XH, Zhang YL, Ge YY (2013). Research on the development and utilization of China's biomass energy, *Agr Sci Technol Equip*, **8**, 40–42.

Hung YT, Aziz HA, Yusoff MS, Kamaruddin MA, Yeh RYL, Liu LH, Huhnke CR, Fu YP (2015). Chemical waste and allied products. *Water Environ Res*, **87**(10), 1312–1359.

Jennelle EM, Gaudy AF (1970). Studies on the kinetics and mechanism of BOD exertion in dilute systems, *Biotech Bioeng*, **12**, 519–539.

Jones Jerry L, Radding SB (1980). Thermal conversion of solid wastes and biomass, *ACS Symposium Series 130*, Washington DC, USA.

Klass Donald L, Emert GH (1981). *Fuels from Biomass and Wastes*, Ann Arbor Science, *Ann Arbor*, MI.

Kobayashi R, Miwa T, Yamamoto S, Nagaski S (1981). Properties and mode of action of F-1, 3-glucanase from *Rhizoctonia* sp, *J Ferment Technol*, **59**, 21–26.

Kobayashi R, Miwa T, Yamamoto S, Nagasaki S (1982). Preparation and evaluation of an enzyme which degrades yeast cell walls, *Eur J Appl Microbiol Biotechnol*, **15**, 14–19.

Kwietniewska E, Tys J (2014). Process characteristics, inhibition factors and methane yields of anaerobic digestion process, with particular focus on microalgal biomass fermentation, *Renew Sust Energ Rev*, **34**, 491–500.

Messing Ralph A (1982). Immobilized Microbes and a high rate, continuous, anaerobic waste processor, Symposium papers *Energy from Biomass and Wastes VI*, IGT, Florida.

Monod J (1949). The growth of bacterial cultures, *Ann Rev Microbiol*, **3**, 371–394.

Monteiro CF (1975). Brazilian experience with the disposal of waste water from cane sugar and alcohol industry, *Proc Biochem*, **10**, 33–41.

Pipyn P, Verstraete W (1979). A Pilot scale anaerobic upflow reactor treating distillery waste waters, *Biotechnol Lett*, **1**, 495–500.

Radhakrishnan I, De SB, Nath B (1969). Evaluation of the loading parameters for anaerobic digestion of cane molasses distillery wastes, *J Water Poll Control Fed*, **41**, R431.

Ramm P, Jost C, Neitmann E, Sohling U, Menhorn O, Weinberger K, Mumme J, Linke B (2014). Magnetic biofilm carriers: the use of novel magnetic foam glass particles in anaerobic digestion of sugar beet silage, *J Renew Energy*, **2014**, 208718.

Pipyn P, Verstraete W (1979). A pilot scale anaerobic upflow reactor treating distillery waste waters, *Biotechnol Lett*, **1**, 495–500.

Report of the committee of technical experts on alcohol and alcohol based industries (1980). Ministry of Petroleum, Chemical & Fertilizers, Govt. of India, New Delhi.

Riera F, Valz-Gianinet S, Callieri D, Sineriz F (1982). Use of a packed-bed reactor for anaerobic treatment of stillage of sugar cane molasses, *Biotechnol Lett*, **4**, 127–132.

Roychowdhury AB (1981). Alcohol based industries in India, current status and future prospects, Seminar organised by ICMA & AABIDA, New Delhi.

Sankaran K, Premalatha M, Vijayasekaran M, Somasundaram VT (2014). DEPHY project: distillery wastewater treatment through anaerobic digestion and phycoremediation: a green industrial approach, *Renew Sust Energy Rev*, **37**, 634–643.

Scammel GW (1975). Anaerobic treatment of industrial wastes, *Proc Biochem*, **10**, 34–36.

Sen BP, Bhaskaran TR (1962). Anaerobic digestion of liquid mollases distillery wastes, *J Water Pollut Control Fed*, **34**, 1015–1025.

Skogman, H (1979). Effluent treatment of molasses based fermentation wastes, *Proc Biochem*, **14**, 5–6.

Stadtman TC, Barkar HA (1951). Studies on the Methane Fermentation, *J Bacteriol*, **61**, 67–80.

Standard Methods for the Examination of Water and Wastewater (1981). APHA, 13th edition.

Statistical Review of World Energy (2013). Workbook (xlsx), London.

Chapter 8

Feedstock for Biohythane

8.1 Introduction

Biohythane production could be envisioned as a renewable source of energy because it produces most from organic biomass. Any organic compound, which is rich in carbohydrates, fats, and proteins, could be considered as a possible substrate for biohythane production. Literature reviews suggest that carbohydrates may be considered as the main source of hydrogen during fermentation. Substrates rich in complex carbohydrates are thus the most suitable feedstock for biohydrogen production. Wide ranges of feedstock are available for hydrogen production (Fig. 8.1).

Most of the reports suggest that 80% hydrogen production from dark fermentation was carried out using pure substrates. Cost-effective hydrogen production can be possible only through renewable feedstock (Show et al., 2012). Some woody plants, aquatic plants, and algae show promise as feedstock. Furthermore, agricultural wastes, wastes from food processing, livestock effluents, and other industrial wastes are important renewable feedstock for hydrogen production. In this section, different types of feedstock are discussed on the basis of their availability, and prospects of these for fermentative hydrogen production are discussed. Different types of feedstock used for fermentative hydrogen production along with the rate of hydrogen production and yield have been shown in Table 8.1.

Biohythane: Fuel for the Future
Debabrata Das and Shantonu Roy

ISBN 978-981-4745-29-1 (Hardcover), 978-981-4745-30-7 (eBook)
www.panstanford.com

Table 8.1 Feedstock used for hydrogen production

Substrate type	Organism	Bioprocess characteristics	HY*	HPR# (L/L d)	References
Monosaccharide					
Glucose	*Caldicellulosiruptor saccharolyticus*	Batch fermentation; pH 7; 70°C	2.8	8.8	Mars et al., 2010
	Thermotoga neapolitana DSM 4359	Batch fermentation; pH 7; 75°C	3	8.5	Ngo et al., 2011
	Thermoanaerobacterium thermosaccharolyticm	Batch fermentation; pH 6.5; 60°C	2.7	8.0	Roy et al., 2014b
	Thermoanaerobacterium thermosaccharolyticum	Batch fermentation; pH 6.5; 60°C	2.5	6.8	Singh et al., 2014
	Clostridium beijerinckii	Batch fermentation; pH 6.5; 37°C	2	15	Taguchi et al., 1994
	Clostridium beijerinckii Fanp 3	Batch fermentation; pH 6.5; 37°C	2.52	9.36	Pan et al., 2008
	Ethanoligenens harbinese	CSTR; pH 5; 35°C	1.93	19.6	Xing et al., 2008
	Clostridium tyrobutyricum	CSTR; pH 5; 35°C	1.18	7.2	Jo et al., 2008
Fructose	*Clostridium butyricum* TM-9A	Batch fermentation; pH 6.5; 37°C	2.4	0.84	Junghare et al., 2012

Substrate type	Organism	Bioprocess characteristics	HY*	HPR# (L/L d)	References
	Thermotoga neapolitana	CSABR ; pH 7; 75°C	3.6	2.66	Ngo et al., 2012
	Thermoanaerobacterium thermosaccharolyticum W16	Batch fermentation; pH 6.5; 60°C	2.62	5.71	Ren et al., 2010
Xylose	*Thermoanaerobacter mathranii* A3N	Batch fermentation; pH 7; 70°C	2.5	1.8	Jayasinghearachchi et al., 2012
	Enterobacter aerogenes IAM 1183	Batch fermentation; pH 6.5; 37°C	2.64	5.79	Ren et al., 2009
Disaccharides					
	Thermoanaerobacter mathranii A3N	Fed batch; pH 7; 70°C	2.69	2.4	Jayasinghearachchi et al., 2012
	Caldicellulosiruptor saccharolyticus	Batch fermentation; pH 7; 70°C	2.96	4.5	Vanniel, 2002
Sucrose	*Clostridium butyricum* CGS5	Batch fermentation; pH 6.5; 37°C	1.39	3.9	Chen et al., 2005
	Clostridium butyricum TISTR 1032	Batch fermentation; pH 6.5; 37°C	1.34	3.11	Plangklang et al., 2012
	Escherichia coli DJT135	Batch fermentation; pH 6.5; 37°C	0.68	–	Ghosh and Hallenbeck, 2009

(*Continued*)

Table 8.1 (*Continued*)

Substrate type	Organism	Bioprocess characteristics	HY*	HPR# (L/L d)	References
Arabinose	*Caldicellulosiruptor owensensis*	Batch fermentation; pH 7; 70°C	2.7	–	Zeidan and van Niel, 2010
	Thermophilic mixed culture	Batch fermentation; pH 6.5; 60°C	1.2	3.6	Abreu et al., 2012
	Clostridium butyricum TM-9A	Batch fermentation; pH 6.5; 37°C	8	0.06	Junghare et al., 2012
Ribose	*Clostridium butyricum* TM-9A	Batch fermentation; pH 6.5; 37°C	8	0.84	Junghare et al., 2012
Cellobiose	*Clostridium thermocellum*	Batch fermentation; pH 6.5; 60°C	1.05	–	Levin et al., 2006
	Sludge	Batch fermentation; pH 6.5; 35°C	2.19	–	Adav et al., 2009
	Thermophilic microflora enriched from cow manure	Batch fermentation; pH 6.5; 75°C	2.68	–	Yokoyama et al., 2007
Lactose	*Clostridium thermolacticum*	Batch fermentation; pH 6.5; 58°C	1.2	0.49	Collet, 2004
Polysaccharides					
Starch	*Thermophilic mixed culture*	UASB; pH 6.5; 60°C	1.17	–	Akutsu et al., 2009
	Thermophilic mixed culture	Batch; pH 6.5; 60°C	1.73	–	Yokoyama et al., 2007
	Clostridium beijerinckii RZF-1108	Batch fermentation; pH 6.5; 37°C	–	8.64	Taguchi et al., 1994
	Enterobacter aerogenes and C. butyricum	Batch fermentation; pH 6.5; 37°C	2.8	–	Yokoi et al., 1998
	Mixed culture	CSTR; pH 6.5; 37°C	0.9	1.14	Lay, 2003

Substrate type	Organism	Bioprocess characteristics	HY*	HPR# (L/L d)	References
Agricultural biomass					
Pretreated wheat straw	*Caldicellulosiruptor saccharolyticus*	Batch fermentation; pH 7; 70°C	3.8	–	Ivanova et al., 2009
Corn stalk	*Clostridium thermocellum*	Batch fermentation; pH 6; 55°C	–	0.44	Li and Liu, 2012
Cassava wastewater	*Clostridium acetobutylicum*	Batch fermentation; pH 6.5; 37°C	2.41	1.32	Cappelletti et al., 2011
Palm oil mill effluent	*Clostridium butyricum* EB6	Batch fermentation; pH 6.5; 37°C	0.22	28.4	Chong et al., 2009
Miscanthus hydrolysate	*Caldicellulosiruptor saccharolyticus*	Batch fermentation; pH 7; 70°C	3.4	6.8	de Vrije et al., 2009
Chlorella vulgaris ESP6 hydrolysate	*Clostridium butyricum* CGS5	Batch fermentation; pH 6.5; 37°C	–	5.8	Liu et al., 2012
Chlorella sorokiniana	Thermophilic mixed culture	Batch fermentation; pH 6; 60°C	2.6	–	Roy et al., 2014a
Chlorella sorokiniana	*Enterobacter cloacae* IITBT08	Batch fermentation; pH 6.5; 37°C	2.54	–	Kumar et al., 2013
Chemical wastewater	Mixed culture	Batch fermentation; pH 6.5; 37°C	1.2	–	Venkatamohan et al., 2007

*Hydrogen yield (mol hydrogen/mol hexose equivalent)
#Hydrogen production rate

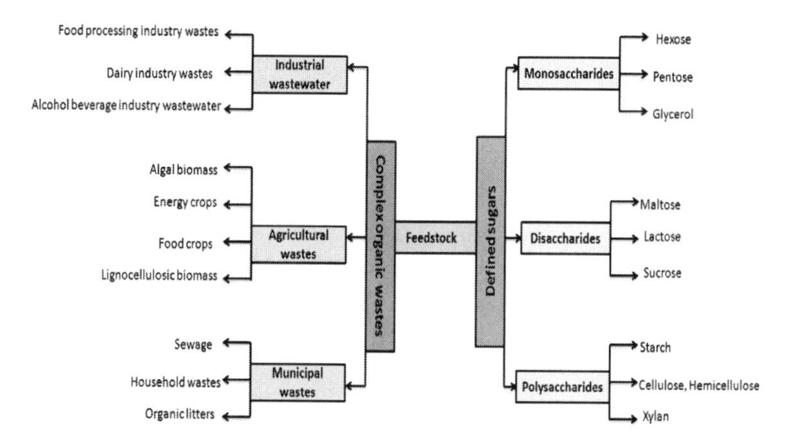

Figure 8.1 Potential feedstock for biohydrogen production.

8.2 Simple Sugars as Feedstock

The mono- and disaccharides are simple sugars, which are mostly used as model substrates for optimization of the fermentative hydrogen production process. Carbon, hydrogen, and oxygen are the backbone of a sugar molecule with an empirical formula of $C_nH_{2n}O_n$. In nature, pentose and hexose sugars are the most abundant, usually found in plant-based and animal-based feedstock. Apart from fructose, fruits also contain xylose and arabinose, which are, in turn, the most abundant pentose sugars. Ribose is available in all plants. Sugarcane or sugar beet is the sources of sucrose, which is a disaccharide. Similarly, maltose is found in germinating grain, corn syrup, and other products; and lactose is mostly found in milk and milk-derived dairy products. Simple sugars are important because of their biodegradability; these help in uncomplicated study of bacterial kinetics, assessment of nutritional requirements, and also optimization of process parameters (Zhao et al., 2011). Moreover, use of simple sugars for fermentation generally takes lesser time compared to complex sugars (Lee et al., 2003). Various studies highlight that fermentative bacteria have certain preferences when different types of sugars are used as substrates. Pure substrates such as glucose, fructose, sucrose, and lactose are a few commonly used sugars for biohydrogen production. However, pure carbohydrate sources are not suitable as feedstock for hydrogen production due to their high costs (Hu and Chen, 2007). Different microbial species have different preferences toward substrates for hydrogen production. For example, it was observed that *Clostridium saccharoperbutylacetonicum* ATCC 27021 grown on disaccharides (lactose, sucrose, and maltose) produced 2.81 mol of hydrogen per mol of sugar, whereas with a monosaccharide, 1.29 mol of hydrogen per mol of hexose was observed, which is twofold lesser yield compared to a disaccharide (Ferchichi et al., 2005). A varied hydrogen yield was also observed with a different array of substrate (2.2 mol of hydrogen per mol of glucose, 6 mol of hydrogen per mol of sucrose, and 5.4 mol of hydrogen per mol of cellobiose) using *Enterobacter cloacae* IIT-BT08 (Kumar and Das, 2000). Moreover, glucose, sucrose, and xylose were used in many experiments to study hydrogen production by the hyperthermophilic microorganism *Caldicellulosiruptor saccharolyticus.* It was observed that during fermentation of 10 g/L glucose and 10 g/L fructose, *Caldicellulosiruptor saccharolyticus* could completely consume all substrates and showed an identical rate of substrate consumption (Panagiotopoulos et al., 2010). In contrast, *Thermotoga neapolitana*, which is also a hyperthermophilic microorganism, showed a higher substrate consumption rate for glucose and fructose, which suggests that glucose is preferred to fructose for hydrogen production. Moreover, higher substrate concentrations led to higher lactate production, thereby decreasing the overall hydrogen yield (de Vrije et al., 2009). To determine the preference of sugar utilization and fermentative behavior of *Thermoanaerobacterium thermosaccharolyticum* W16, various experiments were performed where glucose, xylose, and a mixture of glucose, xylose, and arabinose with a final concentration of 10 g/L were provided. These concentrations of sugars were at par with the sugar content found in the corn stover hydrolysate. It was observed that this microorganism preferred glucose to other types of sugars and sugar mixtures. The bacteria grew well on simple sugars and showed an optical density and the maximum hydrogen production rate at par with media containing hydrolysate (although hydrogen yield was slightly higher on using hydrolysate) (Ren et al., 2010).

8.3 Complex Substrates as Feedstock

There is a relatively high abundance of complex sugars (polysaccharides) in nature. But to provide information about their suitability for hydrogen production, comprehensive research is required targeting the pretreatment and saccharification process (Cheng and Liu, 2011). The biohydrogen production process can be commercialized economically if cheap renewable organic feedstock and low-cost sources such as polysaccharides (starch, cellulose, and xylan) are utilized.

Polymeric sugars such as starch and cellulose consist of a monomeric unit of hundreds or thousands of glucose molecules. In plants, atmospheric CO_2 is fixed into energy storage molecules known as starch. On the other hand, cellulose serves as a rigid building block to maintain cell wall integrity. The cell walls of different parts of the plants (namely leaves, stems, stalks, and woody portions) are made up of cellulose, hemicelluloses, and lignin. Moreover, plant cell walls and some algae also contain xylan. Xylans are polymeric sugars made up of xylose subunits (which is a pentose sugar). Subunits of β-D-xylose are linked via β1,4 linkage to form xylan molecules (structure is analogous to that of cellulose). Lignocellulosic raw materials may be projected as the main component of future feedstock due to their availability. Tightly bound cellulose, hemicelluloses, and lignin form the basic structure of lignocellulic biomass. Cellulose and hemicellulose can be a potential source of sugar for hydrogen fermentation. Lignin is not digested by the anaerobic fermentation process. Moreover, the presence of lignin hinders the accessibility of cellulose and hemicelluloses for enzymatic hydrolysis, thereby resisting mobilization of sugar, and chemically degraded lignin can also inhibit microbial growth (Yokoi et al., 1998).

8.4 Biomass Feedstock

Enhancement of hydrogen production could be achieved through biomass. Properties such as renewability and consumption of atmospheric CO_2 during growth may leave a smaller carbon footprint compared to fossil fuels. However, lower energy content (40% of oxygen content of biomass) and lower hydrogen content of 6% hampers the utility of biomass as a feedstock for thermal gasification. Rather, the biomass could be converted to hydrogen via the biological route by implementing different saccharification technologies that help in converting complex sugars to a simple one. These techniques include thermochemical processes such as acid heat treatment and high-pressure steam explosion and biological processes such as enzymatic hydrolysis and microbial hydrolysis. Biological saccharification processes are slow but are environment friendly compared to thermochemical processes.

8.4.1 Energy Crop

Feedstock used for energy generation is regarded as energy crops (examples include straws from rice, wheat, barley, corn stalk, miscanthus, and cassava). They are commonly referred to as second-generation cellulosic biomass (Elsharnouby et al., 2013). The early 1980s showed the possibility of utilization of whole crops for energy production (Helsel and Wedin, 1981). "Energy crops" are those crops which are grown solely for their further exploitation as feedstock for energy production. The entire biomass or part of it might be used as feedstock. Such feedstock is either directly exploited for its energy content via combustion or biotransformed to biofuels via fermentation.

Sustainable energy production from energy crops could be achieved on the following points (Hawkes, 2002):

- Minimum nutrient and water requirements and thus cost effective
- Immune to stressed environmental conditions
- Higher biomass yields

A high sugar/carbohydrate content and low lignin content are the principal criteria for these plants to be considered suitable for hydrogen production via dark fermentation. Energy crops can be categorized into three broad domains: (i) fermentable simple sugar–rich crops such as sweet sorghum, sugar beet, and sugarcane; (ii) starch-rich crops such as corn, wheat, millets, and rice; and (iii) lignocellulosic biomass, which includes few herbaceous plants such as switch grass, sugarcane bagasse, and fodder grass and few woody plants such as *Miscanthus*, poplar, and conifers.

8.4.1.1 *Miscanthus* sp.

Miscanthus belong to the perennial C_4 grass family. They are rapidly growing plants capable of producing 8 to 15 tons dry weight per hectare with low inputs of nutrients and are found in western European regions. This plant has been extensively studied as a potential feedstock for future energy generation. It needs different types of pretreatment (grinding, ball milling, steam explosion, and alkali treatment) before it can be used as feedstock. The highest hydrogen yield of 3.2 mol/mol glucose was reported using *Miscanthus* hydrolysate fermented by an extreme thermophilic microorganism *Thermotoga elifii* (Devrije, 2002).

8.4.1.2 Sweet sorghum extract

Another perennial C_4 plant is sweet sorghum, also known to have high photosynthetic efficiency. The biomass of sweet sorghum is rich in carbohydrates, and it is a high-yielding crop. Its stalks contain sucrose and glucose, which contribute 55% and 3.2% w/w of dry matter (Antonopoulos et al., 2008). They also contain cellulose (12.4% w/w) and hemicellulose (10.2%). Its high fermentable sugar content makes it an enterprising prospect for fermentative hydrogen production. Overall, out of many "new crops" that are currently being investigated as potential raw materials for energy and industry, sweet sorghum seems to be the most promising one. Using sweet sorghum lysate, the highest methane yield of 78.0 L CH_4/kg and hydrogen yield of 0.86 mol/mol were observed (hexose biomass) (Ntaikou et al., 2008).

8.4.1.3 Sugar beet juice

In the U.K., sugar beet yields around 54 t ha^{-1} (wet extractable sucrose content of about 170 g kg^{-1} wet beet). Reports suggest that a dry mass of the pulp after extraction contains mostly cellulose, hemicelluloses, and pectin, with a small amount of lignin and variable nutrient content. Sugar beet can be cultivated alongside fermentation facilities to be used as an energy crop (Hussy et al., 2005). Hydrogen yields using sugar beet pulp were from 1.7 ± 0.2 to 1.9 ± 0.2 mol/mol hexose converted at a retention time of 14–15 hours (16 kg total sugar m^{-3} d^{-1} organic loading rate). Large-scale cultivation of such crops for biofuel production has led to raised voices against such practices. The requirement of an excessive amount of water, competition with dietary crops for fertile land, use of pesticides, and decrease in fertility of soil due to less crop rotation are the pertaining issues regarding energy crop cultivation. These issues have led to the food versus fuel debate, thereby questioning the use of energy crops as feedstock for biofuels (Srinivasan, 2009). Even in the case of nonfood crop cultivation, the sustainability issue is under question. As a solution, second-generation biofuels have come into play, that is, use of wastes and residues for biofuel production.

8.4.2 Algal Biomass

The rise of lignocellulosic biomass as a source of feedstock is due to its abundant but harsh pretreatment and saccharification processes that hamper its utility to a great extent. Moreover, limitations in resources for water and agriculture are the main disadvantages of using agricultural wastes as a source of bioenergy production (Li et al., 2008). Algal biomass has been considered third-generation feedstock for biofuel production. The cellular structure of algae is simpler compared to higher photosynthetic plants, as algae do not have lignin and hemicelluloses in their cell walls. Due to their large surface-to-volume body ratio, algae can assimilate more nutrients compared to higher plants. Moreover, algal cultivation has certain fundamental advantages (Singh et al., 2008):

- Growth rate: Algae generally have a higher growth rate and a higher rate of biomass production (dry cell weight per unit time and volume) compared to other terrestrial autotrophic plants.
- Carbohydrate content and quality: In harvested biomass, the carbohydrate content varies from 20% w/w to 40% w/w. Lack of a recalcitrant polymer such as lignin makes it ideal feedstock.
- Nutrient preference and rate of substrate utilization: During growth, algae have minimal nutritional requirements.
- Ease of biomass harvesting: Filamentous algae are easier to harvest compared to single-cell algae. Cheap and large-scale harvesting techniques could prove vital for usage of algal biomass as feedstock.

For using algae as feedstock for hydrogen production, various pretreatment methods have been studied. The algal biomass contains complex sugars that were converted to simpler forms by optimizing different physicochemical parameters of the pretreatment. The algal biomass contains complex carbohydrates, which remain attached to the rigid cell wall. Therefore, to gain access to complex sugars and to convert them to simple sugars, it is necessary to break the algal cell wall with harsh pretreatment methods. The overall cost-effectiveness of the hydrogen production process largely depends on the cost of pretreatment of biomass. Several pretreatment methods were explored to break the algal cell wall, which include physical (sonication, bead beating, grinding, milling, and pyrolysis), chemical (acid, alkali, thermal, and H_2O_2), and biological (enzymatic, microbial) methods, resulting in the release of fermentable sugars for biofuel production. All of the methods have their own merits and demerits. Due to higher saccharification efficiency, acid treatment (chemical) methods are preferred over others. Energy-intensive physicochemical methods may be simpler in terms of technology but have limited commercial potential. In the process, formation of furfurals leads to the inhibition of growth of the fermentative microorganisms. Biological methods like co-culture development may be considered an alternative where an organism intensifies the saccharification process, thereby helping in the fermentation process. The costly and time-consuming nature of crude enzymatic techniques is a hurdle for their utility as pretreatment methods. Use of algal biomass as feedstock for hydrogen production has gained interest recently. In a study, two-step hydrogen production was studied where *Clostridium butyricum* was fed with algal biomass hydrolysate to produce hydrogen and short-chain fatty acids produced in the first stage were used for hydrogen production in the second-stage photofermentation, using *Rhodobacter sphaeroides* KD 131 (Kim et al., 2006). In another study, thermophilic, dark-fermentative hydrogen production was reported using *Chlamydomonas reinharditti* biomass (Nguyen et al., 2010). *Anabena variabilis* biomass was hydrolyzed by using amylase, and the hydrolysate was used for thermophilic, dark-fermentative hydrogen production (Nayak et al., 2014). In a similar study, *Chlorella sorokiniana* biomass was treated with acid heat pretreatment and the hydrolysate was used to produce hydrogen by *Enterobacter cloacae* IITBT08 (Kumar et al., 2013).

8.5 Waste as Feedstock

Agricultural wastes as a whole are a bright prospect for use as feedstock for biohydrogen generation because of their vast repertoire and patentability. Various industrial wastewaters could serve a dual purpose in the biohydrogen production process. First, it would lead to generation of clean energy in the form of hydrogen; second, it would also help waste management by reducing the wastewater's chemical oxygen demand (COD) and biological oxygen demand (BOD) content, thereby reducing the toxic effects of wastewater when discharged in receiving land or water bodies. The food manufacturing industry generally produces high-strength organic wastewaters, which are rich in carbohydrates, mainly simple sugars, starch, and cellulose, and could be used as feedstock for biohydrogen production. There is a variety of food industry wastes, such as noodle manufacturing wastes, sugar beet wastewater rice, wheat bran processing by-products, bean curd manufacturing waste, sugar factory wastes, distillery effluents, rice winery wastewater, molasses, dehydrated brewery mixtures, starch manufacturing waste and organic wastewater, and cheese whey (Mizuno et al., 2000). Solid organic wastes are also explored as feedstock for fermentative hydrogen production. Solid wastes generated from household kitchens, food processing, and municipal wastes have a high COD. These wastes are rich in proteins and fats apart from carbohydrates (Lay, 2003).

8.5.1 Municipal Solid Waste

The organic fraction of municipal solid waste (OFMSW) can be considered an important feedstock for hydrogen production because it can lead to a cost-efficient and environment-friendly strategy of production (Noike and Mizuno, 2000). Various value-added by-products may also be derived from conversion of organic wastes to a hydrogen-rich biogas. Municipal solid waste (MSW) poses a major threat to the environment, and its production is globally increasing exponentially at a rate of 6% per year. Nearly 60% of MSW is an organic fraction, which consists of kitchen waste, waste paper, and urban greening waste. OFMSW could be considered a sustainable source of feedstock for biohydrogen production. Its abundant availability at zero cost could improve economic feasibility of biohythane production.

Worldwide biological wastewater treatment processes generate large amounts of sewage sludge. In China, about 4.22 billion tons of municipal wastewater was produced annually in the year 2001. This wastewater is treated in 150 municipal wastewater treatment plants, which generate about 0.55–1.06 million tons of dry sludge (Dunn, 2002).

The waste sludge is treated by anaerobic digestion for methane production, with hydrogen as an intermediate product. Many reports have shown that different wastes with high organic content produce hydrogen by anaerobic fermentation. Polysaccharides and proteins are the major components of sewage sludge and thus can be a potential substrate for fermentative hydrogen production. The raw sewage sludge generally shows inferior potential for hydrogen production (i.e., 0.16 mg of hydrogen [g of dried solids, DS]$^{-1}$). To solubilize the nutrients present in sewage sludge, various pretreatments such as ultrasonication, acid treatment, sterilization, and freezing-thawing were explored to improve fermentative hydrogen production. Sludge pretreated with sterilization or freezing-thawing showed improved hydrogen yields in the range of 1.5–2.1 mmol of hydrogen (g of COD)$^{-1}$ (Wang et al., 2003).

8.5.2 Food Waste

Food waste is considered one of the potential substrates for hydrogen production as it contains about 90% volatile suspended solids (VSS) favoring microbial degradation. Landfill disposal of food waste generates foul smells and pollutes the groundwater. Due to this reason, anaerobic digestion is the most preferred method for treating food wastes. These food wastes include canteen wastes, starchy waste material, and cheese whey, which are potential carbon sources. In the present scenario, researchers are focusing on thermophilic fermentation for the enhancement of biogas production. Institutional food wastes used for thermophilic hydrogen production gave 81 mL hydrogen/g VSS compared to 63 mL hydrogen/g VSS by mesophilic dark fermentation (Chen et al., 2012).

8.6 Industrial Wastewater

8.6.1 Dairy Industry Wastewater

Wastewater from the dairy industry is high in organic content with a high BOD and COD (Orhon et al., 1993). Furthermore, the dairy industry is one of the largest sources of industrial effluents in Europe. Carbohydrates, fats, and proteins from milk contribute to the increase in organic load in natural habitats (Perle et al., 1995). High concentrations of organic matter in dairy waste streams may pose serious problems by increasing organic load on the local municipal sewage treatment facility. Majority of the effluents were generated during cleaning of transport lines, equipment between production cycles, cleaning of tank trucks, washing of milk silos, and equipment malfunctions or operational errors (Danalewich et al., 1998). Various physicochemical and biological treatment methods were employed for treatment of dairy wastewaters. Biological treatment methods were preferred over chemical methods because of high operational costs of the latter (Vidal, 2000). Oxidation ponds, activated sludge plants, and anaerobic treatment are commonly employed biological treatment processes for dairy wastewater treatment (Tawfik et al., 2008). Anaerobic treatment of high-COD dairy effluents is more advisable because of no aeration requirement and low sludge production, and this treatment can be performed in a smaller area compared to the aerobic process.

In dairy waste streams, the organic fraction is contributed by the residues, which are derived from milk and its derivatives. The nitrogen content of such wastewater is contributed mainly from milk proteins, urea, and nucleic acids or in the form of inorganic nitrogen such as NH_4^+, NO_2^-, and NO_3^-. Along with it, orthophosphate (PO_4^{3-}) and polyphosphate ($P_2O_7^{4-}$) are also present (Guillen-Jimenez, 2000). The yardstick generally used for assessment of strength and treatability of wastewater was concentrations of suspended solids (SS) and VSS (Danalewich et al., 1998). SS in dairy wastewater were contributed by coagulated milk solids, cheese curd fines, or flavoring ingredients. The bulk portion of the total COD in cheese-processing wastewater was contributed by lactose, lactate, protein, and fat (Hwang and Hansen, 1998). The main carbohydrate present in dairy wastewater is lactose that could prove to be a potential substrate for anaerobic bacteria. Anaerobic digestion of lactose for biogas production is a mutual biological activity from acidogens, acetogens, and methanogens (Yu and Pinder, 1993). Metabolites namely ethanol, acetate, butyrate, propionate, iso- and normal valerate, caproate, lactate, and formate were produced along with hydrogen during fermentation of lactose. Other components of wastewater were casein protein in milk composition and in dairy effluents and lipids. Lipids are potentially inhibitory compounds for biohydrogen and biomethanation processes. On using dairy wastewater, the maximum hydrogen production of 0.122 mmol hydrogen (g COD)$^{-1}$ was observed when chemical-treated sludge was used as inoculum (Venkata Mohan et al., 2008).

8.6.2 Distillery Wastewater

Distillery wastewater is rich in biodegradable organic material such as sugars, hemicelluloses, dextrin, resins, lignin, and organic acids. For every 1 L of alcohol, 8–15 L of wastewater, which has a high COD (80–160 g L^{-1}), is produced by distillery industries. The distilleries' used technologies such as biomethanation followed concentration and incineration for treatment of such wastewaters. High organic content and availability of a large quantity of wastewater can prove to be a potential feedstock for biohydrogen production by anaerobic fermentation. Many reports were available for biohydrogen production using distillery wastewater in different reactor configurations. In the anaerobic sequencing batch biofilm reactor, maximum hydrogen production of 6.98 mol hydrogen (kg COD_R)$^{-1}$ d^{-1} was observed along with 70% COD removal (Venkatamohan, 2008).

8.6.3 Chemical Wastewaters

These wastewaters were the effluents generated from chemicals, drugs, pharmaceuticals, pesticides, and various chemical processing units. In general, these wastewaters have a low COD but are rich in nitrogen and complex organic compounds. The potentiality of such wastewater for biohydrogen production showed a maximum hydrogen production of 1.25 mmol (g COD added)$^{-1}$ (Venkata mohan et al., 2007). Thus, in the future, such type of chemical industry wastewater can also be considered for feedstock for dark fermentation.

8.6.4 Glycerol

Biodiesel is the clean-burning diesel fuel produced from animal fats, greases, and vegetable oils. Diminishing fossil fuel reserves and environmental consequences have increased the importance of biodiesel. The production of biodiesel is done both chemically and enzymatically. In both cases, glycerol is the major by-product. Treating glycerol waste is a major problem as it increases the overall cost of production of biodiesel. Previously, glycerol waste was treated by mesophilic organisms for hydrogen production. Because glycerol industry effluents are generally released at very high temperatures, these need to be cooled down before treatment. Currently, researchers are moving toward the treatment of this waste by thermophilic organisms to save energy. *Enterobacter aerogenes* HU-101 under mesophilic conditions is able to produce 80 mmol L^{-1} h^{-1} hydrogen using glycerol waste (Ito et al., 2005).

8.6.5 Palm Oil Mill Effluent

Due to its release at high temperature, palm oil mill effluent (POME) can be a potential substrate for the thermophilic dark fermentation process. The major source of POME generation in the palm oil mill is the separator sludge and the sterilizer condensate. On average, 0.9–1.5 m^3 POME is generated from 1 ton of palm oil. It is highly rich in organic content with a BOD of more than 20 g L^{-1} and more than 0.5 g L^{-1} nitrogen. It was reported that with POME using the UASB reactor under thermophilic conditions, a hydrogen production rate of 4.4 L (g POME)$^{-1}$ d^{-1} was achieved (O-Thong et al., 2011).

8.7 Conclusion

The biohydrogen production process is predominated by use of substrates like pure monosaccharides (59%), disaccharides (10%), polysaccharides (11%), and sustainable feedstock (20%). At present, wheat straw, algal biomass, barley straw, and different biomass hydrolysates are explored as sustainable feedstock. Other domains of feedstock are high-strength industrial effluents, agro-based residues, and dairy wastes. Utilization of such feedstock would help not only in clean energy generation but also in waste management.

References

Abreu AA, Karakashev D, Angelidaki I, Sousa DZ, Alves MM (2012). Biohydrogen production from arabinose and glucose using extreme thermophilic anaerobic mixed cultures, *Biotechnol Biofuels*, **5**, 6.

Adav SS, Lee D-J, Wang A, Ren N (2009). Functional consortium for hydrogen production from cellobiose: concentration-to-extinction approach, *Bioresour Technol*, **100**, 2546–2550.

Akutsu Y, Lee D-Y, Chi Y-Z, Li Y-Y, Harada H, Yu H-Q (2009). Thermophilic fermentative hydrogen production from starch-wastewater with bio-granules, *Int J Hydrogen Energy*, **34**, 5061–5071.

Antonopoulou G, Gavala HN, Skiadas IV, Angelopoulos K, Lyberatos G (2008). Biofuels generation from sweet sorghum: fermentative hydrogen production and anaerobic digestion of the remaining biomass, *Bioresour Technol*, **99**, 110–119.

Cappelletti BM, Reginatto V, Amante ER, Antônio RV (2011). Fermentative production of hydrogen from cassava processing wastewater by *Clostridium acetobutylicum*, *Renew Energy*, **36**, 3367–3372.

Chen C-C, Chuang Y-S, Lin C-Y, Lay C-H, Sen B (2012). Thermophilic dark fermentation of untreated rice straw using mixed cultures for hydrogen production, *Int J Hydrogen Energy*, **37**, 15540–15546.

Chen W, Tseng Z, Lee K, Chang J (2005). Fermentative hydrogen production with CGS5 isolated from anaerobic sewage sludg, *Int J Hydrogen Energy*, **30**, 1063–1070.

Cheng X-Y, Liu C-Z (2011). Hydrogen production via thermophilic fermentation of cornstalk by *Clostridium thermocellum*, *Energy Fuels*, **25**, 1714–1720.

Chong M, Rahim R, Shirai Y, Hassan M (2009). Biohydrogen production by *Clostridium butyricum* EB6 from palm oil mill effluent, *Int J Hydrogen Energy*, **34**, 764–771.

Collet C (2004). Hydrogen production by *Clostridium thermolacticum* during continuous fermentation of lactose, *Int J Hydrogen Energy*, **29**, 1479–1485.

Danalewich JR, Papagiannis TG, Belyea RL, Tumbleson ME, Raskin L (1998). Characterization of dairy waste streams, current treatment practices, and potential for biological nutrient removal, *Water Res*, **32**, 3555–3568.

De Vrije T, Bakker RR, Budde MA, Lai MH, Mars AE, Claassen PA (2009). Efficient hydrogen production from the lignocellulosic energy crop Miscanthus by the extreme thermophilic bacteria *Caldicellulosiruptor saccharolyticus* and *Thermotoga neapolitana*, *Biotechnol Biofuels*, **2**, 12.

Devrije T (2002). Pretreatment of Miscanthus for hydrogen production by *Thermotoga elfii*, *Int J Hydrogen Energy*, **27**, 1381–1390.

Dunn S (2002). Hydrogen futures: toward a sustainable energy system, *Int J Hydrogen Energy*, **27**, 235–264.

Elsharnouby O, Hafez H, Nakhla G, El Naggar MH (2013). A critical literature review on biohydrogen production by pure cultures, *Int J Hydrogen Energy*, **38**, 4945–4966.

Ferchichi M, Crabbe E, Gil G-H, Hintz W, Almadidy A (2005). Influence of initial pH on hydrogen production from cheese whey, *J Biotechnol*, **120**, 402–409.

Ghosh D, Hallenbeck PC (2009). Fermentative hydrogen yields from different sugars by batch cultures of metabolically engineered *Escherichia coli* DJT135, *Int J Hydrogen Energy*, **34**, 7979–7982.

Guillen-Jimenez E (2000). Bio-mineralization of organic matter in dairy wastewater, as affected by pH. The evolution of ammonium and phosphates, *Water Res*, **34**, 1215–1224.

Hawkes F (2002). Sustainable fermentative hydrogen production: challenges for process optimisation, *Int J Hydrogen Energy*, **27**, 1339–1347.

Helsel ZR, Wedin W (1981). Direct combustion energy from crops and crop residues produced in Iowa, *Energy Agric*, **1**, 317–329.

Hu B, Chen S (2007). Pretreatment of methanogenic granules for immobilized hydrogen fermentation, *Int J Hydrogen Energy*, **32**, 3266–3273.

Hussy I, Hawkes F, Dinsdale R, Hawkes D (2005). Continuous fermentative hydrogen production from sucrose and sugarbeet, *Int J Hydrogen Energy*, **30**, 471–483.

Hwang S, Hansen CL (1998). Characterization of and bioproduction of short-chain organic acids from mixed dairy-processing wastewater, *Trans Am Soc Agric Eng*, **41**, 795–802.

Ito T, Nakashimada Y, Senba K, Matsui T, Nishio N (2005). Hydrogen and ethanol production from glycerol-containing wastes discharged after biodiesel manufacturing process, **100**, 260–265.

Ivanova G, Rákhely G, Kovács KL (2009). Thermophilic biohydrogen production from energy plants by *Caldicellulosiruptor saccharolyticus* and comparison with related studies, *Int J Hydrogen Energy,* **34**, 3659–3670.

Jayasinghearachchi HS, Sarma PM, Lal B (2012). Biological hydrogen production by extremely thermophilic novel bacterium *Thermoanaerobacter mathranii* A3N isolated from oil producing well, *Int J Hydrogen Energy,* **37**, 5569–5578.

Jo JH, Lee DS, Park JM (2008). The effects of pH on carbon material and energy balances in hydrogen-producing *Clostridium tyrobutyricum* JM1, *Bioresour Technol,* **99**, 8485–8491.

Junghare M, Subudhi S, Lal B (2012). Improvement of hydrogen production under decreased partial pressure by newly isolated alkaline tolerant anaerobe, *Clostridium butyricum* TM-9A: optimization of process parameters, *Int J Hydrogen Energy,* **37**, 3160–3168.

Kim M, Baek J, Yun Y, Junsim S, Park S, Kim S (2006). Hydrogen production from *Chlamydomonas reinhardtii* biomass using a two-step conversion process: anaerobic conversion and photosynthetic fermentation, *Int J Hydrogen Energy,* **31**, 812–816.

Kumar K, Roy S, Das D (2013). Continuous mode of carbon dioxide sequestration by *C. sorokiniana* and subsequent use of its biomass for hydrogen production by *E. cloacae* IIT-BT 08, *Bioresour Technol,* **145**, 116–122.

Kumar N, Das D (2000). Enhancement of hydrogen production by *Enterobacter cloacae* IIT-BT 08, *Proc Biochem,* **35**, 589–593.

Lay J (2003). Influence of chemical nature of organic wastes on their conversion to hydrogen by heat-shock digested sludge, *Int J Hydrogen Energy,* **28**, 1361–1367.

Lee K-S, Lo Y-S, Lo Y-C, Lin P-J, Chang J-S (2003). H_2 production with anaerobic sludge using activated-carbon supported packed-bed bioreactors, *Biotechnol Lett,* **25**, 133–138.

Levin D, Islam R, Cicek N, Sparling R (2006). Hydrogen production by *Clostridium thermocellum* 27405 from cellulosic biomass substrates, *Int J Hydrogen Energy,* **31**, 1496–1503.

Li Q, Liu C-Z (2012). Co-culture of *Clostridium thermocellum* and *Clostridium thermosaccharolyticum* for enhancing hydrogen production via thermophilic fermentation of cornstalk waste, *Int J Hydrogen Energy,* **37**, 10648–10654.

Liu C-H, Chang C-Y, Cheng C-L, Lee D-J, Chang J-S (2012). Fermentative hydrogen production by *Clostridium butyricum* CGS5 using carbohydrate-rich microalgal biomass as feedstock, *Int J Hydrogen Energy,* **37**, 15458–15464.

Mars AE, Veuskens T, Budde MAW, van Doeveren PFNM, Lips SJ, Bakker RR, de Vrije T, Claassen PAM (2010). Biohydrogen production from untreated and hydrolyzed potato steam peels by the extreme thermophiles *Caldicellulosiruptor saccharolyticus* and *Thermotoga neapolitana, Int J Hydrogen Energy,* **35**, 7730–7737.

Nayak BK, Roy S, Das D (2014). Biohydrogen production from algal biomass (Anabaena sp. PCC 7120) cultivated in airlift photobioreactor, *Int J Hydrogen Energy,* **39**, 7553–7560.

Ngo TA, Kim M-S, Sim SJ (2011). Thermophilic hydrogen fermentation using Thermotoga neapolitana DSM 4359 by fed-batch culture, *Int J Hydrogen Energy,* **36**, 14014–14023.

Ngo TA, Nguyen TH, Bui HTV (2012). Thermophilic fermentative hydrogen production from xylose by *Thermotoga neapolitana* DSM 4359, *Renew Energy,* **37**, 174–179.

Nguyen T-AD, Kim K-R, Nguyen M-T, Kim MS, Kim D, Sim SJ (2010). Enhancement of fermentative hydrogen production from green algal biomass of *Thermotoga neapolitana* by various pretreatment methods, *Int J Hydrogen Energy,* **35**, 13035–13040.

Noike T, Mizuno O (2000). Hydrogen fermentation of organic municipal wastes, *Water Sci Technol,* **42**, 155–162.

Ntaikou I, Gavala H, Kornaros M, Lyberatos G (2008). Hydrogen production from sugars and sweet sorghum biomass using Ruminococcus albus, *Int J Hydrogen Energy,* **33**, 1153–1163.

Mizuno O, Ohara T, Shinya M, Noike T (2000). Characteristics of hydrogen production from bean curd manufacturing waste by anaerobic microflora, *Water Sci Technol,* **41**, 25–32.

Orhon D, Görgün E, Germirli F, Artan N (1993). Biological treatability of dairy wastewaters, *Water Res,* **27**, 625–633.

O-Thong S, Mamimin C, Prasertsan P (2011). Effect of temperature and initial pH on biohydrogen production from palm oil mill effluent: long-term evaluation and microbial community analysis, *Electron J Biotechnol,* **14**, 9.

Pan C, Fan Y, Zhao P, Hou H (2008). Fermentative hydrogen production by the newly isolated *Clostridium beijerinckii* Fanp3, *Int J Hydrogen Energy,* **33**, 5383–5391.

Panagiotopoulos IA, Bakker RR, de Vrije T, Koukios EG, Claassen PAM (2010). Pretreatment of sweet sorghum bagasse for hydrogen production by *Caldicellulosiruptor saccharolyticu*, *Int J Hydrogen Energy*, **35**, 7738–7747.

Perle M, Kimchie S, Shelef G (1995). Some biochemical aspects of the anaerobic degradation of dairy wastewater, *Water Res*, **29**, 1549–1554.

Plangklang P, Reungsang A, Pattra S (2012). Enhanced bio-hydrogen production from sugarcane juice by immobilized *Clostridium butyricum* on sugarcane bagasse, *Int J Hydrogen Energy*, **37**, 15525–15532.

Ren N-Q, Cao G.-L, Guo W-Q, Wang A-J, Zhu Y-H, Liu B, Xu J-F (2010). Biological hydrogen production from corn stover by moderately thermophile *Thermoanaerobacterium thermosaccharolyticum* W16, *Int J Hydrogen Energy*, **35**, 2708–2712.

Ren Y, Wang J, Liu Z, Ren Y, Li G (2009). Hydrogen production from the monomeric sugars hydrolyzed from hemicellulose by *Enterobacter aerogenes*, *Renew Energy*, **34**, 2774–2779.

Roy S, Kumar K, Ghosh S, Das D (2014a). Thermophilic biohydrogen production using pre-treated algal biomass as substrate, *Biomass Bioenergy*, **61**, 157–166.

Roy S, Vishnuvardhan M, Das D (2014b). Improvement of hydrogen production by newly isolated *Thermoanaerobacterium thermosaccharolyticum* IIT BT-ST1, *Int J Hydrogen Energy*, **39**, 7541–7552.

Show KY, Lee DJ, Tay JH, Lin CY, Chang JS (2012). Biohydrogen production: current perspectives and the way forward, *Int J Hydrogen Energy*, **37**, 15616–15631.

Singh S, Kate BN, Banerjee UC (2005). Bioactive compounds from cyanobacteria and microalgae: an overview, *Crit Rev Biotechnol*, **25**, 73–95.

Singh S, Sarma PM, Lal B (2014). Biohydrogen production by *Thermoanaerobacterium thermosaccharolyticum* TERI S7 from oil reservoir flow pipeline, *Int J Hydrogen Energy*, **39**, 4206–4214.

Srinivasan S (2009). The food v. fuel debate: a nuanced view of incentive structures, *Renew. Energy*, **34**, 950–954.

Taguchi F, Mizukami N, Hasegawa K, Saito-Taki T, Morimoto M (1994). Effect of amylase accumulation on hydrogen production by *Clostridium beijerinckii* strain AM21B, *J Ferment Bioeng*, **77**, 565–567.

Tawfik A, Sobhey M, Badawy M (2008). Treatment of a combined dairy and domestic wastewater in an up-flow anaerobic sludge blanket (UASB) reactor followed by activated sludge (AS system), *Desalination*, **227**, 167–177.

Vanniel E (2002). Distinctive properties of high hydrogen producing extreme thermophiles, *Caldicellulosiruptor saccharolyticus* and *Thermotoga elfii*, *Int J Hydrogen Energy*, 27, 1391–1398.

Venkata Mohan S, Lalit Babu V, Sarma PN (2008). Effect of various pretreatment methods on anaerobic mixed microflora to enhance biohydrogen production utilizing dairy wastewater as substrate, *Bioresour Technol*, **99**, 59–67.

Venkata Mohan S (2008). Simultaneous biohydrogen production and wastewater treatment in biofilm configured anaerobic periodic discontinuous batch reactor using distillery wastewater, *Int J Hydrogen Energy*, **33**, 550–558.

Venkata Mohan S, Vijayabhaskar Y, Muralikrishna P, Chandrasekhararao N, Lalitbabu V, Sarma P (2007). Biohydrogen production from chemical wastewater as substrate by selectively enriched anaerobic mixed consortia: influence of fermentation pH and substrate composition, *Int J Hydrogen Energy*, **32**, 2286–2295.

Vidal G (2000). Influence of the content in fats and proteins on the anaerobic biodegradability of dairy wastewaters, *Bioresour Technol*, **74**, 231–239.

Wang CC, Chang C, Chu CP, Lee DJ, Chang B-V, Liao CS (2003). Producing hydrogen from wastewater sludge by *Clostridium bifermentans*, *J Biotechnol*, **102**, 83–92.

Xing D, Ren N, Wang A, Li Q, Feng Y, Ma F (2008). Continuous hydrogen production of auto-aggregative *Ethanoligenens harbinense* YUAN-3 under non-sterile condition, *Int J Hydrogen Energy*, **33**, 1489–1495.

Yokoi H, Tokushige T, Hirose J, Hayashi S, Takasaki Y (1998). H_2 production from starch by a mixed culture of *Clostridium butyricum* and *Enterobacter aerogenes*, *Biotechnol Lett*, **20**, 143–147.

Yokoyama H, Moriya N, Ohmori H, Waki M, Ogino A, Tanaka Y (2007). Community analysis of hydrogen-producing extreme thermophilic anaerobic microflora enriched from cow manure with five substrates, *Appl Microbiol Biotechnol*, **77**, 213–222.

Yu J, Pinder KL (1993). Intrinsic fermentation kinetics of lactose in acidogenic biofilms, *Biotechnol Bioeng*, **41**, 479–488.

Zeidan AA, van Niel EWJ (2010). A quantitative analysis of hydrogen production efficiency of the extreme thermophile *Caldicellulosiruptor owensensis* OLT, *Int J Hydrogen Energy*, **35**, 1128–1137.

Zhao X, Xing D, Fu N, Liu B, Ren N (2011). Hydrogen production by the newly isolated *Clostridium beijerinckii* RZF-1108, *Bioresour Technol*, **102**, 8432–8436.

Chapter 9

Effect of Physicochemical Parameters on the Biohythane Process

9.1 Introduction

Biohythane production processes are greatly influenced by complex biochemical and physical parameters. The parameters such as the age and size of the inoculum, complexity of substrates, reactor configurations, pH, temperature, alkalinity, hydraulic retention time (HRT), etc., needed to be studied intensely for a better understanding of the system. Moreover, it also helps in scaling up of the process for commercial application. The performance of microorganisms in converting the reactant to the product is dependent on the efficiency of its enzymatic machinery. The optimum output from microorganisms could be expected at a defined pH, temperature, and substrate concentration. The bioreactor in which the microorganisms are grown also plays a crucial role. The design and configuration of the fermentor help in improvement of mixing characteristics, manipulation of overhead gas partial pressure, etc. Parameters like HRT and recycle ratio are influenced the bioreactor configuration.

An ample number of reports was available regarding the improvement of biohydrogen and biomethane production via optimization of various physicochemical factors. This chapter illustrates the role of the above-mentioned parameters on hydrogen

Biohythane: Fuel for the Future
Debabrata Das and Shantonu Roy

ISBN 978-981-4745-29-1 (Hardcover), 978-981-4745-30-7 (eBook)
www.panstanford.com

production by dark-fermentative bacteria and biomethanation production.

9.2 Factors Influencing Biohythane Production

The efficiency of dark fermentation process involved in biohythane production is governed by the performance of the bioreactor, various physicochemical parameters, etc. These factors help in the manipulation of various biochemical pathways for improvement of product formation.

The important physicochemical parameters which influence the biohythane production pathways are mentioned below:

- Temperature
- pH
- Alkalinity
- Inoculum
- Effect of micronutrient's HRT
- Hydrogen and CO_2 partial pressure

9.2.1 Role of Temperature in Biohythane Production

All microorganisms prosper at a temperature where they function optimally. This temperature is regarded as the optimal temperature. The temperature required for their growth is quite varied. The operation temperature influences the growth rate of bacteria by influencing the biochemical reaction responsible for maintenance homeostasis and metabolism. Thus, it might also influence the nutritional requirement, metabolic end-product formation, and characteristics of microbial cells. Most of the literature reports on biohydrogen and biomethane production are based on mesophilic dark fermentation (Li et al., 2001). Temperature also plays a vital role in biohythane production. Operation temperature governs metabolism via mediating the enzymatic reactions. Every enzyme has an optimum temperature range at which maximum productivity is observed. Extreme temperature would lead to denaturation of metabolic and life-supporting enzymes. The shift in the metabolic pathways and microbial community is significantly influenced by operation temperature. Therefore, several studies on understanding the effect of temperature variance on microbial community distribution were reported.

The choice of microorganisms decides the operational temperature of the fermenter. Fermentative microorganisms capable of producing hydrogen were reported at different temperatures, viz., psychrophilic (0°C–20°C), mesophilic (20°C–42°C), and thermophilic (42°C–75°C). Many mesophilic bacterial isolates (such as *Clostridium* and *Enterobacter* strains) showed optimal hydrogen production in the temperature range of 37°C–45°C (Vindis, 2009). The growth rate of the organism was affected severely as the temperatures deviate from the optimal ranges.

On working with mixed consortia, variation in temperature leads to a change in microbial population dynamics. Mixed consortia could harbor mesophilic, thermotolerant, thermoduric, thermophilic microorganisms. Depending upon the operation temperature, the respective microorganisms might become dominant compared to another habitant of the consortia. Thermophilic hydrogen-producing microbes showed a higher hydrogen yield compared to the mesophilic counterparts (Groenestijn et al., 2002). Higher temperatures make the hydrogen production process thermodynamically favorable. It increases the entropy of the system; thus the overall free energy increases, making the process spontaneous.

Roy et al. (2012) studied the effect of temperature on thermophilic dark fermentation (Fig. 9.1). It was found that hydrogen production increased linearly up to 60°C. An increase in temperature beyond 60°C leads to a decrease in hydrogen production. The decrease in hydrogen production could be due to denaturation of the enzymatic machinery and changes in membrane fluidity of the microorganisms.

High temperatures increase the kinetic mobility of H^+ and e^-, which makes the hydrogen production thermodynamically favorable. High temperatures also help in solubilization of complex feedstock and improve hydrolysis of the substrate (Veeken and Hamelers, 1999). Operation at high temperatures also decreases the risk of methanogenic bacterial contamination (Egorova and Antranikian, 2005). The enzymatic activity doubles with every 10°C increase in temperature till it reaches an optimal temperature, and with further increase in temperature enzymatic activity decreases.

The effect of temperature on the hydrogen production rate and yield can be illustrated using a case study by Lee et al. in which they used a carrier-induced granular sludge bed (CIGSB) reactor with sucrose-based artificial wastewater (Lee et al., 2006). When the temperature was increased from 30°C to 40°C, the hydrogen production increased; however, with further increase in temperature hydrogen production decreased. This observation can be attributed to the positive kinetic effect of the enzymatic processes, which increases up to the threshold temperature and beyond this temperature thermal deactivation of biocatalyst occurs (Slininger et al., 1990). Thus, it can be concluded that the decrease in efficiency of the CIGSB was due to the thermal deactivation of enzymatic reactions that control the metabolic pathway. To compute the thermal deactivation mechanism of hydrogen production, the activation-deactivation enthalpy can be calculated using the modified Arrhenius equation (Eq. 9.1) (Fabiano and Perego, 2002):

$$r_{\mathrm{m}} = AX_{\mathrm{g}}Y_{\frac{H_2}{X_{\mathrm{g}}}}\exp\left(\frac{-\Delta H^*}{RT}\right) \quad (9.1)$$

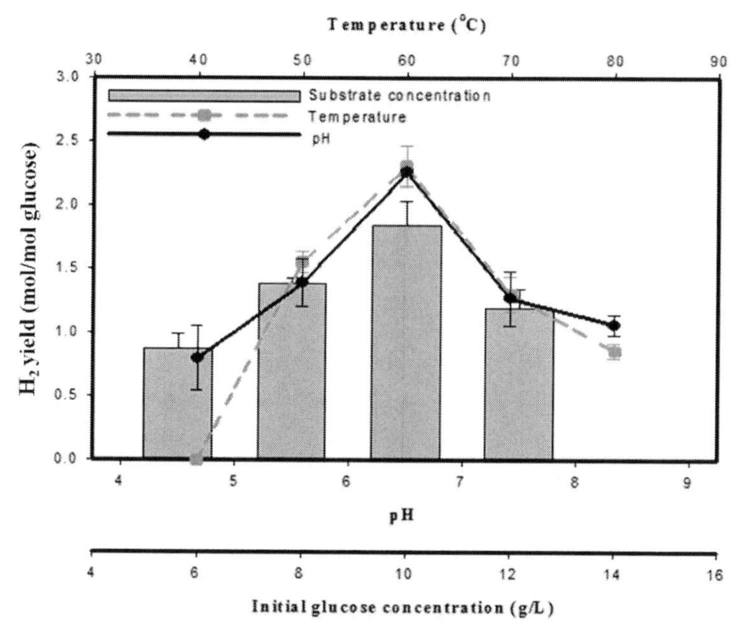

Figure 9.1 Single-parameter optimization of physicochemical parameters in a thermophilic mixed culture. Experimental conditions: temperature, 40°C to 90°C; pH, 4.5–7.5; glucose concentration, 6 g L^{-1} to 14 g L^{-1} (Roy et al., 2012).

By modifying Eq. 9.1, the activation and deactivation enthalpies could be expressed distinctly as

$$\ln r_{\mathrm{m}} = \ln\left(AX_{\mathrm{g}} Y_{\frac{H_2}{X_{\mathrm{g}}}} \right) - \frac{\Delta H^*}{RT} \tag{9.2}$$

$$\ln r_{\mathrm{m}} = \ln\left(BX_{\mathrm{g}} Y_{\frac{H_2}{X_{\mathrm{g}}}} \right) - \frac{\Delta H^{**}}{RT} \tag{9.3}$$

where r_{m} is the maximum specific hydrogen production rate, X_{g} (g L^{-1}) is the cell mass concentration, $Y_{\mathrm{H2}/X}$ (mL hydrogen g^{-1} cell mass) is the yield, ΔH^* (kJ mol^{-1}) is the activation enthalpy of fermentation, and A is the Arrhenius pre-exponential factor (Nath et al., 2006).

Moreover, the influence of temperature on activation entropy and thermal deactivation entropy (ΔS^* and ΔS_{d^*}, respectively) could be expressed as

$$\Delta S^* = R \ln \frac{Ah}{k_{\mathrm{B}}T} \tag{9.4}$$

$$\Delta S_{\mathrm{d}}^* = R \ln \frac{Bh}{k_{\mathrm{B}}T} \tag{9.5}$$

where h and k_{B} are Planck's and Boltzmann's constants, respectively.

The activation enthalpy of the biohydrogenation process using a potent mesophilic microorganism (*Enterobacter cloacae* IITBT08) was found to be 47.34 kJ mol^{-1} K^{-1} (Nath et al., 2006). In another report, on using *E. aerogenes* NCIMB 10102, the activation energy of 67.3 kJ mol^{-1} K^{-1} was observed. This shows that the thermal activation energy varies with the characteristic microorganisms (Fabiano and Perego, 2002).

Exploitation of microorganisms growing at high temperatures, that is, thermophiles, has great potential in hydrogen production. Till date, the highest rate and yield of hydrogen have been observed in thermophilic microorganisms. Thermophilic hydrogen-producing microbes were explored from natural geysers, anaerobic digester sludge (Uneo et al., 1995), hot springs (Koskinen et al., 2008), cow dung (Yokoyama et al., 2007), and wastewater treatment plants (Thong et al., 2007). High-temperature hydrogen production has certain operational advantages:

- Low solubility of hydrogen and CO_2 at high temperatures
- Hydrogen production least influenced by the partial pressure of hydrogen at high temperature
- Better solubility of the substrate at high temperatures
- Improved hydrolysis and thermodynamic efficiency

Caldicellulosiruptor owensensis strain OLT is a well-known thermophilic microorganism which has been reported to be the highest-hydrogen-yielding microorganism so far. It shows a hydrogen yield of the theoretical maximum, that is, 4 mol of hydrogen per mol of glucose consumed (Zieden and Neil 2010). On the other hand, there are many disadvantages of thermophilic dark-fermentative hydrogen production, such as requirement of increased power for the maintenance of high temperatures and maintenance of a strict anaerobic environment. These factors would eventually influence the economic feasibility of the process. A probable solution of the requirement of high electrical power could be the usage of thermal insulation to minimize the heat loss from the reactor. There is a plethora of materials that have better insulation properties and that have been explored, mainly polymeric materials shown in Table 9.1. Glass wool is generally used for insulation of instruments related to temperature control. The high cost of glass wool makes it least feasible as an insulating material for scale-up applications. Cheaper insulating materials like coconut coir and concrete can be used as insulating materials. It has been reported that heat loss in the thermophilic process is about 57% higher compared to the mesophilic process (Kelly and Gibson, 2008). Thus cheap and effective insulating materials can prevent this heat loss and can make the process more viable. Therefore, in the future, many technological breakthroughs would be required in order to make thermophilic hydrogen production economically feasible.

Temperature is the most influential parameter for methanogenesis. For psychrophiles, the optimal temperature range is between 10°C–15°C while for mesophiles and thermophiles it is 35°C–40°C and 58°C–68°C, respectively. The methane production rate is slowest with psychrophiles compared to mesophiles and thermophiles, and thus, they are seldom used for scale up of methane production. It is observed that methanogens are extremely subtle to change in temperature and even a small temperature variation (of 2°C to 3°C) can lead to volatile fatty acid (VFA) accumulation (Speece, 1983). This leads to a decrease in the methane generation rate for all anaerobic microorganisms, especially at thermophilic conditions. Nevertheless, by maintaining a suitable temperature, these microorganisms can recover.

Table 9.1 Properties of different types of insulation materials

Type of insulation and thickness	Maximum heat flux (Wm^{-2})	Thermal conductivity ($Wm^{-1}K^{-1}$)	References
Glass wool (25 mm)	1000	0.04	Nahar, 2003
Coconut coir (50 mm)	1000	0.074	Yandoh et al., 2007
Polyurethane (30 mm)	1140	0.040	Chuawittayawuth and Kumar, 2002
Polystyrene (15 mm)	880	0.030	Abdullah and Abou-Ziyan, 2003

9.2.2 Effect of pH on Biohythane Production

Among all the chemical factors influencing dark fermentation, pH is considered the most influential. It influences the stability of the acid-producing fermentative bacteria and acetoclastic methane-producing bacteria. It plays a major role in the oxidation-reduction potential of the anaerobic process. Thus, it directly impacts the metabolic pathway. As fermentation takes place inside the reactor, the pH of the system changes. This results in variation of the enzymatic machinery. In an archetypal dark anaerobic process, hydrogen is produced only during the exponential growth phase, that is, it's a growth-associated product. In the stationary phase, the accumulated VFAs were converted to solvents via a process called solventogenesis.

The hydrogen production is governed by the initial pH of the medium. In most of the literature reports, a pH of 5.8 has been considered the optimum pH for hydrogen evolution. The redox potential of hydrogen could be given as

$$
\left.
\begin{aligned}
E &= E_0 + \frac{RT}{2F}\ln\frac{\left[H^+\right]^2}{P_{H_2}} = \frac{2.303RT}{F}\text{pH} - \frac{RT}{2F}\ln P_{H_2} \\
&= 0 + \frac{RT}{2F}\left[\ln\left(H^+\right)^2 - \ln P_{H_2}\right] \\
&= 0 + \frac{2RT}{2F}\ln\left[H^+\right] - \frac{RT}{2F}\ln P_{H_2} \\
&= 0 + \frac{2.303\,RT}{F}\log\left[H^+\right] - \frac{RT}{2F}\ln P_{H_2} \\
&= -\frac{2.303\,RT}{F}\times\text{pH} - \frac{RT}{2F}\ln P_{H_2} \\
&\therefore \frac{2.303\,RT}{F} = 0.0592\text{ V at }25^\circ\text{C} \\
&\therefore \text{At pH} = 6 \text{ and } P_{H_2} = 1 \quad \therefore \ln 1 = 0 \\
E &= -0.0592\times 6 - 0 \\
E &= -0.3552 = 0.3552\text{ V}
\end{aligned}
\right\} \quad (9.6)
$$

Carbon dioxide and hydrogen are the gaseous products produced by *E. aerogenes*. They are evolved at a ratio of 2:1. The total gas evolved is quite a few times the gas present in the overhead space. This leads to increment of the partial pressure of hydrogen (up to 0.33 atm). The redox potential of hydrogen (E) was found to be 0.340 V when a pH of 6.0 and a pressure of 0.3 atm were maintained. The correlation between redox potential and pH on hydrogen production is as follows: During hydrogen production, NADH is oxidized to NAD^+. This reaction is favored at a pH below 6. A pH above it might lead to shifting in the metabolic pathway toward solventogenesis. In the case of *E. cloacae,* this pH was found to be 4.8.

$$\text{NADH} + \text{H}^+ \leftrightarrow \text{NAD}^+ + \text{H}^+ \quad (9.7)$$

The enzymes complexes such as NADH:ferredoxin oxidoreductase, ferredoxin, and hydrogenase utilize NADH as an electron source for catalysis in much obligates such as *Clostridium* sp.

9.2.2.1 The role of redox potential of NAD on intracellular pH

A model organism such as *Escherichia coli* has an intracellular pH of 8 approximately. Under such conditions, the extracellular pH was varied from 5.5 to 9.0. A pH of 7 is considered most suitable for most of the biochemical reactions in many bacteria. However, a few bacteria favor a slightly alkaline pH inside the cell.

The redox potential of NAD at pH 8.00 could be calculated as

$$\left.\begin{aligned} E &= E_0 + \frac{RT}{2F}\ln\frac{\left[\mathrm{NAD^+}\right]\left[\mathrm{H^+}\right]}{\left[\mathrm{NADH}\right]} \\ &= E_0 - \frac{2.303\,RT}{2F}\mathrm{pH} + \frac{RT}{2F}\ln\frac{\left[\mathrm{NAD^+}\right]}{\left[\mathrm{NADH}\right]} \end{aligned}\right\} \tag{9.8}$$

where E_0 is the midpoint potential ([NADH] = [NAD^+]) at pH 0 and 25°C. Since the midpoint potential at pH 7 is –320 V, E_0 is postulated as –0.113 V. When the intracellular pH value is 8, a midpoint potential of –0.349 V was observed. This value of midpoint potential was found to be higher than the midpoint potential of hydrogen (–0.474 V). Moreover, under a pH of 6.0 and a pressure of 0.3 atm, the E value was lower than the potential of hydrogen (–0.340 V). Depending on the operational pH, the membrane-bound hydrogenase in *E. coli* could take part for either hydrogen production or its uptake activity. The optimum pH for hydrogen evolution was 6.5, while that for uptake was 8.5. The enzyme is a homodimer with each subunit having a molecular weight of 113,000 Da.

The importance of these results lies in the fact that the optimum pH of hydrogen evolution was same as the optimum culture pH of *E. aerogenes* for hydrogen production, and this was also the same as the internal pH of the cells. The hydrogenase enzyme is a membrane-bound quaternary protein. Each subunit of hydrogenase enzyme has an active site. Moreover, at different pH values ranging from 5.5 to 6.5, the two active sites might be facing a different side of the membrane, that is, one active site might be interacting with NADH on the cytoplasmic side at pH 8, and the other might be interacting with protons on the periplasmic space.

Considering all the assumptions mentioned above, hydrogen production could be related with culture pH and NADH concentration as

$$\left.\begin{aligned}
\Delta E &= -\frac{2.303\,RT}{F}\text{pH} - \frac{RT}{2F}\,\text{In}\,\text{P}_{\text{H}_2} - E_{\text{o}} - \frac{2.303\,RT}{2F}\text{pH} + \frac{RT}{2F}\,\text{In}\,\frac{[\text{NAD}^+]}{[\text{NADH}]} \\
&= -\frac{2.303\,RT}{F}(0.5\text{pH}) - \frac{RT}{2F}\,\text{In}\,\text{P}_{\text{H}_2} = -59.2\,\text{pH} - 113 - 29.6\,\text{pH} \\
&= -0.113V - \frac{2.303\,RT}{F}\text{pH} - \frac{2.303RT}{2F}\text{pH} \\
&= -0.113V - \text{pH}\left(\frac{2.303\,RT}{F} - \frac{2.303\,RT}{2F}\right) \\
&= -0.113V - \text{pH}\left(\frac{4x(a) - (2a)}{4F}\right) \\
&= -0.113 - \text{pH}\,x\,.02996
\end{aligned}\right\} \tag{9.9}$$

When P_{H2} = 0.1 atm (under partial vacuum around the cells and NADH/NAD$^+$ = 20 within the cells, the potential differences are

$$\nabla = -59.2\ (\text{pH } 7.1)\ (\text{at } P_{H2} = 0.1 \text{ atm, NADH/NAD}^+ = 20) \tag{9.10}$$

From the above observations, it could be said that the potential difference becomes larger as the culture pH becomes smaller, and larger the potential difference, the easier the hydrogen evolution. But this hydrogen evolution decreases if the culture pH becomes smaller than the certain minimum value (e.g., 5.5 in this case); this could be explained by taking into account ATP productivity of the cells.

By considering that the ATP yield (Y_{ATP}) is 10.0 g dry cells per mole of ATP, the ATP productivity (P_{ATP}) of the cell is described from the productivity of cell mass, as a function of the culture pH:

$$P_{ATP} = 3.8(\text{pH} - 4.4) \text{ for pH} < 7.0 \tag{9.11}$$

Many facultative and obligate anaerobes metabolize glucose through the Embden–Meyerhof–Parnas (EMP) pathway to generate ATP. In this pathway, NAD$^+$ is reduced to NADH in proportion to the production of ATP. Therefore, hydrogen production activity could be mathematically represented as

$$\int a\ \alpha(\Delta E)(P_{\text{ATP}}) \tag{9.12}$$

$$\int a = -52\,(\text{pH} - 5.75)^2 + 100 \tag{9.13}$$

This equation shows the relationship between hydrogen evolution activity and pH. In the case of dark-fermentative hydrogen production, the pH falls below 4.5 (Khanal et al., 2004). Another study shows that when the intracellular short-chain fatty acid concentration exceeds 440 mM, the solventogenesis process gets induced. In support of this fact, this switching to solventogenesis could be prevented if the fermentation is performed under controlled pH. In view of this, maintenance of controlled extracellular pH and its effect on hydrogen production and glucose uptake was studied in *E. cloacae* IIT-BT 08 (Khanna et al., 2011)

This study showed that on controlling the initial pH, the hydrogen yield as well as biomass increased significantly. Moreover, substrate conversion efficiency also improved (Fig. 9.2a,b). Short-chain fatty acids such as acetate, butyrate, propionate, etc., and solvents such as ethanol are a concomitant of the fermentative hydrogen production process. Accumulation of VFAs causes a drop in pH of the extracellular medium compared to the initial pH of the medium. Butyrate and acetate are two main products that are produced during the fermentation reaction. The accumulated VFAs are the feedstock to acetoclastic methanogens. Before channelizing the spent media toward the methanation process, the pH of the spent media needs to be raised to 7 to 8.2. Biomethanogenesis is favored at alkaline pH. Acidogens exhibit maximum activity at pH 5.5–6.5, while methanogens exhibit maximum activity at pH 7.8–8.2. Thus anaerobic digesters are usually maintained in the range of 7–8 to enhance methanogenesis. At a pH higher than 8, dissociation of NH_4^+ to neutral NH_3 occurs, which inhibits the growth of methanogens (Hansen et al., 1998). An important marker of pH persistence in anaerobic digesters is alkalinity. Usually, pH fluctuations caused by the generation of VFAs and carbon dioxide are buffered by bicarbonate alkalinity buffers at a pH close to neutral. A stable ADP is characterized by the bicarbonate alkalinity in the range of 1000 to 5000 mg L^{-1} as $CaCO_3$. The ratio between VFAs and alkalinity should be in the range of 0.1–0.25. A further increase of the VFA-to-alkalinity ratio indicates possible process deterioration and requires the organic loading rate (OLR) to decrease in order to lower the VFA formation rate.

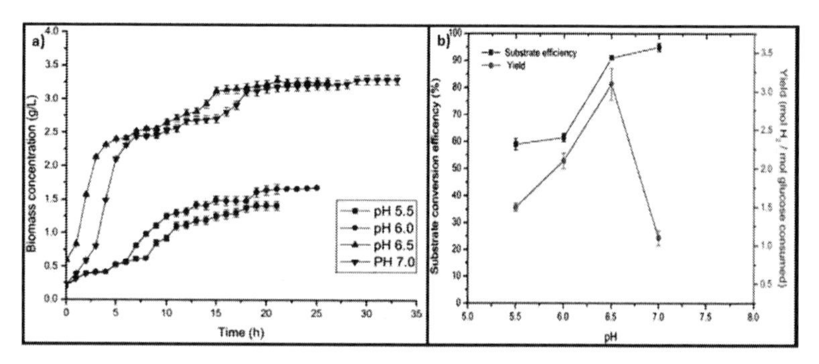

Figure 9.2 (a) Variation in biomass production in response to different regulated pH values. Medium used, MYG; working volume, 2 L; operational pH, 5.5–7.0; temperature, 37°C; agitation speed, 200 rpm; bar represents the SD for $n = 3$. (b) Variation in substrate conversion efficiency in response to regulated pH. Medium used, MYG; working volume, 2 L; operational pH, 5.5–7.5; temperature, 37°C; agitation speed, 200 rpm; bar represents the SD for $n = 3$.

9.2.3 Effect of the Inoculum on Biohythane Production

The reported studies on dark-fermentative hydrogen production mostly use simple sugars/soluble fermentable sugars as substrates. With the dawn of the concept of converting organic wastes to energy, the need for the development of a mixed microbial consortium comprising a symbiotically allied group of bacteria came to existence. Developing an enriched inoculum is very important for obtaining the desired product. A single group of bacteria might not have all the hydrolytic enzymes required for hydrolysis of complex organic compounds like cellulose; however, several bacteria in an enriched mixed consortium have the ability to produce hydrolytic enzymes. These hydrolytic enzymes thus help in solubilization of complex carbohydrates present in the organic waste. The soluble fermentable sugars could be then utilized for hydrogen production. The natural microbial consortia comprise different microbes like hydrogen producers, hydrogen consumers, methanogens, acetogens, electrogens, etc. To select hydrogen-producing microbes amongst the mixed microbial population is regarded as an enrichment of the culture. In the enrichment process, an artificial selection procedure was applied that would selectively promote hydrogen-producing bacteria and eliminate hydrogen producers.

Various pretreatment processes had been explored for enrichment processes. Different pretreatment methods, that is, physical (heat, freezing/thawing, ultrasonic, ultraviolet) and chemical (acid, alkali, organic), were commonly used (Fang et al., 2002; Lin and Lay, 2004). In heat shock pretreatment, the non-spore-formers get eliminated. Most of the hydrogen-consuming microorganisms are non-spore-formers. Moreover, methanogens belong to non-spore-forming archaebacterial, which get eliminated with heat shock treatment. Adverse conditions such as high temperature, acidity, and alkalinity promote endospore formation in a certain group of bacteria which also include hydrogen-producing bacteria. Thus, when favorable conditions are provided, the endospores germinate and the hydrogen-forming bacteria dominate the system. Chemical pretreatment such as bromoethane sulfonate (BES) treatment also helps to inhibit methanogens. BES acts as a competitive inhibitor of coenzyme M, responsible for methane production. One of the advantages of the development of a functional microbial consortium is the presence of symbiotically associated microbes. These microbes might produce hydrolytic enzymes which were otherwise absent in the principle hydrogen-producing microorganisms.

The natural habitat, sources of microorganisms, has been discussed in the previous chapter (Chapter 2). The sources of inoculums such as sludge, leachate, hot springs, and cow dung slurry possess a plethora of microorganisms having vivid characteristics. When these sources are used as the inoculum, it may harbor either a pure microbe or a mixed/constructed consortia. Obligate anaerobes such as *Clostridium* sp. and facultative anaerobes such as *Enterobacter* sp. are well-known groups of microorganisms that are most used as inoculums for mesophilic dark-fermentative hydrogen production. The hydrogen-producing inoculum might thus consist of sporulating bacteria like *Bacillus* or *Clostridium* or nonsporulating bacteria, for example, *Enterobacter, Escherichia, Klebsiella* sp., etc. On the other hand, for thermophilic hydrogen production species of *Thermoanaerobacter* sp., *Thermoanaerobaterium* sp., *Caldicellulosiruptor* sp., and *Thermotogales* sp. were the commonly used microorganisms. The preparation of a microbial population which would be suitable for introduction in the final production stage may be regarded as inoculum development. During preparation of the inoculum the following things were taken into consideration:

- The revival of microorganisms from stock
- Successive subculturing to activate all the cellular machinery
- Optimization of the strength of the active inoculum so as to minimize the lag phase
- Quantification of active cells in the microbial population

The revival of cells from dormant stock leads to the development of seed culture. This seed culture is further subcultured for activation of the vital cellular machinery. A number of active cells to be introduced in the fresh media are regarded as inoculum strength. In pilot- and bench-scale operations, small reactors are used for the purpose of inoculum development. During inoculum development, a sequential subculturing technique was used to increase the volume of the inoculum to the desired level (Fig. 9.3). For typical dark-fermentative hydrogen production, a varied range of inoculum size of 0.5%–20% v/v has been used. This strategy increases the inoculum volume by 20–200-fold at each step of inoculum buildup. The source of a potential inoculum could be either isolate or mixed consortia. According to the temperature requirements, the inoculums are incubated (mesophiles or thermophiles).

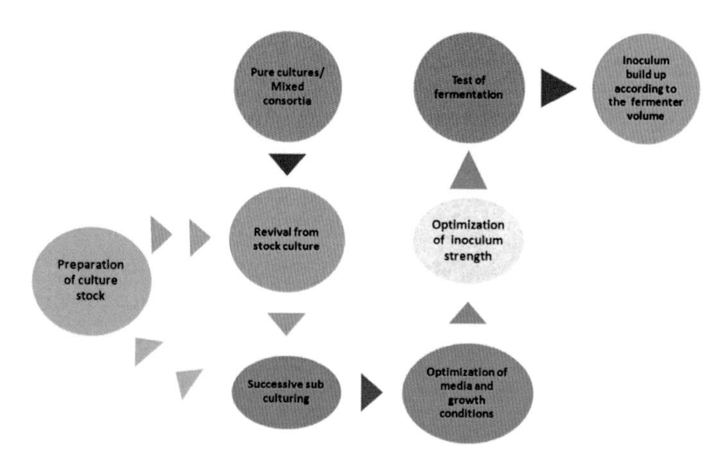

Figure 9.3 Strategy of inoculum builds up for biohythane production.

For optimum product yield, the physiological state of the inoculum and inoculum percentage are vital factors. Dark-fermentative hydrogen production is a growth-associated product; thus the infusion of a seed inoculum in the midlog phase would result in an expected improvement of yield. The inoculum volume varies from 5%–20% (v/v). This depends upon the characteristics of the species and medium used. Obligate anaerobes produce very less amount of biomass; thus, a larger inoculum strength is required for production. The inoculum age also matters during fermentation. Cells growing at the exponential phase have the entire enzymatic machinery active, which is required for hydrogen and methane production.

Properties of an enriched mixed consortium could be established with the help of the metagenomics approach (José and Thorsten, 2007). It is a novel genomics tool that could provide vital information on the presence of potential hydrogen producers in an enriched mixed consortium. Techniques such as ribotyping followed by denaturation gradient gel electrophoresis (DGGE) could help in mining the information regarding the microbial profile of the mixed culture (Fig. 9.4). Thus, using a suitably enriched culture could prove handy in utilizing complex organic wastes (proteins, lipids, complex cellulosic and starchy materials) for maximum energy extraction. On using the polymerase chain reaction (PCR)-DGGE technique, the alkali treatment shifted the microbial profile from non-hydrogen-producing acidogens to hydrogen-producing microbes (Kim and Shin, 2008).

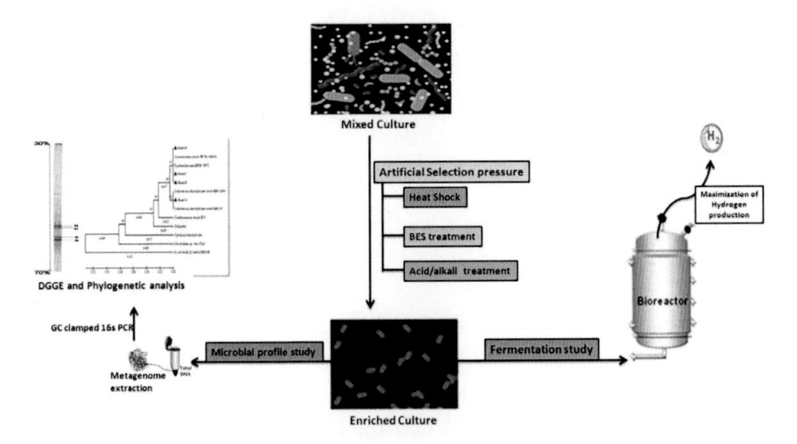

Figure 9.4 Conceptualization of enrichment process for maximization of hydrogen production and molecular approaches for microbial community analysis.

A thermophilic mixed culture was enriched from anaerobic digester sludge by using heat shock treatment and BES treatment, which comprised of long rod-shaped microorganisms dominated by *Thermoanaerobacterium* sp. (Roy et al., 2012). This enriched mixed culture could utilize different types of carbon sources (glucose, sucrose, L-arabinose, D-xylose, D-cellobiose, and starch) to produce hydrogen (Table 9.2), depicting the broad spectrum of the enzymatic machinery of the mixed consortia which helps in utilizing complex carbon sources like cellobiose and starch. Apart from inoculum development, the inoculum age is observed to be an equal influential factor for hydrogen production. Generally, for the production media an inoculum in its midlog phase is used for inoculation.

Table 9.2 Hydrogen production potential of enriched mixed consortia using different substrates

Carbon source	Cumulative hydrogen production ($mL\ L^{-1}$)	Volatile fatty acid concentration ($g\ L^{-1}$) Ethanol concentration	Acetate concentration	Butyrate concentration
Glucose	2285	0.22 ± 0.10	1.25 ± 0.34	1.76 ± 0.36
Fructose	1714	0.19 ± 0.04	0.56 ± 0.10	1.60 ± 0.42
L-arabinose	2142	0.20 ± 0.08	0.90 ± 0.10	1.82 ± 0.76
D-xylose	2860	0.28 ± 0.10	1.28 ± 0.12	1.88 ± 0.78
Sucrose	3482	0.4 ± 0.12	1.36 ± 0.20	2.30 ± 0.54
Cellobiose	2142	0.20 ± 0.10	1.20 ± 0.12	1.80 ± 0.48
Starch	1430	0.18 ± 0.04	0.43 ± 0.12	1.66 ± 0.32

9.2.4 Role of Medium Alkalinity in Biohythane Production

It is already known that for effective biohydrogen production, the optimal pH range is 5.0–6.8, and as the fermentation proceeds, the pH drops below this optimum range. If this drop in pH is compensated, a substantial improvement in hydrogen yields can be expected. The buffering capacity of the media is also regarded as alkalinity (Mohan et al., 2007), and its strength is directed by the presence of divalent ions like Ca^{2+} and Mg^{2+} or other ions such as phosphates, carbonates, citrates, etc.

In a fascinating study by Venkat Mohan et al. (2007) the importance of alkalinity in negating the effect of accumulated organic acids was reported. They showed that by maintaining the buffering capacity, the pH level was balanced inside the reactor, which helped in the improvement of hydrogen production (Fig. 9.5). By varying the organic rate from 2.4 kg COD m^{-3} d^{-1} the alkalinity concentration was varied from a maximum of 2900 mg L^{-1} to a minimum of 300 mg L^{-1}. During stable hydrogen production, the operation alkalinity values varied from a maximum of 1600 mg L^{-1} to a minimum of 125 mg L^{-1}.

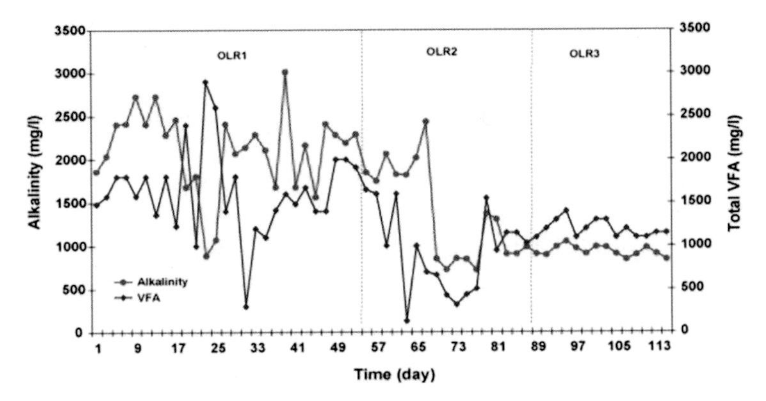

Figure 9.5 Variation of VFA and alkalinity during the total operation phase of the reactor (OLR_1, 2.4 kg COD/m^3-day; OLR_2, 3.5 kg COD/m^3-day; OLR_3, 4.7 kg COD/m^3-day).

Continuous hydrogen production was performed using a continuous stirred tank reactor (CSTR), where a buffered system was used instead of controlling the pH and a substantial influence of alkalinity was observed on hydrogen production (Shi et al., 2009). In this case, the maximum methane production rate was observed when the alkalinity was increased in the range from 500 mg L^{-1} to 1000 mg L^{-1}. Thus, it can be concluded that for steady hydrogen production, the range of alkalinity must be optimized. After hydrogen production is completed, the accumulated VFAs in the system are subsequently used for the biomethanation process. For this process, the ratio between VFAs and alkalinity should be in the range of 0.1–0.25. In general, for the biomethanation process, the alkalinity ranges from 1000 to 5000 mg L^{-1} as $CaCO_3$.

9.2.5 Effect of Hydraulic Retention Time on Biohydrogen Production

The total time that the cells and soluble nutrients reside in the reactor is called the HRT. Unlike methanogenesis, hydrogen production occurs at lower HRTs and is dependent on the volume of the reactor and the flow rate of the feed (HRT = Volume of reactor/Feed flow rate). For continuous hydrogen production, the HRT should be optimized as at a suitable HRT, the hydrogen production rates can be maximized, while at a very low HRT washout of the active cells might occur. This dictates that HRT optimization has a direct effect on the specific growth rate of the organism. During hydrogen production using mixed microflora comprising methanogens and acidogenic hydrogen producers, a shift in the microbial profile is observed with a change in the HRT. Lower HRTs lead to the enrichment of acidogenic hydrogen producers inside, while the methanogens are washed out (Lo et al., 2009). Thus, it was concluded that acidic pH (6–6.5) and a low HRT might totally suppress methanogens in mixed consortia. The HRT also plays a crucial role in the end-metabolite formation, which is concomitantly associated with a change in the microbial profile with respect to the HRT. When the HRT was decreased from 10 h to 6 h, the hydrogen production was increased in *Clostridium* sp. with a subsequent decrease in propionate concentration (Zhang et al., 2006). Similarly, when anaerobic sludge was used as the inoculum with an HRT of 24 h, hydrogen production increased with a concomitant increase in the butyrate/acetate (B/A) ratio (Mariakakis et al., 2012). Also no methane was observed in the total biogas produced. HRT studies helped in designing the experiment and reactor for treatment of industrial wastewater, where lower HRTs improved hydrogen production and chemical oxygen demand (COD) removal (Şentürk et al., 2013). Since the OLR is a function of the HRT, the OLR is also reflected as one of the important factors to be considered for continuous hydrogen production.

The HRT could also help in enrichment of microbial consortia since it directly affects the specific growth rate of bacteria. By manipulating the HRT slow-growing microbes like methanogens and other hydrogen-consuming microbes can be expelled out of the reactor, thus leading to selective enrichment of hydrogen-producing microbes (Zhang et al., 2006). This approach of using a short HRT for suppressing methanogens led to improvement in hydrogen production (Kim et al., 2004). This demonstrates that microbial consortia inside a bioreactor are very delicate toward a change in HRTs. In another study, lowering the HRT from 8 to 6 h prompted the increase in hydrogen production. In addition, propionate concentration of the soluble metabolite diminished with abatement in the HRT (Zhang et al., 2006). The issue with shorter HRTs is the danger of cell washout, which is otherwise called "bleeding" of the reactor. The mean cell residence time in a CSTR is the same as the HRT. This limits the hydrogen production at a shorter HRT. Of late it has been accounted that there is self-granulation or flocculation even in a CSTR, which results in a longer mean cell residence time. This decoupled mean cell residence time from the HRT prompts higher biomass at low HRTs (Fang et al., 2002). For keeping up a higher biomass concentration inside the reactor, immobilized/granulated reactors are more proficient. Hydrogen production by utilizing immobilized cells has several benefits compared to suspended cells (Azbar et al., 2009). The cell washout issue in a suspended cells reactor is overcome by utilizing immobilized cells, and in this manner a high rate of hydrogen production can be accomplished, even at short HRTs (Hoist et al., 1999). The increment in hydrogen production by an immobilized whole-cell reactor is because of higher substrate conversion effectiveness and an increment in the mean cell residence time (Roy et al., 2014). The common whole-cell immobilization techniques depend on adsorption or entrapment methods. The disadvantages of gel entrapment of cells are degradation of the gel during long-term operation and restriction in nutrient and metabolite mass transport. Despite what might be expected, natural adsorption of cells on the matrix is a straightforward and economical strategy. Such a system outlines lower internal mass transfer and is less expensive to implement (Rattanapan et al., 2011). Different reactor configurations are used for hydrogen production (Fig. 9.6).

Immobilized whole-cell systems are considered for hydrogen production in granular reactors, fixed-bed reactors, fluidized-bed reactors, and up-flow anaerobic sludge blanket (UASB) reactors (O-Thong et al., 2008). There are a few impediments in utilizing UASB reactors, for example, low stability of hydrogen-framing granules and mass transfer resistance.

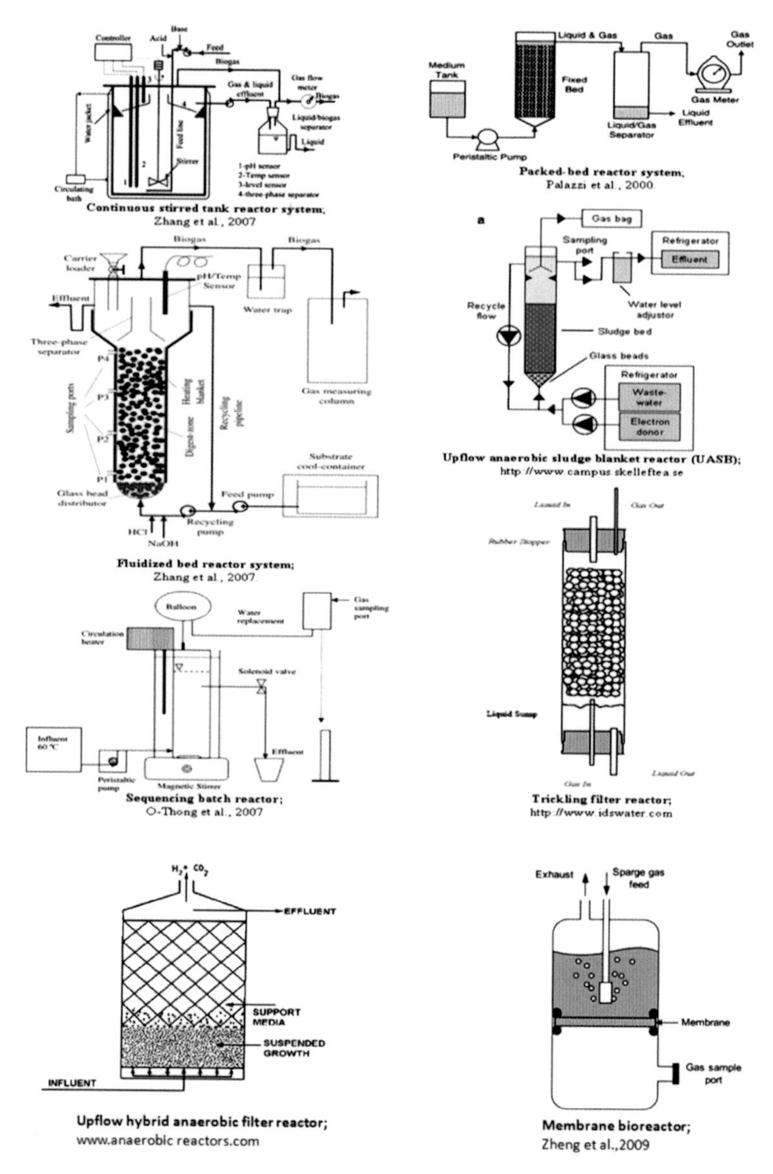

Figure 9.6 Different bioreactor configurations that could be considered for biohythane production.

During methanogenesis, the HRT should be kept twofold greater than the generation time of the slowest-growing microbes (Dohanyos and Zabranska, 2001). The HRT should be held for a suitable duration so that the dead zones get eliminated, and it would also help in promoting an efficient syntrophy amongst the microorganisms present in the mixed culture.

9.2.6 Effect of Metal Ions on Biohythane Production

Biohydrogen and biomethane production requires various types of metal ions as micronutrients. These metal ions play a critical role in the metabolism of microorganisms. Metal ions such as Fe^{+2}, Zn^{+2}, Ni^{+2}, Na^{+1}, Mg^{+2}, and Co^{+2} play a pivotal role in both hydrogen and CH_4 production. These micronutrients might be required in trace amounts but they have an influential role as cofactors, transport processes facilitators, and structural skeletons of many enzymes (Fe-Fe hydrogenase, Ni-Fe hydrogenase, etc.).

9.2.6.1 Effect of iron and nickel on fermentative hydrogen production

Iron plays a vital role in dark fermentation. Mainly in fermentative hydrogen production, Fe is a major component of the hydrogenase enzyme. The hydrogenase enzyme has two categories, that is, either a metallic iron–iron center in [FeFe] hydrogenase or a nickel–iron center in [NiFe] hydrogenase. These enzymes catalyze the reaction, which leads to the formation of molecular hydrogen from the proton. This enzyme cannot directly transfer an electron to the proton (H^+ ion). It needs a mediator known as ferredoxin. Ferredoxin is involved in pyruvate oxidation to acetyl-coenzyme A, CO_2, and hydrogen. As is clear from the name, ferredoxin contains an iron–sulfur cluster. The improvement of hydrogen production on supplementing media with Fe^{2+} and its effect on expression of both Fe–S and non-Fe–S proteins components of the hydrogenase enzyme was first observed by Vanacova et al. (2001). This observation has led many researchers to explore the effect of Fe^{2+} on hydrogen production in wild-type and hydrogenase-overexpressed systems. In one such report, the [NiFe] hydrogenase was homologous overexpressed and further improvement of hydrogen production was observed with Fe^{2+} ion and Ni^{2+} ion supplementation (Kim et al., 2010). In another study, hydrogen production was studied using an obligate anaerobic microorganism *Clostridium saccharoperbutylacetonicum* ATCC 27021 where the media was supplemented with Fe^{2+} ions (Ferchichi et al., 2005). Some report suggested that supplementation with Fe^{2+} ions leads to a shift in the metabolic profile. It was observed that supplementation with a Fe^{2+} ion concentration of 12 mg L^{-1} showed a metabolic shift from lactic acid fermentation to butyric acid fermentation. Iron also plays a critical role in the biomethanation process. Most members of the order *Methanosarcinales* are known to contain cytochromes b and c and other proteins with protoheme-derived prosthetic groups (Kamlage and Blaut, 1992; Kuhn and Gottschalk, 1983). The iron is also required by methanogenic archaea like *Methanosarcina barkeri* to synthesize protoheme via precorrin-2, which is formed from uroporphyrinogen III in two consecutive methylation reactions utilizing S-adenosyl-L-methionine (Buchenau et al., 2006).

In biohythane production, methanogenesis (second stage) also involves two major pathways:

- Reduction of carbon dioxide
- Fermentation of acetate

On investigating the above two pathways, it was observed that the demethylation of methyl-coenzyme M to methane was the common reaction. In addition, methyl-coenzyme M and heterodisulfide reductases reduce the heterodisulfide of coenzyme M and coenzyme B. The above observation indicates that the methanogenesis process requires the involvement of various enzymes, novel cofactors, and metal ions. Nickel is an essential metal which plays a critical role in functioning of many enzyme systems that are responsible for methane production. Enzymes such as carbon monoxide dehydrogenase, hydrogenase, and methyl-coenzyme M reductases require nickel in their active site.

The majority of methane is produced by the conversion of the methyl group of acetate to methane. This process accounts for the origin of 70% of biologically produced methane. However, the presence of metabolic groups of anaerobes governs methane production via the methyl group of acetate or reduction of CO_2. The pathway for methane generation from acetate includes three major steps:

1. Activation of acetate to acetyl-CoA
2. Carbon–carbon and carbon–sulfur bond cleavage of acetyl-CoA

3. Methyl transfer to HS-CoM and reductive demethylation of CH_3-S-CoM to methane

In this pathway, acetate is activated to acetyl-CoA, followed by cleavage of the carbon–carbon and carbon–sulfur bonds (decarbonylation) catalyzed by the nickel/iron–sulfur component of the CO dehydrogenase (CODH) enzyme complex. The nickel/iron–sulfur component oxidizes the carbonyl group to CO_2 and reduces a ferredoxin. The methyl group is transferred to the corrinoid/iron–sulfur component within the complex and finally to coenzyme M (HS-CoM) catalyzed by at least two methyl transferases. The CH_3-S-CoM is reductively demethylated to methane with electrons derived from the sulfur atoms of CH_3-S-CoM and HS-HTP, which results in the formation of the heterodisulfide CoM-S-S-HTP. Reduction of the heterodisulfide to the corresponding sulfhydryl forms of the cofactors is accomplished with electrons derived from reduced ferredoxin.

Methane-producing bacteria like *Methanobacterium thermoautotrophicum*, *Methanobrevibacter smithii,* and *Methanosarcina barkeri* are reported to require nickel for growth as they contain a nickel-tetrapyrrole designated factor F430 (Dikert et al., 1981). F430 is the prosthetic group of the enzyme methyl-coenzyme M reductase. This enzyme catalyzes the release of methane in the final step of methanogenesis:

$$CH_3\text{-S-CoM} + \text{HS-CoB} \rightarrow CH_4 + \text{CoB-S-S-CoM} \quad (9.14)$$

F430 is the most reduced tetrapyrrole in nature with only five double bonds. This particular tetrapyrrole derivative is called a corphin. Because of its relative lack of conjugated unsaturation, it is yellow, not the intense purple-red associated with more unsaturated tetrapyrroles. It is also the only tetrapyrrole derivative found in nature to contain nickel.

9.2.6.2 Effect of magnesium on fermentative biohydrogen production

Magnesium is one of the most abundant elements within any microorganism. Magnesium is principally required by ribosomes. It also functions as a cofactor of many crucial enzymes such as kinases and synthetases. In glycolysis, many enzymes require magnesium ions as a cofactor. The activation of these enzymes (hexokinase, phosphofructokinases, glyceraldehyde-3-phosphate dehydrogenases, and enolases) helps bacteria to metabolize substrates and produce ATP (Wang et al., 2007).

9.2.6.3 Effect of other heavy metals on fermentative biohydrogen production

In certain industrial wastewaters and municipal sludge, substantial concentrations of heavy metals are present, which lead to upsetting of the wastewater treatment process (Stronach et al., 1986; Fang and Chan, 1997). These heavy metals have an inhibitory effect on metabolic processes, depending upon the concentrations at which they are present. It is well known that acidogenesis and methanogenesis of anaerobic processes are negatively affected by cobalt, chromium, copper, and zinc ions (Hickey et al., 1989). Very few reports are available on the effect of heavy metals on biohythane production. In one such report, the effect of six different heavy metals (that were used for electroplating) on hydrogen production was investigated (Li and Fang, 2007). An anaerobic sludge blanket reactor was used, which was fed with sucrose containing wastewater with varying concentrations of heavy metals. The order of toxicity was Cu>Ni>Zn>Cr>Cd>Pb (least toxic) for hydrogen.

On the contrary, in the biomethanation process, metals like cobalt (Co) or zinc (Zn) play a stimulatory effect. Other than nickel-tetrapyrrole coenzyme F430, *Methanosarcinales* also contain the cobalt tetrapyrrole hydroxy benzimidazolyl carbamide, which is a vitamin B12 homologue (Sauer et al., 1999). These tetrapyrroles are involved in different steps of methanogenesis. Few methane-producing organisms like *Methanosarcina barkeri* have been reported to contain carbonic anhydrase (Karrasch et al., 1989). Carbonic anhydrase (EC 4.2.1.1) is a zinc-containing enzyme that catalyzes the following reaction:

$$CO_2 + H_2O \rightarrow HCO_{3-} + H^+ \tag{9.15}$$

9.2.6.4 Effect nitrogen and phosphate on fermentative biohydrogen production

For stable fermentation, a carbon/nitrogen (C/N) ratio greater than 10 is recommended (Fig. 9.7). The dark-fermentative microorganisms showed improvement in hydrogen production when they were grown in a fermentation media having a C/N ratio greater than 20. Lower nitrogen content favors acidogenesis. However, the majority of studies on the effect of the C/N ratio on hydrogen production were observed in batch fermentation. No study so far is available on the role of the C/N ratio in a continuous mode of hydrogen production. The spent media generated after dark-fermentative hydrogen production still has a high amount of organic acids and unutilized nitrogen sources. The C/N ratio of such spent media generally ranges from 20 to 30 (FAO, 1996). There are pros and cons of the C/N ratio on methanogenesis. A very high or very low C/N ratio shows an adverse effect on methane production. If the C/N ratio of the system is very high, then it would lead to rapid utilization of nitrogen by methanogens for meeting their protein requirements. Thus, the leftover carbon would remain unutilized, leading to a decrease in methane production. Similarly, if the C/N ratio is very low then there would be an accumulation of ammonia in the system due to the liberation of nitrogen.

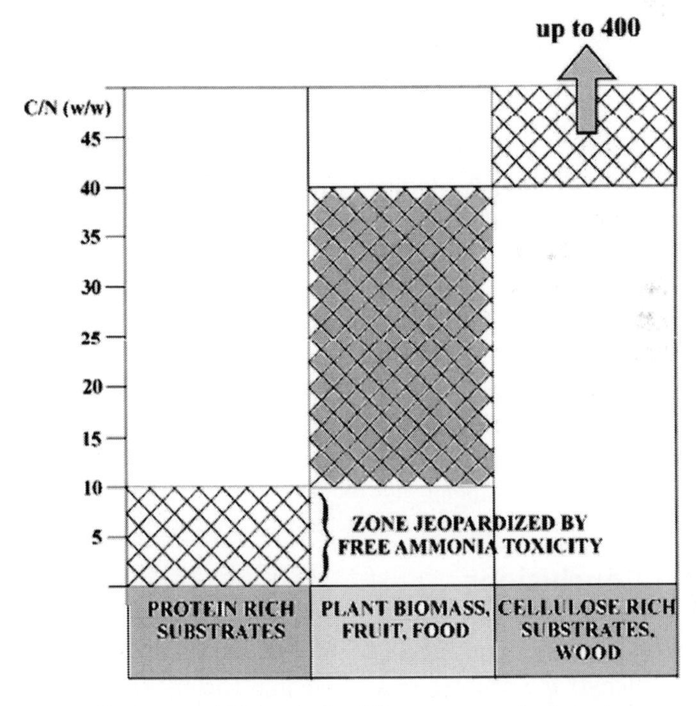

Figure 9.7 Variation of C/N ratio for different types of substrates.

This would lead to an increase in pH. A pH value greater than 8–8.5 is considered unfavorable for methanogenic bacteria (Fig. 9.8) (FAO, 1996). Thus for optimal second-stage methane production, feedstock (spent media) with a high C/N ratio could be mixed with that with a low C/N ratio to bring the average ratio of the composite input to a desirable level.

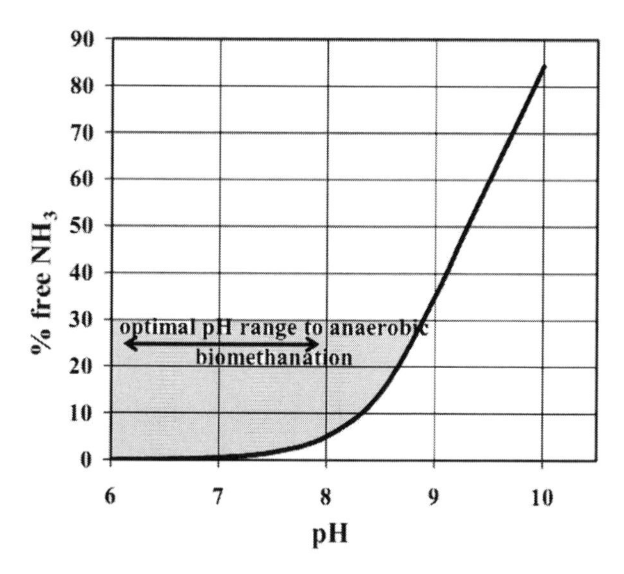

Figure 9.8 Percentage ratio of free nondissociated ammonia (in system NH^+/NH_3) at various pH values (Straka et al., 2007).

The requirement of phosphates is also important in dark fermentation. Phosphates not only help in maintaining buffered conditions during fermentation but also provide the building blocks of nucleic acids and ATPs. In dark fermentation, an increase in phosphate concentration leads to enhancement of the hydrogen production potential. No reports are available on the role of phosphates in methanogenesis.

9.3 Conclusions

Many bioprocess parameters play a crucial role in the operation of the biohythane process in a fermenter. Factors such as mixing characteristics, manipulation of overhead gas partial pressure,

HRT, and recycle ratio influence the process significantly. Chemical parameters such as temperature, pH alkalinity, etc., also play a defining role in the performance of a reactor, which are the prerequisite of any microbial growth. They influence the biochemical reactions that are responsible for maintaining homeostasis and metabolism. Many metal ions also found to be important for the biohythane process, viz., Fe^{+2}, Zn^{+2}, Ni^{+2}, Na^{+1}, Mg^{+2}, and Co^{+2}. Nitrogen (in terms of the C/N ratio) and phosphates are required as building blocks of important cellular components and biomass production. Moreover, the study of the effect of heavy metals on the biohythane process reveals the robustness of the process.

References

Abdullah AH, Abou-Ziyan HZ (2003). Thermal performance of flat plate solar collector using various arrangements of compound honeycomb, *Energy Convers Manage*, **44**, 3093–3112.

Azbar N, Çetinkaya Dokgöz FT, Keskin T, Korkmaz KS, Syed HM (2009). Continuous fermentative hydrogen production from cheese whey wastewater under thermophilic anaerobic conditions, *Int J Hydrogen Energy*, **34**, 7441–7447.

Buchenau B, Kahnt J, Heinemann IU, Jahn D, Thauer RK (2006). Heme biosynthesis in *Methanosarcina barkeri* via a pathway involving two methylation reactions, *J Bacteriol*, **188**, 8666–8668.

Chuawittayawuth K, Kumar K (2002). Experimental investigation of temperature and flow distribution in a thermosiphon solar water heating system, *Renew Energy*, **26**, 431–448.

Diekert G, Konheiser U, Piechulla K, Thauer RK (1981). Nickel requirement and factor F430 content of methanogenic bacteria, *J Bacteriol*, **148**(2), 459–464.

Dohányos M, Zábranská J (2001). Anaerobic digestion. In *Sludge Into Biosolids: Processing, Disposal, and Utilization*. IWA Publishing. Chapter 13, 223–241.

Egorova K, Antranikian G (2005). Industrial relevance of thermophilic Archaea, *Curr Opin Microbiol*, **8**, 649–655.

Fabiano B, Perego P (2002). Thermodynamic study and optimization of hydrogen production by *Enterobacter aerogenes*, *Int J Hydrogen Energy*, **27**, 149–156.

Fang HH, Liu H, Zhang T (2002). Characterization of a hydrogen-producing granular sludge, *Biotechnol Bioeng*, **78**, 44–52.

Fang HHP, Chan OC (1997). Toxicity of electroplating metals on benzoate-degrading granules. *Environ Technol*, **18**, 93–99.

FAO (1996). Report on the Meeting for the Development of a National Biogas Policy Framework and Celebration of the 10,000th Biogas Plant Construction with BSP Support. http://www.fao.org/sd/EGdirect/EGre0022.htm.

Ferchichi M, Crabbe E, Gil GH, Hintz W, Almadidy A (2005). Influence of initial pH on hydrogen production from cheese whey, *J Biotechnol*, **120**, 402–409.

Hansen KH, Angelidaki I, Ahring BK (1998). Anaerobic digestion of swine manure: inhibition by ammonia, *Water Res*, **32**, 5–12.

Hickey RF, Vanderwielon J, Switzenbaum, MS (1989). The effect of heavy metals on methane production and hydrogen and carbon monoxide level during batch anaerobic sludge digestion, *Water Res*, **23**, 207–218.

Hoist O, Manelius A, Krahe M, Miirkl H, Rawen N, Sharp R (1997). Thermophiles and fermentation technology, *Camp Biochem Physiol*, **118A**(3), 415–422.

José LS, Thorsten K (2007). Molecular biology techniques used in wastewater treatment: an overview, *Proc Biochem*, **42**, 119–133.

Kamlage B, Blaut, M (1992). Characterization of cytochromes from *Methanosarcina* strain Göl and their involvement in electron transport during growth on methanol, *J Bacteriol*, **174**(12), 3921–3927.

Khanal SK, Chen WH, Sung LLS (2004). Biological hydrogen production: effects of pH and intermediate products, *Int J Hydrogen Energy*, **29**, 1123–1131.

Khanna N, Kotay SM, Gilbert JJ, Das D (2011). Improvement of biohydrogen production by *Enterobacter cloacae* IIT-BT 08 under regulated pH, *J Biotechnol*, **152**, 15–30.

Kim JYH, Jo BH, Cha HJ (2010). Production of biohydrogen by recombinant expression of [NiFe]-hydrogenase 1 in *Escherichia coli*, *Microb Cell Fact*, **9**, 1–10.

Kim SH, Han SK, Shin HS (2004). Feasibility of biohydrogen production by anaerobic co-digestion of food waste and sewage sludge, *Int J Hydrogen Energy*, **29**, 1607–1616.

Kim SH, Shin HS (2008). Effects of base-pretreatment on continuous enriched culture for hydrogen production from food waste, *Int J Hydrogen Energy*, **33**, 5266–5274.

Koskinen PEP, Lay CH, Puhakka JA, Lin PJ, Wu SY, Örlygsson J (2008). High efficiency hydrogen production by an anaerobic, thermophilic enrichment culture from Icelandic hot spring, *Biotechnol Bioeng*, **101**, 665–678.

Kühn W, Gottschalk G (1983). Characterization of the cytochromes occurring in *Methanosarcina* species, *Euro J Biochem*, **135**(1), 89–94.

Lee KS, Lo YC, Lin PJ, Chang JS (2006). Improving biohydrogen production in a carrier-induced granular sludge bed by altering physical configuration and agitation pattern of the bioreactor, *Int J Hydrogen Energy*, **31**, 1648–1657.

Li CL, Fang HHP (2007). Fermentative hydrogen production from wastewater and solid wastes by mixed cultures, *Crit Rev Environ Sci Technol*, **37**, 1–39.

Lin CY, Lay CH (2005). A nutrient formation for fermentative hydrogen production using anaerobic sewage sludge microflora, *Int J Hydrogen Energy*, **30**, 285–292.

Lo YC, Su YC, Chen CY, Chen WM, Chang JS (2009). Biohydrogen production from cellulosic hydrolysate produced via temperature-shift-enhanced bacterial cellulose hydrolysis, *Bioresour Technol*, **100**, 5802–5807.

Mariakakis I, Meyer C, Steinmetz H (2012). Fermentative hydrogen production by molasses; effect of hydraulic retention time, organic loading rate and microbial dynamics. In: Minic D, (ed), *Hydrogen Energy - Challenges and Perspectives*, InTech.

Nahar NM (2003). Year round performance and potential of a natural circulation type of solar water heater in India, *Energy Buildings*, **35**, 239–247.

Nath K, Kumar A, Das D (2006). Effect of some environmental parameters on fermentative hydrogen production by *Enterobacter cloacae* DM11, *Can J Microbiol*, **52**, 525–532.

O-Thong S, Prasertsan P, Karakashev D, Angelidaki I (2008). High-rate continuous hydrogen production by *Thermoanaerobacterium thermosaccharolyticum* PSU-2 immobilized on heat-pretreated methanogenic granules, *Int J Hydrogen Energy*, **33**, 6498–6508.

Rattanapan A, Limtong S, Phisalaphong M (2011). Ethanol production by repeated batch and continuous fermentations of blackstrap molasses using immobilized yeast cells on thin-shell silk cocoons, *Appl Energy*, **88**, 4400–4404.

Roy S, Ghosh S, Das D (2012). Improvement of hydrogen production with thermophilic mixed culture from rice spent wash of distillery industry, *Int J Hydrogen Energy*, **37**(21), 5867–15874.

Roy S, Vishnuvardhan M, Das D (2014). Improvement of hydrogen production by newly isolated *Thermoanaerobacterium thermosaccharolyticum* IIT BT-ST1, *Int J Hydrogen Energy*, **39**(14), 7541–7552.

Sauer K, Thauer RK, Banerjee R (1999). The role of corrinoids in methanogenesis, *Chem Biochem B*, **12**, 655–679.

Şentürk E, İnce M, Engin GO (2013). The effect of transient loading on the performance of a mesophilic anaerobic contact reactor at constant feed strength, *J Biotechnol*, **164**, 232–237.

Slininger PJ, Bothast RJ, Ladisch MR, Okos MR (1990). Optimum pH and temperature conditions for xylose fermentation by *Pichia stipitis*, *Biotechnol Bioeng*, **35**, 727–731.

Speece RE (1996). *Anaerobic Biotechnology for Industrial Wastewaters. Archae Press*, Nashvillee, TN, USA.

Straka F, Jenicek P, Zabranska J, Dohanyos M, Kuncarova M (2007). Anaerobic fermentation of biomass and wastes with respect to sulfur and nitrogen contents in treated materials, In *Proceedings of the 11th International Waste Management and Landfill Symposium*.

Stronach SM, Rudd T, Lester JN (1986). Anaerobic digestion process. In *Industrial Wastewater Treatment*. Springer Verlag, Berlin.

Thong SO, Prasertsan P, Intrasungkha N, Dhamwichukorn S, Birkeland NK (2007). Improvement of biohydrogen production and treatment efficiency on palm oil mill effluent with nutrient supplementation at thermophilic condition using an anaerobic sequencing batch reactor, *Enz Microb Technol*, **41**, 583–590.

Ueno Y, Otsuka S, Morimoto M (1996). Hydrogen production from industrial wastewater by anaerobic microflora in chemostat culture, *J Fermen Bioeng*, **82**, 194–197.

van de Werken, Harmen JG, Verhaart Marcel RA, Van Fossen, Amy L, Willquist Karin, Lewis, Derrick (2008). Hydrogenomics of the extremely thermophilic bacterium *Caldicellulosiruptor saccharolyticus*, *Appl Environ Microbiol*, **74**(21), 6720–6729.

Van Groenestijn JW, Hazewinkel JHO, Nienoord M, Bussmann PJT (2002). Energy aspects of biological hydrogen production in high rate bioreactors operated in the thermophilic temperature range, *Int J Hydrogen Energy*, **27**, 1141–1147.

Vanacova S, Rasoloson D, Razga J, Hrdy I, Kulda J, Tachezy J (2001). Iron-induced changes in pyruvate metabolism of *Tritrichomonas foetus* and involvement of iron in expression of hydrogenosomal proteins, *Microbiology*, **147**, 53–62.

Veeken A, Hamelers B (1999). Effect of temperature on hydrolysis rates of selected biowaste components, *Bioresour Technol*, **69**, 249–254.

Venkata Mohan S, Lalit Babu V, Sarma PN (2007). Anaerobic biohydrogen production from dairy wastewater treatment in sequencing batch reactor (AnSBR): effect of organic loading rate, *Enz Microb Technol*, **41**, 506–515.

Vindis P (2009). The impact of mesophilic and thermophilic anaerobic digestion on biogas production, *Int J Hydrogen Energy*, **36**, 192–198.

Wang, XJ, Ren NQ, Xiang, WS, Guo WQ (2007). Influence of gaseous end-products inhibition and nutrient limitations on the growth and hydrogen production by hydrogen producing fermentative bacterial B49, *Int J Hydrogen Energy*, **33**, 1153–1163.

Yandoh H, Gbaha P, Toure S (2007). Experimental study on the comparative thermal performance of a solar collector using coconut coir over the glass wool thermal insulation for water heating system, *J Appl Sci*, **7**(21), 3187–3197.

Yokoyama H, Waki M, Ogino A, Ohmori H, Tanaka Y (2007). Hydrogen fermentation properties of undiluted cow dung, *J Biosci Bioeng*, **104**, 82–85.

Zeidan AA, van Niel, WJ (2010). A quantitative analysis of hydrogen production efficiency of the extreme thermophile *Caldicellulosiruptor owensensis* OLT, *Int J Hydrogen Energy*, **35**, 1128–1137.

Zhang ZP, Show KY, Tay JH, Liang DT, Lee DJ, Jiang WJ (2006). Effect of hydraulic retention time on biohydrogen production and anaerobic microbial community, *Proc Biochem*, **41**, 2118–2123.

Chapter 10

Policy of Biohythane Technology

10.1 Environmental Impact

Air quality has been deteriorating for a long time now and is a major source of concern all over the world. It is estimated that in the United States, around 50% of the population lives in areas which are highly polluted by at least one pollutant (Kojima and Lovei, 2001). These pollution levels are potent enough to affect the environment and human health. The major contributors to the higher pollution levels in air are power plants and commercial vehicles. Strategies are being developed today by different countries to reach acceptable levels of air pollution and improve the air quality by focusing on metro cities and making sure that proper conduct and air control practices are being followed; this has been done to reinforce the Clean Air Act. The improvement of commercial buses by using biohythane as a fuel is one such step for improving the air quality.

New renewable energy technologies are generally more costly than fossil fuels like wind energy. This has become cost-competitive over time. However, there are numerous advantages associated with renewable energy resources compared to conventional nonrenewable fuels. A few benefits of biohythane are listed below (Sims et al., 2003):

Biohythane: Fuel for the Future
Debabrata Das and Shantonu Roy

ISBN 978-981-4745-29-1 (Hardcover), 978-981-4745-30-7 (eBook)
www.panstanford.com

- Reduced greenhouse gas (GHG) emissions
- Stimulation of rural economies
- Less dependence on declining fossil fuel supplies
- Better energy security (through reduced dependence on imported energy)
- Potential reduction in air pollutant emissions
- Improved water quality through better manure management

To scale up the production of biohythane, it is important for the government to provide support for the development of the industry. The present chapter discusses various environmental policy drivers, some of which could be used to promote the biohythane industry. It also focuses on specific government policies and incentives in three areas related to the use of biogas and biohythane: renewable energy (electricity), alternative vehicle fuels, and programs that could be tapped for financial support. Finally, it discusses why public support of this industry is not only necessary but also justified. If the biohythane industry is to prosper, the government must launch policy initiatives to give direct financial incentives or tax credits to the biohythane industry similar to renewable electricity, ethanol, and biodiesel.

Recent research on how human health is affected by air pollution has revealed that lungs suffer serious damage from the combustion of fossil fuels. It has been speculated that substituting conventional fossil fuel sources with biohythane will result in improved physical health (Zweig, 2009). One of the major advantages of combustion of biohythane is that the process is not accompanied by release of volatile organic compounds (VOCs), SO_2, CO_2, CO, and particulate matter (PM). However, it does release NO_x and vapor into the atmosphere, among which emission of NO_x depends majorly on flame temperature and duration. One of the significant properties of biohythane that can be used to curtail NO_x emissions is the wide flammability region that allows us to modify the engine design so that less NO_x is released during its combustion (Hordeski, 2011). If the current circumstances and CO_2 impact on the atmosphere are taken into consideration, it seems inevitable that biohythane would be introduced as a source of fuel. The introduction of biohythane would dramatically affect the carbon dioxide content in the atmosphere, which would reach a maximum of 520 ppm before the year 2050. The total economic activity and energy consumption would be higher in comparison to the no-biohythane scenario but much lower than in the case if biohythane had been introduced in the year 2000. The net carbon dioxide would continue to increase until approximately 2070, reaching 620 ppm (Barbir et al., 1993). If this transition starts at 2050 there would be almost no positive effects. The sooner we incorporate the use of a solar biohythane energy system in our economy, the more beneficial it would be for overall sustainable development (Barbir et al., 1993).

In view of this, the automotive industry has seen a significant change in the last few years, where companies are now targeting more energy-efficient and environment-friendly vehicles. Taking the detrimental effects of pollution into consideration, the use of biohythane is being promoted in several countries for common public transport and other vehicular transportation. A detailed analysis of the overall energy efficiency has revealed that the fuel cell drive for city buses offers significant environmental improvements compared to diesel internal combustion (IC) engines. This study takes into consideration both local emissions of trace gases as well as emissions of GHGs. Given the global energy crisis today, the problem of global warming can be tackled only if we are able to commercialize renewable energy systems and offer more tax waivers and other economic benefits (Wurster et al., 1998).

A major advantage of fuel cell vehicles (and all fuel cells) is that they represent an inherently clean, efficient, and quiet technology and can optimize use of fuels from environmentally benign energy sources and feedstocks such as solar energy, wind, geothermal energy, and biomass. The calculated CO_2 emissions for an average car with electrical propulsion as well as with different biohythane technologies are presented in this chapter (Jorgensen, 1998).

10.2 Biohythane Policy

In accordance with the reduced emissions target set up by the Kyoto Protocol, many research groups have performed studies that have focused on biohythane as a means of meeting the demand for clean energy. The biohythane policy is covered under the energy policy for most countries, owing to its numerous advantages. The dairy manure management system and biogas production are influenced by the environmental policy of the country. Environmental emissions contributing to global warming (as a result of methane), ozone depletion (VOCs), etc., are profoundly associated with the release of biogas into the atmosphere. Ammonia, which is a precursor for PM, may also be released from undigested dairy wastes. Public policies are gradually shifting to address emissions from dairy biogas. Public agencies can respond to concerns over dairy gas emissions in the same manner that they respond to other emissions of environmental concern (Gerber et al., 2010):

- Regulate the criteria for air pollutants and GHG emissions.
- Control and reduce emissions through market incentives such as a carbon-trading market or an emission reduction credit (ERC) market.
- Develop and promote technologies that assist the dairies to directly or indirectly reduce their emissions. This might also include setting up subsidies to dairies to help them reduce the creation of biogas or limit its release into the environment.

Another alternative approach that can be taken up is that the dairies can capture the naturally occurring biogas or enhance its production and storage by improved engineering of the system. The stored biogas can then be flared, combusted to generate heat or electricity, or upgraded into pipeline-quality gas through further downstream processing (biohythane) for use in vehicles or other applications.

10.3 Environmental Regulation

The regulation of air quality has been addressed by federal and state policies. However, application and extension of these policies to agricultural activities such as dairy farming has been more or less nonexistent or highly minimalistic. Vehicular emissions have been under the radar for stringent control, especially in the past decade. However, the control strategy has gradually shifted to the use of biofuels such as ethanol and biodiesel to help control emissions. Regulatory requirements related to vehicle emissions could impact the alternative fuel industry.

For smooth transition to a biohythane-based economy, energy policy matters need to be discussed. It calls for strong cooperation between the government and industry. Energy policy attributes to production, consumption (efficiency and emission standards), taxation and other public policy techniques, energy-related research and development, energy economy, general international trade agreements and marketing, energy diversity, and risk factors contrary to the possible energy crisis. There are two key areas of focus, (1) research, development, and demonstration of biohythane technologies by industries and (2) incentives to encourage investment in biohythane infrastructure by the government. The entire plan can be broken down into four significant phases: the technology development phase, followed by establishment of an initial market, the infrastructure investment phase, and a completely developed open market.

The US Department of Energy (DOE) has developed a four-stage roadmap for the biohythane technology to be successfully implemented. The plan states that in the first phase, research will be undertaken by private organization and technologies will be developed at the level of simulation and prototype developments for demonstrations so that it can be followed by major investments in infrastructure. Raising public awareness about the technology and educating people about its potential advantages would accompany this phase. Making the technology available in the market on a wide scale constitutes the second phase of the plan. However, the market penetration capacity of the technology still remains a factor to be considered, but this can be improved by raising awareness among the public so that it is more perceptive to the product and would purchase it as end consumers. The role of the government will be critical here in providing incentives to customers in order to increase sales and outreach. The long-term returns on provision of incentives in terms of environmental and socioeconomic benefits would determine the feasibility of these. To favor biohythane production, the government will also have to change tax policies to impose favorable taxation, carbon penalties, etc. Industry standards, particularly in the sector of automobile and equipment manufacturing, will have to be restructured and modified for designing, building, testing, and ultimately marketing biohythane-related equipment. Collaborations with fuel providers and other related businesses would play a crucial role in biohythane's success. After a successful hold is established in the market, phase III can be initiated, which aims at holistic growth by patronizing the technology to stimulate the market. There exist numerous examples where the government has taken proactive measures and shifted the energy use pattern to favor developing technologies. In central Europe, the policy of preferential taxes was introduced by the government to encourage the development of a natural gas transmission and distribution network. In the final phase of the plan, the technology may be commercially integrated into the national infrastructure and expanded to be available in international markets as well (Jamasb and Pollitt, 2005).

Further, to the benefit of the industry and stakeholders, the policy interests in moving toward a biohythane-based economy are rising. However, due to a lack of the right technological advancements and nonfeasibility of setting up industries, the practical applications of biohythane energy are very limited irrespective of the economy of the country. In most countries, the technology has not yet crossed the R&D phase. The developed countries are slightly ahead and are facilitating the introduction of new biohythane technologies through collaborative international programs as the technology is advancing. Milestones in biogas-related technologies have only been reached because of the heavy investments in research, especially from the member states of the European Union (EU), the United States, Canada, and Japan. These countries undertake about two-thirds of the total public spending in R&D. However, it is important that international organizations extend financial and technological to support research in developing countries for successful transition to a biohythane economy.

For advances in this technology and demonstration of the prototype, British Petroleum is providing the infrastructure for delivery of biohythane for demonstration in transport projects in around 10 cities around the world, including the Clean Urban Transport for Europe (CUTE) bus project in London (Hughes, 2007). This particularly aims at shifting the focus from research to the commercialization of the technology; especially the current EU policies on alternative fuels focus greatly on promotion of biofuels. Policies with respect to bioenergy are very important, and they should be applied in such a way that cross-subsidies between classes of consumers are not created (Balat, 2007). In a proposed biofuels directive from 2009, minimum blending shares of biofuels with conventional fuels has been made mandatory (Demirbas, 2008). However, to introduce biohythane in the system and promote its usage, concrete steps need to be taken. Stepwise transition to implementation of technology can be a possible development path for biohythane. For example, in the beginning when demand for biohythane energy is low, it can be delivered by trucks instead of laying extensive pipelines. The centralization of delivery in such a situation would be beneficial. Utilizing the excess capacity in the merchant biohythane systems can help in setting up the prototype demonstration of the project. Production of biohythane or mobile refuelers can be used for such purpose. Pipelines could be installed later with increase in demand, which would be more economical for heavy consumption and demands. The way biohythane will evolve in the future depends greatly on the existing infrastructure. Therefore, the need of the hour is to introduce changes which will ensure sustainable use of biohythane in the future, and this can only be done by the policy makers and people in the academia and industry. Moreover, it is important to educate the public as its understanding of biohythane will determine the current as well as future policy initiatives for stationary and portable applications as well as vehicle policies. Numerous studies have been performed to analyze the public's current perception and understanding of biohythane. Both positive and negative stigmas exist in the general public opinions when it comes to most energy-producing technologies that depend on individual perspectives, too. For example, individuals who are aware of the environmental effects of a possible nuclear meltdown may deem nuclear power a negative entity, whereas individuals who are more environmentally sensitive and are aware of the increasing carbon emissions may find nuclear power a positive entity. The future of biohythane policies depends greatly on educating the people, which needs to be a combination of scientific understanding with common associated social themes anchored by pre-existing knowledge.

The carbon footprint of biohydrogen and biomethane production processes is still less compared to chemical processes (Korres et al., 2010). Gaseous energy recovery in terms of only hydrogen might not be sufficient to make this process commercially viable. Only 20% to 30% of the total energy can be recovered through hydrogen production. On integration with photofermentation, theoretically, 12 mol mol^{-1} of glucose could be recovered. Many challenges are associated with such integrated systems such as the scaling-up problem and the shading effect of pigments produced by photofermentative organisms (Miyake et al., 1998). Thus, biohythane production provides an encouraging opportunity for converting organic residues rich in carbohydrates, fats, and proteins for clean energy generation. In biohythane, two-stage anaerobic processes for biohydrogen and biomethane were performed using organic waste (Kvesitadze et al., 2011).

In one such study, two-stage biohythane production was studied using starchy wastewater (Roy et al., 2015). The volumetric rate of hydrogen production was found to be 12 L L^{-1} d^{-1}. The calorific value of hydrogen is known to be 12.64 kJ L^{-1}. Therefore, in the first stage, total energy recovery as hydrogen per day was observed as 75.84 kJ (12 L L^{-1} × 12.64 kJ L^{-1} × 0.5 L = 75.84 kJ). Similarly, in the case of biomethanation, the rate of volumetric CH_4 production was 1.15 L L^{-1} d^{-1}. The calorific value of CH_4 is 35 kJ L^{-1}. Therefore, the total energy recovery as CH_4 per day in the second stage was 68.7 kJ (1.5 L L^{-1} × 35.0 kJ L^{-1} × 1.7 L = 68.4 kJ). The calorific value of the substrate was 268.8 kJ. The total gaseous energy recovery for the two-stage process was found to be 53.6%. From single-stage hydrogen production, gaseous energy recovery was only 28%. Improvement in gaseous energy recovery by 57% was also reported by Pawar et al. (2013) on using wheat straw hydrolysate for thermophilic biohydrogen, followed by biomethanation. In another study, desugared molasses was used as the substrate, and the total gaseous energy recovery was improved by 15% (Kongjan et al., 2013).

The prospect of using lingocellulosic biomass for the biohythane process was also explored (Kumari and Das, 2015). Dry sugarcane bagasse was delignified by mixed it with aqueous NaOH solution (solid loading of 5% w/v) at 50°C for 30 min, followed by enzymatic hydrolysis. For enzymatic hydrolysis, the delignified biomass was resuspended in a citrate buffer (0.05 M, pH 5) and cellulase from *Trichoderma reesei* (0.8 units/mg solid) was added to it. After 48 h incubation, the hydrolysate was collected by centrifugation. Alkali pretreatment followed by enzymatic hydrolysis of bagasse gave a substantial amount of reducing sugars, which were further used for the biohythane process. Thus a two-stage system (biohythane) might help in improving gaseous energy recovery from organic wastes to make the process commercially viable.

10.4 Issues and Barriers

- Strong government support to help the technology penetrate the market.
- Moreover, the rates of biohythane produced by various biohythane systems are expressed in different units, making it difficult to assess and compare the rates and amounts of biohythane synthesized by different biohythane technologies.
- Public awareness must be developed about the safety and use. To this end, it is important to develop safety procedures and codes for use of biohythane in energy applications.
- The major issues facing the fast development of a biohythane-based economy is the lack of interaction between the developers of the technology and the large-scale buyers who can put it to end use. This causes a lack of biohythane infrastructure for introduction of fuel cell vehicles.
- For biohythane renewables, the issue at hand is primarily the high cost of the project. In particular matching supply and demand during transition at low cost is a key issue.
- Development of a consistent energy policy to address societal problems of climate change, air pollution, and national security. A strong consensus stand is required to cut carbon emissions.
- The current yield and rates of production, especially from the microbial process, may not be economically feasible.
- Even in a scenario of technical success and strong policy, it may still be probably 10–15 years before biohythane energy technologies start to enter the market. Therefore, they may have no immediate effect on the current oil usage and/or carbon dioxide emissions
- Another crucial drawback of using biohythane as a transportation fuel is the huge onboard storage tanks which are required because of biohythane's extremely low density, as already discussed before. However, the low ignition temperature is one of the major advantages for biohythane

to be used directly as a fuel. Hence it can be used as a fuel indirectly by making fuel cells for producing electricity.

10.5 Status of Biohythane in Developed and Developing Countries

Worldwide biofuels are attracting great attention, with some governments seeing it as an opportunity to reduce GHG emissions and dependence on petroleum-based fuels, leading to their commitment toward biofuel programs. Among all the bio-based fuels, biohythane has gained a lot of popularity. It is speculated that over the next 10–20 years huge investments would flow into government-supported R&D programs and cutting-edge research in biohythane and fuel cell technology will be pursued. However, the use of biohythane will range widely across the world, depending on the technology breakthroughs, the policy adopted, and the cost analysis. It is estimated that biohythane-based vehicles are projected to reach shares of 30%–70 % of the global vehicle stock by 2050, resulting in a biohythane demand between 7 EJ and 16 EJ. The current speculations state that most of this demand will be from central Europe, North America, and China, and the net effect in the reduction of oil consumption would be in the range of 7–16 million barrels per day (Ball and Weitschel, 2009).

Although most biohythane research is taking place in industrialized countries, it is important that the developing countries also take up this research because their economies are more likely to be affected due to the political instability in the oil-producing countries. However, it is difficult for them to carry out full-fledged research due to a lack of necessary funds. It is speculated that engaging developing countries early on in the process may help speed up the transition to biohythane. Statistics also indicate that the highest energy demand is to come from developing countries in the near future. This would mean more utilization of fossil fuels if a breakthrough in energy policy is not endorsed. Therefore, it is critical for the countries to adopt this transition to biohythane as quickly as possible for attaining early energy stability. So far, the United States, Japan, and the EU have been carrying out the largest ongoing projects related to biohythane. The challenge is to establish a link between the regional and international activities using common methodologies and tools, augmenting analysis for all countries, and supporting development efforts. Some countries have also launched integrated R&D programs that cover all elements of biohythane supply and end users. Primarily, the approach adopted by the governments to implement the biohythane program reflects their own needs and resources, for example, Australia focuses on biohythane production from coal because it has large coal reserves. On the other hand, Germany focuses on fuel cell for vehicles as Germany is a leader in vehicle manufacturing. Accordingly, the amount of budget distributed and the resources used by different countries vary within their frameworks.

10.5.1 United States

In the United States, biohythane received an enormous boost during the second Bush administration. Much interest was shown for the development of biohythane fuel cell technologies within the transportation sector by the then government. The reason behind this interest was the reduction of the environmental impact caused due to the burning of fossil fuels and the desire to decrease the United States' dependence on foreign oil. The former president Bush in his State of the Union address in 2003 announced the initiative to promote fuel cell technology. He stated, "With a new national commitment, our scientists and engineers will overcome obstacles . . . so that the first car driven by a child born today could be powered by biohythane and pollution free. Join me in this important innovation to make our air significantly cleaner and our country much less dependent on foreign sources of energy." To begin to map the future of biohythane technology research, development, and demonstration, a new national commitment, a Biohythane Posture Plan, was drafted, and to accelerate research, development, and demonstration, appropriate $1.2 billion was announced for the research by Bush.

Most of the biohythane and fuel cell research carried out by the US government is funded by the DOE. The government follows the strategy of concentrating funding in the early stages of development in high-risk applied research on technologies and leverage private sector funding through partnerships. Several programs to promote alternative fuels, including biohythane, have been created by the US federal government. Some such programs are Clean City and Clean Construction USA. The main focus of the Clean City program is the reduction of GHG emissions by the transport sector. A network of more than 80 offices that develop public–private partnerships to promote alternative fuels, advanced vehicles, fuel blends, hybrid vehicles, and idle reduction are part of the Clean City program. They also provide information about financial opportunities, newsletters, and related technical and informational material; coordinate technical assistance projects; publish fact sheets; and update and maintain energy databases. Clean Ports USA is an incentive-based program that encourages port authorities to redesign and replace older diesel engines with new technologies and cleaner fuels to reduce emissions. The US Environmental Protection Agency's National Clean Diesel Campaign offers funding to port authorities to overcome obstacles in preventing the adoption of cleaner diesel technologies. These are only two of many such examples of ongoing programs run by the DOE to effectively reduce dependence on fossil fuels and GHG emissions in the coming decade.

10.5.2 Europe

The EU has funded some 200 projects on biohythane and fuel cell energy technologies since 1986, with a total contribution of over 550 million euros. The main focus of these projects was advancements in research in all the basic areas of biohythane research, including production, delivery, storage, biohythane-fueled vehicles, use of biohythane in cost-effective fuel cells, and other related policies aimed at transition of biohythane. Long-term collaborations are fostered among different organizations that are active in the same field. They exchange experience, create links, and might continue cooperation even after the project has finished by working together in projects. Also, research is channeled toward marketable solutions as universities and businesses cooperate and partners are found, creating supply chains. Further, the EU has joined forces with European industry and research institutes in a public–private partnership, the Fuel Cells and Biohythane (FCH) Joint Technology

Initiative (JTI), to accelerate the development and deployment of biohythane as a fuel in the most efficient way.

To accelerate the commercialization of FCH technologies in a number of application areas, the partners will together implement a program of research, technological development, and demonstration. Additionally, recently a new Energy Plan 2020 has been announced by the Danish government, which includes establishing a range of initiatives for biohythane infrastructure and fuel cell electric vehicles (FCEVs), with an aim of reaching 100% fossil independence by 2050. The government initiatives take up the recommendations of a recent Danish industry coalition analysis and roadmap on "biogas for transport in Denmark onwards 2050."

The German government has planned to take biohythane research to all new levels. Initiatives have been taken by the Ministry of Transport, Building and Urban Development to solve the classic "chicken and egg" problem. Fifty new public biohythane fuel stations have been announced. The estimated expenditure is more than $50 million in the project in the next year. Germany, having 15 biohythane fuel stations, has enough to power the 5000 biohythane-powered vehicles currently operating in the country.

10.5.3 Asia Pacific

The pace of funding for the design and rollout of biohythane-refueling stations was expected to grow in 2013, especially in Asia Pacific and Europe. Strong interest was apparent in 2012 in biohythane-refueling stations. For example, Japan released subsidies in 2013 to start the building program of biohythane-refueling stations, called "Subsidy for Biohythane Supply Facility Preparation." The government of Japan has set aside a "war chest" of $0.5 billion for 2013. The program will provide half funding of the construction cost of the station. Japan was among the first countries investing in biohythane research in a ten-year project that was completed in the year 2002, and soon after its completion the Japanese government initiated a second project in 2003. The Japanese government is confident that an economical biohythane-based fuel may soon be a reality with continued funding in this area.

10.6 Future Outlook

Biohythane's role in today's society is inevitable if one wants realization in both energy security as well as the control the pollution. In fact, the demand for energy is growing globally in concurrence with socioeconomic living standards. The world energy demand will increase by half around the year 2030, with more than two-thirds of this increase coming from developing and emerging countries according to a survey by the International Energy Agency. Alternative energy plays an important role in socioeconomic development. The majority of experts believe that biohythane works as an important energy carrier in the future energy sector.

Biological production of biohythane may play a key role, though present contributions might seem to be insignificant. To realize the role of biological biohythane, the yields will have to be improved and the economics of the process looked into by analyzing the reactor sizes and efficiencies of production. As the sizes of bioreactors required are huge, these systems may be considered impractical for our hypothetical application in the near future. Concerted efforts would bring about a greater contribution of this technology to the existing biohythane production technologies. Although the research on biohythane production has come a long way, concerted efforts are a prerequisite for industrial-scale production. For economically feasible, realistic applications, the biohythane yields and production rates must surpass considerably the present achievements. More efforts on the research on pilot-scale productions are warranted.

Significant work has already been reported by the group in terms of biohythane production and yields. This initial success could prove to be a stepping stone toward the more significant path to a biohythane-based economy, but there still remains much work to be done for commercial application. Thus, complete transition to a biohythane-based economy from a fossil-based economy is a daunting task and the following points need to be considered and taken care of immediately. For establishing a new setup or organization, a common platform needs to be set up so that communication gaps can be avoided. Therefore, organizations need to be developed so that research on hydrogen production, storage, and use in fuel cells can be done together.

10.6.1 Advent of Fuel Cell Technology

One of the reasons of the underperformance of biohydrogen production was also the absence of technologies related to its direct application. With the advent of technology such as fuel cells, the hydrogen produced by dark fermentation could be directly converted to electricity. This has infused new life into the implementation of a hydrogen-based economy. A fuel cell is a device that is similar to a continuously recharging battery that generates electricity by the low-temperature electrochemical reaction of hydrogen and oxygen (Winter and Brodd, 2004). The contrasting difference with batteries that store energy is that a fuel cell can produce electricity continuously as long as hydrogen and oxygen are supplied to it. Hydrogen-powered fuel cells produce water as a by-product and virtually no pollutant. Fuel cells operate at temperatures much below the IC engine. Fuel cells are not bound by the limitations of the Carnot cycle; thus they can efficiently convert fuel to electricity compared to IC engines. The operating temperature, the type of fuel, and a range of applications of fuel cells depend on the electrolyte they use. The electrolyte can be an acid, a base, a salt, or a solid ceramic or polymeric membrane that conducts ions. However, at present, fuel cells cannot compete with conventional energy conversion technologies in terms of cost and reliability. High-temperature solid oxide fuel cells (SOFCs) and molten carbonate fuel cells (MCFCs) are ideal for the distributed energy supply operating today with natural gas, which enables the development and use of this technology independently from the establishment of a hydrogen infrastructure. Indeed, they offer an interesting transition to the hydrogen economy. It has given a fresh breath to biofuel research, where gaseous energy or ethanol can be converted to electricity directly. The types of fuel cell technologies available are:

- Low-temperature proton exchange membrane fuel cells (Ralph, 1997): The highest power density is provided by proton exchange membrane fuel cells (PEMFCs) and alkaline fuel cells (AFCs). The requirement of a costly platinum catalyst and the need of highly pure hydrogen are the major bottlenecks in using such fuel cells. PEMFCs are most favored for mass market automotive and small-scale combined heat and power (CHP) applications, and there is a massive global effort to develop commercial systems.

- Phosphoric acid fuel cells (PAFCs) (Suzuki, 1994): These types of fuel cells are more tolerant toward impurities that remain as contaminants with hydrogen. PAFCs could be potentially used for stationary power generation and large vehicles. They are commercially available today but have a relatively high cost. Direct methanol fuel cells (DMFCs) are powered by methanol and are considered for a number of applications, particularly those based around replacing batteries in consumer applications such as mobile phones and laptop computers. Thus a hydrogen economy has a great potential for creating employment. Moreover, its environmental benefits also give it an upper hand in considering it as a future fuel.

10.7 Conclusions

Pollution and global warming are the "two faces of the same coin." Awareness among governments and citizens regarding these "evils" has given the required impetus to the renewable energy industry in recent times. Various policies and roadmaps have been discussed in this chapter that might drive or promote the biohythane industry. A paradigm shift in strategies of using renewable energy sources for electricity, alternative vehicle fuels, and alternative fuel vehicles and examining programs that could be promoted for financial support has been discussed in detail. A variety of microorganisms and their metabolic interplay during biohythane production are truly a complex mechanism. Waste management, waste disposal, socioeconomic issues, etc., may be safeguarded by using the biohythane process. The advent of fuel cell technologies has made a great impact on conversion of hydrogen to electricity. This has also reduced the need for sophisticated and expensive storage systems for hydrogen. Rather, low-cost storage and real-time conversion of hydrogen to electricity have been proposed. Improved gaseous energy recovery in terms of calorific value not only showed promise in preliminary studies but also seemed encouraging in large-scale application. So, encouraging government policies and socioeconomic assessment of the biohythane process could help in realizing the goal of "waste to energy."

References

Balat M (2007). Hydrogen in fueled systems and the significance of hydrogen in vehicular transportation, *Energy Source, Part B*, **2**, 49–61.

Ball M, Wietschel M (2009). The future of hydrogen opportunities and challenges, *Int J Hydrogen Energy*, **34**, 615–627.

Barbir F, Plass Jr HJ, Veziroglu TN (1993). Modeling of the solar hydrogen energy system, *Proceeding of the DOE/NREL Hydrogen Program Review*, 151.

Demirbas, A (2008). Biofuels sources, biofuel policy, biofuel economy and global biofuel projections, *Energy Convers Manage*, **49**(8), 2106–2116.

Gerber P, Key N, Portet F, Steinfeld H (2010). Policy options in addressing livestock's contribution to climate change, *Animal*, **4**(3), 393–406.

Hordeski MF (2011). *Megatrends for Energy Efficiency and Renewable Energy*, The Fairmont Press, Inc., Lilburn, GA.

Hughes AN (2007). Organizations and institutions relating to the development of hydrogen and fuel cell activities in the UK, *UKSHEC Social Science Working Paper No 34*, London Policy Studies Institute, London.

Jamasb T, Pollitt M (2005). Electricity market reform in the European Union review of progress toward liberalization and integration, *Energy J*, **26**, 11–41.

Jorgensen K (1998). Hydrogen energy progress, *Proceedings of the 12th World Hydrogen Energy Conference*, Buenos Aires, Argentina.

Kojima M, Lovei M (2001). Urban air quality management coordinating transport, environment, and energy policies in developing countries, Vol 508, World Bank Publications, Washington DC.

Kongjan P, O-Thong S, Angelidaki I (2013). Hydrogen and methane production from desugared molasses using a two-stage thermophilic anaerobic process, *Eng Life Sci*, **13**, 118–125.

Kumari S, Das D (2015). Improvement of gaseous energy recovery from sugarcane bagasse by dark fermentation followed by biomethanation process, *Bioresour Technol*, **194**, 354–363.

Kvesitadze G, Sadunishvili T, Dudauri T, Zakariashvili N, Partskhaladze G, Ugrekhelidze V (2011). Two stage anaerobic process for biohydrogen and biomethane combined production from biodegradable solid wastes, *Energy*, **37**, 94–102.

Miyake M, Sekine M, Vasilieva LG, Nakada E, Wakayama T, Asada Y, Miyake J (1998). Improvement of bacterial light dependent hydrogen production by altering the photosynthetic pigment ratio. In: Zaborsky OR, Benemann JR, Matsunaga T, Miyake J, Pietro AS (eds), *BioHydrogen*, Springer US, 81–86.

Pawar SS, Nkemka VN, Zeidan AA, Murto M, van Niel EWJ (2013). Biohydrogen production from wheat straw hydrolysate using *Caldicellulosiruptor saccharolyticus* followed by biogas production in a two step uncoupled process, *Int J Hydrogen Energy*, **38**, 9121–9130.

Ralph TR (1997). Proton exchange membrane fuel cells, *Platinum Metals Rev*, **41**, 102–112.

Sims RE, Rogner HH, Gregory K (2003). Carbon emission and mitigation cost comparisons between fossil fuel, nuclear and renewable energy resources for electricity generation, *Energy Policy*, **31**(13), 1315–1326.

Stambouli AB, Traversa E (2002). Solid oxide fuel cells (SOFCs) a review of an environmentally clean and efficient source of energy, *Renew Sust Energy Rev*, **6**(5), 433–455.

Suzuki N (1994). US Patent No 5,346,780, Washington, DC, US Patent and Trademark Office.

Winter M, Brodd RJ (2004). What are batteries fuel cells and supercapacitors, *Chem Rev*, **104**(10), 4245–4270.

Wurster R, Altmann M, Sillat D, Drewitz HJ, Kalk KW, Hammerschmidt A, Stuhler W, Holl E (1998). Hydrogen energy progress XII, *Proceedings of the 12th World Hydrogen Energy Conference*, Buenos Aires, Argentina.

Zweig JS, Ham JC, Avol EL (2009). Air pollution and academic performance evidence from California schools, *Work Pap, Dep Econ, Univ, Md by VALE Trial Account on*, **5**(20), 14.

Index